Melanin-Concentrating Hormone and Sleep

S. R. Pandi-Perumal • Pablo Torterolo •
Jaime M. Monti
Editors

Melanin-Concentrating Hormone and Sleep

Molecular, Functional and Clinical Aspects

Springer

Editors
S. R. Pandi-Perumal
Somnogen Canada Inc.
Toronto, Ontario, Canada

Pablo Torterolo
Department of Physiology, School of Medicine
University of the Republic
Montevideo, Uruguay

Jaime M. Monti
Department of Pharmacology and Therapeutics, School of Medicine Clinics Hospital
University of the Republic
Montevideo, Uruguay

ISBN 978-3-030-09322-8 ISBN 978-3-319-75765-0 (eBook)
https://doi.org/10.1007/978-3-319-75765-0

Printed on acid-free paper

This Springer imprint is published by the registered company Springer International Publishing AG part of Springer Nature.
The registered company address is: Gewerbestrasse 11, 6330 Cham, Switzerland

To our families...
for their abundant support, for their patience and understanding, and for their everlasting love and affection.

Preface

The melanin-concentrating hormone (MCH) is a neuropeptide that was initially isolated from the teleost pituitary as a chromophore modulator by Kawauchi et al. (1983). Twenty years later, Verret et al. described that this neuropeptide has a strong relationship with the generation and maintenance of sleep. Since then, several authors have described the anatomy and physiology of the MCHergic system. Regarding sleep, the MCHergic system is now considered a critical player for both REM and NREM sleep generation. In addition, there is preclinical and clinical evidence suggesting a role of MCH in human pathology. Finally, synthetic MCH-receptor ligands are under study for the treatment of different medical conditions.

The first two chapters in this volume deal with anatomy: Bittencourt and Dinitz review the neuroanatomical structure of the MCHergic system, while Lee gives an elaborate account of the projections of the MCHergic neurons toward the structures involved in the regulation of sleep and wakefulness.

The next three chapters focused on the role of MCH in sleep physiology. Gao summarized the evidence supporting the role of MCH in the regulation of sleep homeostasis and also discussed the physiological implications of MCH as a central node of the overall brain circuitry controlling physiological functions and complex behaviors. Monti et al. described the effects of MCH on the neurotransmitter systems involved in the generation and maintenance of wakefulness. Finally, Blanco-Centurion et al. performed a careful analysis of the optogenetic studies that focused on understanding the functions of the MCHergic neurons.

In chapters "Cannabinoids, Sleep, and the MCH System", "MCH and Thermoregulation", "MCH, Sleep, and Neuroendocrine Functions" and "Melanin-Concentrating Hormone: Role in Nursing and Sleep in Mother Rats", the role of the MCHergic system was analyzed in the frame of different physiological functions. Murillo-Rodriguez et al. underlined the interactions between endocannabinoids and MCH in the control of sleep, while Luppi reviewed the role of MCH in thermoregulation, highlighting the hypothesis that the regulation of temperature, metabolism, and sleep is closely linked. D'Almeida and coworkers provide an integrative discussion of the role of MCH in sleep and neuroendocrine functions. Thereafter, Benedetto et al. evaluated the function of MCH in the postpartum rat, stressing the modulation by MCH of maternal behavior and sleep, functions that are integrated within the preoptic area.

The scope of the last three chapters was to integrate basic with clinical research. Costa et al. examined the state of the art of MCH in medical conditions. Then, Scorza and coworkers discussed the preclinical findings in relation to MCH and depression, a condition where sleep is highly affected. Finally, Chaki reviewed the current knowledge regarding the MCH antagonists for the treatment of depression and anxiety disorders.

We consider that understanding the role of MCH in the control of sleep and related functions is highly relevant for human's health. Investigation in this field is advancing at a rapid pace, and this book reveals novel findings and presents questions for future research.

Montevideo, Uruguay — Jaime M. Monti
Toronto, ON, Canada — S. R. Pandi-Perumal
Montevideo, Uruguay — Pablo Torterolo

References

Kawauchi et al (1983) Characterization of melanin-concentrating hormone in chum salmon pituitaries. Nature 305(5932):321–323

Verret et al (2003) A role of melanin-concentrating hormone producing neurons in the central regulation of paradoxical sleep. BMC Neurosci 4:19

Credits and Acknowledgements

This volume owes its final shape and form to the assistance and hard work of many talented people. Creating a book, which surveys a broad interdisciplinary field such as sleep and neuropharmacology, involves the collaborative scholarship of many individuals. We express our profound gratitude to the many people who have helped and also to some who have contributed without realizing just how helpful they have been.

The editors wish to express their sincere appreciation and owe endless gratitude to all our distinguished contributors for their scholarly contribution that facilitated the development of this volume. Our greatest debt is obviously to our outstanding authors who, regardless of how busy they were, managed to find time for this project. They, in a most diligent and thoughtful way, have brought a wide range of interests and disciplines to this volume entitled *Melanin-Concentrating Hormone and Sleep: Molecular, Functional and Clinical Aspects*. They accepted our submission deadlines and tolerated with great patience our repeated reminders, our frequent phone calls, and our bombardment with high-priority e-mail messages.

We would like to thank our secretarial and administrative staffs of our respective institutions, for helping us to stay on task and for their attention to detail.

No volume can be completed without the untiring efforts of many publishing professionals. Producing a volume such as this is a team effort, and we acknowledge with gratitude the work of the editorial department of Springer International Publishing AG, Switzerland. We are especially indebted to Dr. Beatrice Menz, Senior Editor, Biomedicine, Springer International Publishing AG, who was an enthusiastic and instrumental supporter from the start to the end. Our profound gratitude is also offered to Hannah Dean, Editorial Assistant, Biomedicine, whose equally dedicated efforts promoted a smooth completion of this important project. Both Beatrice and Hannah provided unflagging dedication, invaluable help, and encouragement. We appreciate their intellectual rigor and personal commitment to our project.

We also thank the Springer International Publishing AG production department colleagues for their meticulous work. They all gave unstintingly of their time, energy, and enthusiasm. This talented and dedicated team of copy and production editors strengthened, polished, trimmed, and conscientiously checked the text for errors.

The editors would also like to acknowledge the close cooperation we have received from each other. We think we made a good team, even if we say it ourselves!

Every effort has been made by the authors, editors, and publishers to contact the copyright holders to obtain their permission for the reproduction of borrowed material. Regrettably, it remains possible that this process could have been incomplete. Thus, if any copyrights have been overlooked, the publisher will ensure correction at the first opportunity for a subsequent reprint of this volume.

Last, but certainly not least, we are most grateful to our wonderful wives and families, who provided love and support too valuable to measure. We owe everything to them. Their understanding and patience, wisdom, creativity, constant support, and encouragement while the book was being developed are immeasurably appreciated. Without the love and support of our families and friends, we could not have completed this project. Being able to spend more time with them is our chief reward for finishing. They saw the work through from the conception of an idea to the completion of an interesting project with unwavering optimism and encouragement. They were the source of joy and inspiration for us, and we thank them for their continuing support and for understanding the realities of academic life!

To all the people who contributed to this project, we want to say "thank you!." Their willingness to contribute their time and expertise made this work possible, and it is to them that the greatest thanks are due. They make our work possible and pleasurable.

For this, and for so much else, we are ever grateful.

Montevideo, Uruguay Jaime M. Monti
Toronto, ON, Canada S. R. Pandi-Perumal
Montevideo, Uruguay Pablo Torterolo

Contents

Editors and Contributors

About the Editors

Jaime Monti is Doctor of Medicine at the University of the Republic, Uruguay. He has completed postdoctoral studies at the National Institutes of Health, USA. He is full professor of Pharmacology and Therapeutic and the School of Medicine of the Republic University, professor of the Basic Sciences Development Program (PEDECIBA), and researcher of the National system of Researchers. His research has been awarded the Claude Bernard Prize (France government) and the Schering Award for Basic Sleep Research. He is an internationally recognized expert in the neuropharmacology of sleep and wakefulness, with more than a hundred of papers in this topic.

S. R. Pandi-Perumal is the president and chief executive officer of Somnogen Canada Inc, a Canadian Corporation. He is a well-recognized sleep researcher, both nationally and internationally, and has authored over 200 publications. His general area of research interest includes sleep and biological rhythms. He is a well-known editor in the field of sleep medicine and has edited over 25 volumes dealing with various sleep-related topics. He received an honorable mention in the *New York Times* in 2004. The India International Friendship Society awarded him the prestigious Bharat Gaurav award on January 12, 2013. The Bharat Gaurav Award is usually given in recognition of a person who has made achievements in a particular field, which in turn will have a positive impact on society at large.

Pablo Torterolo is Doctor of Medicine and Doctor of Neurosciences at the University of the Republic, Montevideo, Uruguay. He has completed postdoctoral studies at the University of Bologna and at the University of California, Los Angeles. He is currently associate professor of the Department of Physiology of the School of Medicine, University of the Republic. He is also professor of the Basic Sciences Development Program (PEDECIBA) and researcher of the National System of Researchers. His research has been awarded the Prize of the National

Academy of Medicine and the Elio García-Austt Award. His studies focus on the neurobiology and pathophysiology of sleep.

Contributors

Luciana Benedetto Department of Physiology, School of Medicine, University of the Republic, Montevideo, Uruguay

Jackson C. Bittencourt Laboratory of Chemical Neuroanatomy, Department of Anatomy, Institute of Biomedical Sciences, University of Sao Paulo, Brazil

Centre for Neuroscience and Behavior, Institute of Psychology, University of Sao Paulo, Brazil

Carlos Blanco-Centurion Department of Psychiatry & Behavioral Sciences, Medical University of South Carolina, Charleston, SC, USA

Shigeyuki Chaki Research Headquarters, Taisho Pharmaceutical Co., Ltd, Saitama, Japan

Alicia Costa Department of Physiology, School of Medicine, University of the Republic, Montevideo, Uruguay

Miriel de-la-Cruz Laboratorio de Neurociencias Moleculares e Integrativas, Escuela de Medicina, División Ciencias de la Salud, Universidad Anáhuac Mayab, Mérida, Yucatán, Mexico

Grupo de Investigación en Envejecimiento, División Ciencias de la Salud, Universidad Anáhuac Mayab, Mérida, Yucatán, Mexico

Intercontinental Neuroscience Research Group, Mérida, Yucatán, Mexico

Clelia De-la-Peña Centro de Investigación Científica de Yucatán A.C., Unidad de Biotecnología, Mérida, Yucatán, México

Giovanne B. Diniz Laboratory of Chemical Neuroanatomy, Department of Anatomy, Institute of Biomedical Sciences, University of Sao Paulo, Brazil

Vânia D'Almeida Department of Psychobiology, Universidade Federal de São Paulo, São Paulo, Brazil

Annabel Ferreira Facultad de Ciencias, Sección de Fisiología y Nutrición, Universidad de la República, Montevideo, Uruguay

Xiao-Bing Gao Department of Comparative Medicine, Program on Integrative Cell Signaling and Neurobiology of Metabolism (ICSNM), Yale University School of Medicine, New Haven, CT, USA

Carlos Eduardo Neves Girardi Department of Psychobiology, Universidade Federal de São Paulo, São Paulo, Brazil

Master Program in Psychology/Psychosomatics, Universidade Ibirapuera, São Paulo, Brazil

Débora Cristina Hipólide Department of Psychobiology, Universidade Federal de São Paulo, São Paulo, Brazil

Patricia Lagos Department of Physiology, School of Medicine, University of the Republic, Montevideo, Uruguay

Hyun Sook Lee Department of Anatomy, College of Medicine, Konkuk University, Seoul, Republic of Korea

Meng Liu Department of Psychiatry & Behavioral Sciences, Medical University of South Carolina, Charleston, SC, USA

Ximena López Department of Experimental Neuropharmacology, Instituto de Investigaciones Biológicas Clemente Estable, Montevideo, Uruguay

Marco Luppi Dipartimento di Scienze Biomediche e NeuroMotorie, Università degli Studi di Bologna, Bologna, Italy

Lorena Macías-Triana Laboratorio de Neurociencias Moleculares e Integrativas, Escuela de Medicina, División Ciencias de la Salud, Universidad Anáhuac Mayab, Mérida, Yucatán, Mexico

Grupo de Investigación en Envejecimiento, División Ciencias de la Salud, Universidad Anáhuac Mayab, Mérida, Yucatán, Mexico

Intercontinental Neuroscience Research Group, Mérida, Yucatán, Mexico

Jaime M. Monti Department of Pharmacology and Therapeutics, School of Medicine Clinics Hospital, University of the Republic, Montevideo, Uruguay

Daniela Morales-Lara Laboratorio de Neurociencias Moleculares e Integrativas, Escuela de Medicina, División Ciencias de la Salud, Universidad Anáhuac Mayab, Mérida, Yucatán, Mexico

Grupo de Investigación en Envejecimiento, División Ciencias de la Salud, Universidad Anáhuac Mayab, Mérida, Yucatán, Mexico

Intercontinental Neuroscience Research Group, Mérida, Yucatán, Mexico

Eric Murillo-Rodríguez Laboratorio de Neurociencias Moleculares e Integrativas, Escuela de Medicina, División Ciencias de la Salud, Universidad Anáhuac Mayab, Mérida, Yucatán, Mexico

Grupo de Investigación en Envejecimiento, División Ciencias de la Salud, Universidad Anáhuac Mayab, Mérida, Yucatán, Mexico

Intercontinental Neuroscience Research Group, Mérida, Yucatán, Mexico

Seithikurippu R. Pandi-Perumal Somnogen Canada Inc., Toronto, ON, Canada

José Carlos Pastrana-Trejo Laboratorio de Neurociencias Moleculares e Integrativas, Escuela de Medicina, División Ciencias de la Salud, Universidad Anáhuac Mayab, Mérida, Yucatán, Mexico

Grupo de Investigación en Envejecimiento, División Ciencias de la Salud, Universidad Anáhuac Mayab, Mérida, Yucatán, Mexico

Intercontinental Neuroscience Research Group, Mérida, Yucatán, Mexico

Karen Romero-Cordero Laboratorio de Neurociencias Moleculares e Integrativas, Escuela de Medicina, División Ciencias de la Salud, Universidad Anáhuac Mayab, Mérida, Yucatán, Mexico

Grupo de Investigación en Envejecimiento, División Ciencias de la Salud, Universidad Anáhuac Mayab, Mérida, Yucatán, Mexico

Intercontinental Neuroscience Research Group, Mérida, Yucatán, Mexico

Cecilia Scorza Department of Experimental Neuropharmacology, Instituto de Investigaciones Biológicas Clemente Estable, Montevideo, Uruguay

Priyattam Shiromani Department of Psychiatry & Behavioral Sciences, Medical University of South Carolina, Charleston, SC, USA

Ralph Johnson Medical Center, US Department of Veterans Affairs, Charleston, SC, USA

Pablo Torterolo Department of Physiology, School of Medicine, University of the Republic, Montevideo, Uruguay

Jessika Urbanavicius Department of Experimental Neuropharmacology, Instituto de Investigaciones Biológicas Clemente Estable, Montevideo, Uruguay

Neuroanatomical Structure of the MCH System

Jackson C. Bittencourt and Giovanne B. Diniz

Abstract The mammalian melanin-concentrating hormone (MCH) system has been studied for almost 30 years as of the writing of this chapter. The understanding of any neural system starts with its structure, as the anatomical organization of its peptides and receptors provides important clues about its physiological activity. With that in mind, the MCH system will be reviewed as a whole in the first segment of this chapter, including the genes and proteins that comprise this system. In the following segment, a description of the morphological and neurochemical characteristics of MCH neurons will be provided, including the areas in the central nervous system where these neurons are found and their projection pathways and targets. In the last segment, the receptors for MCH will be briefly discussed, and the available anatomical information about these receptors will be presented to the reader. By the end of this chapter, the reader should be able to identify the sites of MCH synthesis, the major targets of MCH action, and the areas responsive to MCH owing to the presence of its receptor.

List of Abbreviations

3V	Third ventricle
3Vdc	Third ventricle dorsal cap
aa	Amino acid(s)
AHA	Anterior hypothalamic area

J. C. Bittencourt (✉)
Laboratory of Chemical Neuroanatomy, Department of Anatomy, Institute of Biomedical Sciences, University of Sao Paulo, Sao Paulo, Brazil

Centre for Neuroscience and Behavior, Institute of Psychology, University of Sao Paulo, Sao Paulo, Brazil
e-mail: jcbitten@icb.usp.br

G. B. Diniz
Laboratory of Chemical Neuroanatomy, Department of Anatomy, Institute of Biomedical Sciences, University of Sao Paulo, Sao Paulo, Brazil
e-mail: giovanne.diniz@usp.br

S. R. Pandi-Perumal et al. (eds.), *Melanin-Concentrating Hormone and Sleep*,
https://doi.org/10.1007/978-3-319-75765-0_1

ANP	Atrial natriuretic peptide
AROM	Antisense-RNA-overlapping-MCH gene
AVP	Arginine vasopressin
BNST	Bed nucleus of the *stria terminalis*
CART	Cocaine- and amphetamine-regulated transcript
CNS	Central nervous system
CRF	Corticotropin-releasing factor
f	Fornix
GAD67	Glutamate decarboxylase 67 kDa
GPCR	G protein-coupled receptor
ic	Internal capsule
IHy	Incertohypothalamic area
IRE	Interferon-gamma responsive element
LDT	Laterodorsal tegmental nucleus
LHA	Lateral hypothalamic area
MCH	Melanin-concentrating hormone (peptide)
MCHR1	Melanin-concentrating hormone receptor 1 (peptide)
MCHR1	Melanin-concentrating hormone receptor 1 (primate gene)
Mchr1	Melanin-concentrating hormone receptor 1 (rodent gene)
MCHR2	Melanin-concentrating hormone receptor 2 (peptide)
MCHR2	Melanin-concentrating hormone receptor 2 (primate gene)
ME	Median eminence
mfb	Medial forebrain bundle
mt	Mammillothalamic tract
MGOP	MCH-gene-overprinted-polypeptide
MPOA	Medial preoptic area
NEI	Neuropeptide glutamic acid-isoleucine
NGE	Neuropeptide glycine-glutamic acid
NPY	Neuropeptide Y
opt	Optic tract
ORF	Open reading frame
ORX	Orexin/hypocretin
PACAP	Pituitary adenylate cyclase-activating peptide
PC	Prohormone convertase(s)
PH	Posterior hypothalamic area
Pmch	Pro-melanin-concentrating hormone (rodent gene)
PMCH	Pro-melanin-concentrating hormone (primate gene)
PMCHL1	Pro-melanin-concentrating hormone linked gene 1 (chimeric gene)
PMCHL2	Pro-melanin-concentrating hormone linked gene 2 (chimeric gene)
pmPRt	Paramedian pontine reticular formation
ppMCH	Prepro-melanin-concentrating hormone
PVH	Paraventricular nucleus of the hypothalamus
REM	Rapid eye movement sleep
SDL	Subcoeruleus/sublaterodorsal nucleus
SLC-1	Somatostatin-like coupled receptor 1, same as MCHR1

SP	Substance P
SSTR	Somatostatin receptor(s)
TM	Transmembrane
TMdm	Tuberomammillary nucleus, dorsomedial part
Tu	Olfactory tubercle
X-ir	Immunoreactive to X
ZI	*Zona incerta*
α-MSH	α-Melanocyte-stimulating hormone

1 Introduction

The melanin-concentrating hormone (MCH) system is a complex vertebrate neuropeptidergic system, composed of several neuropeptides and multiple receptor subtypes that span several clades. Although MCH was initially isolated from the teleost pituitary as a chromophore modulator (Kawauchi et al. 1983), less than a decade was necessary for the description of mammalian orthologues of its gene (*Pmch/PMCH*) (Nahon et al. 1989, 1993; Presse et al. 1990). Simultaneously, the amino acid (aa) sequence of the peptides encoded by these genes (Vaughan et al. 1989) was described, and their distribution in the mammalian brain soon followed (Bittencourt et al. 1992). The description of the neuroanatomical structure of the MCH system was then followed by a profusion of functional works, which slowly started to unveil the numerous roles played by its peptides in mammalian physiology. Among these functions, and of particular interest to this work, are the modulation of motivated behaviors (reviewed in Diniz and Bittencourt 2017) and of the sleep–wake cycle (reviewed in Ferreira et al. 2017a).

It is necessary, however, to clarify the definitions of "motivated behaviors" and "sleep–wake cycle" employed in this chapter, as these concepts are often used in the literature without further consideration. The most common conception of motivated behaviors is as *behaviors that the animal performs to meet homeostatic needs*. This description is true of ingestive behavior, perhaps the most thoroughly studied motivated behavior. However, this definition fails to include, for example, mating and maternal behaviors; both of these types of behavior may even result in a worsening of the animal's metabolic needs, but it performs these actions nevertheless because they are necessary for the maintenance of the species. Motivated behaviors could therefore be redefined as *actions that the animal executes that are necessary to its survival or the survival of its species*. Although this definition can be applied to all behaviors discussed so far, it opens up the question of whether sleeping can be considered a motivated behavior. For the purposes of this work, it will not be. While other behaviors require the animal to interact with exogenous elements (e.g., food, mate, pups) for their full completion, sleep can occur regardless of exogenous elements. Therefore, we will define motivated behaviors in this chapter as *actions that the animal executes toward exogenous agents that are necessary for its survival or the survival of its species.* Sleep, on the other hand, will be defined as "a periodic, complex intrinsic state alteration that encompasses synaptic alterations, hormone secretion, and consciousness suspension and that is necessary for the support of several other physiological processes."

As a complex physiological process, the sleep–wake cycle is sustained in the central nervous system (CNS) by an equally complex network of nuclei, connections, neurotransmitters, and neuromodulators. Central in this network is the hypothalamus and, in particular, the lateral hypothalamic area (LHA), which has been long known to be an integrative center of homeostatic controls and motivated behaviors, receiving inputs from and sending outputs to the whole CNS (Diniz and Bittencourt 2017). Although the importance of the LHA for several behaviors has been recognized for a long time, most of its inner workings have remained a black box for decades, mainly owing to its remarkable cyto- and chemoarchitecture and the complexity of its connectivity (Simerly 2015). The study of the LHA was further complicated by the extensive presence of the medial forebrain bundle (*mfb*) in this area, as the LHA has been considered the bed nucleus of the *mfb* (Nieuwenhuys et al. 1982). It was only after the discovery of LHA-specific neurochemical markers (e.g., MCH, orexin/hypocretin [ORX] and cocaine- and amphetamine-regulated transcript [CART]) that this area could be explored in a more detailed way (Bittencourt et al. 1992; Sakurai et al. 1998; de Lecea et al. 1998; Douglass et al. 1995). The discovery of these markers, in addition to the development of new methodologies, such as optogenetics, chemogenetics, in vivo electrophysiology, polysynaptic viral tracing, calcium imaging, and miniaturized fluorescence microscopes, forever changed the way we understand the LHA and other hubs related to the control of the sleep–wake cycle.

The first attempts to anatomically describe the modulatory centers of sleep and wakefulness, however, predate these techniques by far and can be traced to the work of von Economo, in 1930. After attending to patients during the encephalitis epidemic of 1917–1926, von Economo noticed that some of his patients displayed a subset of signs and symptoms related to sleep, which he called *encephalitis lethargica*. Interestingly, while some patients displayed increased sleep time, others had a decrease, even though they were all part of the same viral epidemic. The correlation between symptoms and the location of the lesions in these patients led von Economo to propose a dual model of sleep control: the brain stem as the wakefulness-promoting center and the anterior hypothalamus as the sleep-promoting center (Von Economo 1930). Although simple, von Economo's model was surprisingly accurate and paved the way for many other nuclei to be included in this network in the following decades (Saper 2013). Likewise, several neuromodulators, including MCH and other lateral hypothalamic neuropeptides, were also implicated in this control system as new evidence was incorporated into the model. In the next part of this chapter, the basic aspects of the MCH system and its anatomical structure in the mammalian CNS will be reviewed.

2 The MCH System

In broad strokes, the mammalian MCH system can be split into two components: a peptidergic element and a receptor element. The peptidergic component is composed of a single gene, *Pmch/PMCH*, which encodes a precursor that originates three mature peptides: MCH, neuropeptide glycine-glutamic acid (NGE), and neuropeptide glutamic acid-isoleucine (NEI). While substantial anatomical and functional data have been obtained for MCH and NEI (as reviewed in Diniz and Bittencourt 2017; Bittencourt and Celis 2008), the status of NGE as a neuroactive substance has remained controversial. It has been suggested that NGE may modulate cAMP production (Hintermann et al. 2001), but further studies are needed to give us a better understanding of this putative peptide. The receptor component consists of two genes, *Mchr1/MCHR1* and *MCHR2*, each encoding one receptor subtype (MCHR1 and MCHR2, respectively). The subtype 2 receptor is not ubiquitous among vertebrates, however, as a functional MCHR2 cannot be found in species of the *Glires* superorder. No receptor is known for NEI or NGE. The discovery of their receptors or their mechanisms of action, if existent, could represent the next breakthrough in the understanding of the MCH system. In this section, the general characteristics of each component of the MCH system will be explored.

2.1 MCH, NEI, and NGE

Since there is substantial complexity in the MCH system across different vertebrate clades, which goes beyond the scope of this work, the focus of this section will be on mammalian species, more specifically on data from rats, mice, and humans. As the *Pmch/PMCH* gene is remarkably conserved among mammals, it is likely that observations made on the cited animals can be extrapolated to other species.

The evidence collected so far point to the existence of a single authentic *Pmch* gene in mammals, spanning approximately 1.4 kb. In rats (*rPmch—gene ID, 24659*), this gene is located on chromosome 7q13 and overlaps with the PARP1 binding protein-coding gene (*Parpbp*) (Nahon et al. 1992). In the same *locus* are the genes coding for insulin-like growth factor 1 (*Igf1*), nucleoporin 37 (*Nup37*), and WASH complex subunit 3 (*Washc3*). The mouse gene (*mPmch—gene ID, 110312*) is located on chromosome 10 C1 10 43.7 cM, and, as in the rat, it overlaps with the *Parpbp* gene and is flanked by *Igf1*, *Nup37*, and *Washc3* genes. Finally, the human gene (*PMCH—gene ID*, 5367) is located on chromosome 12q23.2 and has a similar synteny to the murine species (Pedeutour et al. 1994).

In all these species, the *Pmch/PMCH* gene follows the same basic structure, i.e., it is formed by three exons (Exon 1 coding sequence, 249 base pairs [bp]; Exon 2, 199 bp; Exon 3 coding sequence, 47 bp), in addition to two introns and untranslated 5′ and 3′ sequences whose lengths differ between species (Nahon et al. 1989; Breton et al. 1993a; Thompson and Watson 1990; Presse et al. 1990). Figure 1 illustrates the

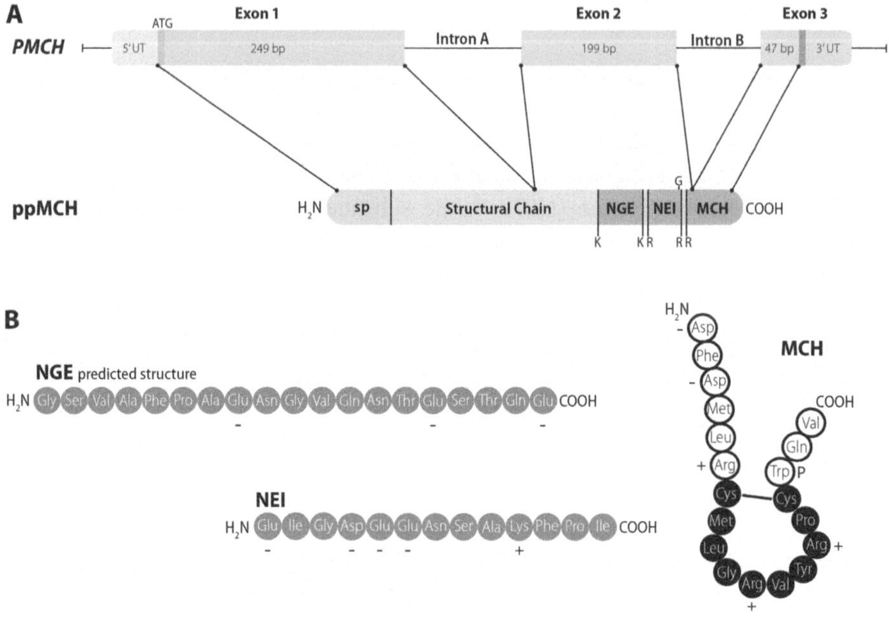

Fig. 1 Structure of the *PMCH* gene and its family of peptides. Schematic representation of the human *PMCH* gene and the MCH, NEI, and NGE peptides. (**a**) The *PMCH* gene is composed of three exons interrupted by two intronic sequences. The first exon corresponds to the signal peptide (sp) and the first part of the structural chain of prepro-MCH (ppMCH). The second exon codes for the remaining segment of the structural chain, NGE, NEI, and the first three amino acids of MCH. The third exon will complete the fourth codon corresponding to MCH and the final 15 amino acids. The sites of cleavage by prohormone convertases are indicated in the ppMCH structure by "K" and "R." The amidation site of NEI is indicated by "G." (**b**) The residue sequences for MCH, NEI, and the predicted structure of NGE. Plus and minus signs indicate the polarity of electrically charged amino acids. The letter "P" indicates a residue that is important for binding potentiation of MCH (Trp^{17}). White circles denote residues that are not essential for MCH function, while black circles denote the necessary amino acids for MCH binding. Gray circles are used for peptides without known binding properties

PMCH gene structure. In the upstream flanking region of the *Pmch/PMCH* gene, several promoter sequences are found, including two AP-1-binding sequences, one interferon-γ response element (IRE), and partial sequences of a glucocorticoid response element (Viale et al. 1997). Through AP-1 sites, the transcription of *Pmch* may respond to the FOS/JUN transcription factors, suggesting that FOS may be an adequate marker for the activity of MCH-containing neurons. Through the IRE, MCH neurons may respond to immunological events, such as sickness behavior or hypothalamic inflammation. The glucocorticoid response element may act to couple MCH populations to the defensive stress response/arousal circuits. Between the two AP-1 sequences, there is a CAAT box, and approximately 27 bp upstream of the cap nucleotide is a TATA box. These two sequences are good indicators that the area is the regulatory zone of *PMCH* transcription initiation (Viale et al. 1997).

As mentioned above, Exon 1 contains a 249 bp coding sequence, including the ATG start codon, in addition to the 5′ untranslated region, whose length depends on the species. The 5′ untranslated sequence is highly conserved between humans and mice and presents a palindromic sequence juxtaposed to the start codon, which could act as part of a transcription control mechanism through the formation of a hairpin. The substitutions in the rat sequence do not support the presence of a folding structure in this species (Breton et al. 1993a). There is a second putative ATG start codon four positions downstream of the first codon. Although both are viable starting points, several works support the first ATG codon as the primary initiation *locus* (Nahon et al. 1989; Breton et al. 1993a; Presse et al. 1990). There is high identity among all three species concerning the coding region of Exon 1, with the mouse equidistant from humans and rats and a slightly bigger difference between rats and humans (R vs. M, 86.75%; R vs. H, 82.33%; H vs. M, 86.37%). It is noteworthy that Exon 1 is the least conserved of the three exons.

Exons 1 and 2 are separated by Intron A, a sequence whose length varies depending on the species, ranging from 302 to 350 bp. Exon 2 is 199 bp in length and is slightly more conserved among the three mammalian species, contrasting with the upstream exon by having higher similarity between the murine species than between the murine and human sequences (R vs. M, 94.48%; R vs. H, 87.94%; H vs. M, 89.95%). After an intermission consisting of Intron B, with lengths ranging from 250 to 271 bp, Exon 3 completes the coding portion of the gene with 47 bp. The Exon 3 coding region is remarkably conserved among mammalian species (R vs. M, 96%; R vs. H, 94%; H vs. M, 96%). Interestingly, the intronic regions display a higher degree of identity between rats and mice (86–88%) than the exonic sequences do, but a much lower degree between mice and humans (60–70%) (Thompson and Watson 1990; Breton et al. 1993a, b). Following the stop codon, there is a 3′ untranslated sequence of variable length that shows a high content of AT motifs and a putative polyadenylation site (ATTAAA), possibly responsible for posttranslational regulation of the *Pmch/PMCH* cDNA, as well as two opposite repeat sequences that could contribute to the formation of a hairpin structure (Breton et al. 1993a).

The *Pmch/PMCH* open reading frame (ORF) is a highly conserved sequence of 495 bp that encodes a prepro-MCH (ppMCH) of 165 aa in all three species (Fig. 1). The first 63 nucleotides (Exon 1) encode a signal peptide that is cleaved by signal peptidases at Gly^{21}, generating a pro-MCH of 144 aa. The remaining 186 nucleotides of Exon 1 and the first 75 of Exon 2 combine after Intron A is spliced to form the 87-aa-long structural chain. Also in Exon 2 is the 124 nucleotides that code for NGE (19 aa), NEI (13 aa), and the first three amino acids of MCH. Intron B is inserted in the codon that originates Met^4 of MCH. After this intermission, the 47 nucleotides of the Exon 3 coding sequence complete Met^4 and encode the final 15 residues of MCH (19 aa in total) (Vaughan et al. 1989; Nahon et al. 1989; Breton et al. 1993a; Thompson and Watson 1990). After the formation of pro-MCH, the activity of prohormone convertases (PCs) will be essential for the formation of mature NGE, NEI, and MCH. Neuropeptide G-E is separated from the structural chain by cleavage of ppMCH Lys^{109} and separated from NEI through the cleavage of the dibasic ppMCH residues Lys^{129}–Arg^{130}. Neuropeptide E-I is separated from MCH through cleavage of the dibasic ppMCH residues Arg^{145}–Arg^{146}. Although not completely elucidated, the cleavage of ppMCH appears to be dependent on several PCs. While PC7, PACE4, PC1/3, PC2, PC5/6-A, and PC5/6-B seem to be able to cleave the Arg^{145}–Arg^{146} residues to generate mature MCH, PC2 appears to be exclusively responsible for the cleavage of residues Lys^{129}–Arg^{130} to generate NEI-G, which will be further processed to generate mature NEI (Viale et al. 1999). The importance of PC2 in the maturation of the MCH peptide family is further supported by ample colocalization between *Pmch* mRNA and PC2 immunoreactivity or *Pcsk2* expression in the hypothalamus. On the other hand, only 15–20% of hypothalamic *Pmch* mRNA-expressing cells are immunoreactive to PC1/3 (Viale et al. 1999). Further studies are necessary to fully clarify the action of PCs on the formation of mature MCH and NEI. While some information is available regarding MCH and NEI, virtually nothing is known about NGE synthesis, as no one has provided evidence for its synthesis in vivo or in vitro. One possibility is that, instead of NGE, pro-MCH encodes an neuropeptide PE, spanning residues Pro^{104} to Glu^{128}, as another Lys residue is found in position 103 and could act as a cleavage site, although little support for this theory is found in the literature (Nahon et al. 1989; Breton et al. 1993a).

While NGE has no putative post-translational modification and is predicted to be linear, NEI undergoes C-terminal amidation on ppMCH residue Gly^{144} to reach its mature form but stays linear (Parkes and Vale 1992). On the other hand, MCH has a cyclic structure due to a disulfide bond between residues Cys^7 and Cys^{16} (Nahon et al. 1989; Vaughan et al. 1989) (Fig. 1). This cyclic portion appears to be fundamental for MCH activity, as disturbances of the ring structure strongly decrease its activity (Macdonald et al. 2000). Furthermore, the cyclic region contains two positive residues, Arg^{11} and Arg^{14}, which confer a predominantly positive charge on the ring. While the substitution of Arg^{14} has minimal effects on the binding properties of MCH, the loss of Arg^{11} virtually abolishes its binding to the MCH receptor. While the cyclic structure is necessary for the activity of MCH, Trp^{17}, located outside the ring, appears to potentiate MCH activity. Despite the binding activity being focused on the ring structure, the whole MCH amino acid sequence is conserved in mammals. The same is observed in NEI, whose aa sequence has 100% identity among humans, mice, and rats. Neuropeptide G-E, on the other hand, has a single substitution differing between humans and mice and four substitutions differing between rats and mice, suggesting less conservation pressure on this peptide than on the other peptides in the MCH family (Breton et al. 1993a).

In addition to the "canonical" MCH, NEI, and NGE, there are several other transcription products that may arise from the *Pmch/PMCH* gene. MCH-gene-overprinted-polypeptide (MGOP) is a putative peptide that arises from alternative splicing of the *PMCH* gene. Both in rats and in mice, MGOP originates from the transcription of Exon 1, total absence of Exon 2, and a frameshifted transcription of Exon 3, which generates a protein whose N-terminal region is conserved when compared with ppMCH but whose C-terminal region is entirely different (Toumaniantz et al. 1996, 2000). Immunoreactivity to MGOP is found in MCH-immunoreactive (MCH-ir) neurons of rodents, but its function is unknown at the time (Allaeys et al. 2004). The antisense-RNA-overlapping-MCH (*AROM*) gene is found in the opposite DNA strand corresponding to the locus of

PMCH. This gene has homologues in rats, mice, and humans and spans approximately 34 kb of genomic DNA. It has a complex genomic structure, containing over 12 exons and multiple initiation sites, possibly generating several coding and noncoding transcripts. Some of the proteins that possibly originate from *AROM* display nucleotide strand-binding properties, suggesting that these products may regulate gene transcription (Borsu et al. 2000; Moldovan et al. 2012). Finally, there are two *PMCH*-linked chimeric genes in primates, termed *PMCHL1* and *PMCHL2*, which have partial homology to the authentic *PMCH* gene and are found on chromosomes 5p14 and 5q13, respectively. Preliminary data suggests that the products of these genes are long noncoding RNAs, with a trans-spliced transcript of *PMCHL1* being expressed in the CNS and putatively modulating *PMCH* mRNA expression (Breton et al. 1993b; Courseaux and Nahon 2001; Presse et al. 2014).

2.2 Melanin-Concentrating Hormone Receptor 1

MCHR1 is found in virtually all known vertebrates. It was the first receptor discovered that has a selective binding affinity for MCH, although some years passed between its description and the discovery of its natural ligand. The hunt for an MCH receptor started just 4 years after the description of the mammalian protein, with the development of a tritiated MCH analog that could be used as a radioligand to locate binding sites for MCH (Drozdz et al. 1993). However, a tenfold decrease in the activity of this analog compared with MCH prompted those authors to develop N- and C-terminal-iodinated MCH analogs that retained their full biological activity. Using these iodinated analogs, they were able to detect MCH binding in several cell lines, including G4F-7, B16F-1, G4F, human RE melanoma, COS-7, and PC12. No binding was detected for CHO cells or human fibroblasts. Competition experiments revealed that α-melanocyte-stimulating hormone (α-MSH), neuropeptide Y (NPY), pituitary adenylate cyclase-activating peptide (PACAP), and substance P (SP) were all unable to displace the iodinated MCH analogs, with the exception of rat atrial natriuretic peptide (ANP), which had two- to fourfold lower affinity than MCH (Drozdz and Eberle 1995; Drozdz et al. 1995). Further works expanded the number of MCH-binding cell lines, including SVK14, A-431, and SCC-25 (Burgaud et al. 1997) and reinforced the specificity of MCH binding, as corticotropin-releasing factor (CRF), arginine vasopressin (AVP), NEI, and MGOP were also not able to displace the radioligand.

A significant advance in the identification of the MCH receptor was made by Kolakowski et al. (1996). These authors identified a genomic DNA fragment that matched the orphan G protein-coupled receptor (GPCR) GPR24, which had significant homology to somatostatin receptors (SSTRs) and could be mapped to the human chromosome 22, locus q13.3. Although this receptor shared approximately 40% of its residues with SSTRs in the predicted transmembrane (TM) domains and had several key residues in critical positions for the binding activity of somatostatin, there was no evidence of binding to somatostatin-14 or opioid ligands. For this reason, this receptor was renamed somatostatin-like coupled receptor 1 (SLC-1). Further refinements of the knowledge regarding SLC-1 were provided by Lakaye et al. (1998), who identified the rat orthologue of SLC-1, described its gene sequence, including a previously unidentified intron, and located the consensus sites for glycosylation and phosphorylation. The ligand of SLC-1, however, remained unknown.

In 1999, a major leap in the understanding of the MCH peptidergic system occurred when five groups identified MCH as the ligand of SLC-1 (Lembo et al. 1999; Bachner et al. 1999; Chambers et al. 1999; Saito et al. 1999; Shimomura et al. 1999). These groups employed the method of "reverse pharmacology," transfecting cell lines with SLC-1, exposing those cells to either brain extracts or panels of neuroactive substances and then measuring parameters of receptor activation such as intracellular Ca^{2+} mobilization, cAMP production, or K^+ currents. In all cases, MCH was the only neuroactive substance capable of eliciting a response in the transfected cell lines, with an EC_{50} that ranged from 0.2 nM to approximately 120 nM depending on the cell line, stability of transfection, and response measured. These works also demonstrated that other neuroactive substances have no competitive activity and started to dissect the intracellular pathways of MCHR1 activation.

All those works contributed to a clear picture of MCH receptor 1. In the rat, this receptor is encoded by the *Mchr1* gene (*Gene ID: 83567*), which is located on chromosome 7q34. The coding sequences are composed of two exons of 82 bp and 977 bp, separated by an intron of 1265 bp. The human *MCHR1* gene (*Gene ID: 2847*), on the other hand, is found on chromosome 22q13.2 and has an intronic sequence of 1214 bp (Fig. 2). The intron position is the same in humans and rodents, splitting the codon of Gly^{28} into a G on Exon 1 and a GG (rat) or GA (human) in Exon 2. The resulting protein is 353 aa long in both species, forming the characteristic seven-TM domain structure of GPCRs. The aa sequence of this protein predicts three consensus sites for asparagine-linked glycosylation in the extracellular N-terminus, two potential phosphorylation sites for protein kinase A in the second and third intracellular loops, six for protein kinase C in the second and third intracellular loops and in the C-terminal portion, and one for protein kinase CK2 in the C-terminal portion. The human and rat proteins share 96% identity, indicating high conservation pressure on this receptor. The three-dimensional model of MCHR1 predicts a hydrophilic pocket between TM domains 3 and 7 and another among TM domains 4, 5, and 6, which are separated by a central hydrophobic region (Macdonald et al. 2000) (Fig. 2).

2.3 The Melanin-Concentrating Hormone Receptor 2

Two years after the discovery of MCHR1, six groups were able to identify another receptor for MCH (An et al. 2001; Hill et al. 2001; Mori et al. 2001; Rodriguez et al. 2001; Sailer et al. 2001; Wang et al. 2001). Using the consensus structure of GPCRs and the available sequence of *MCHR1*, these groups were able to search human

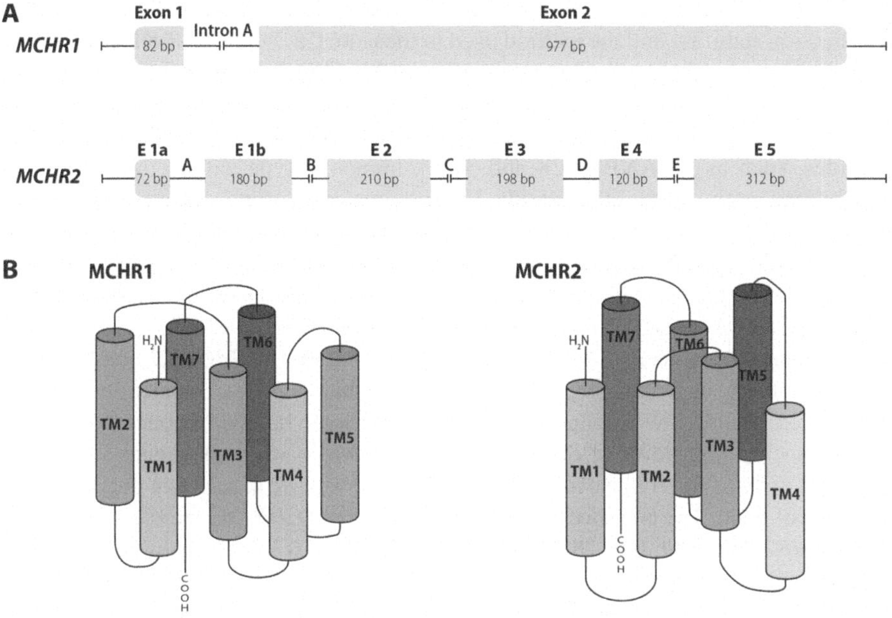

Fig. 2 Structure of the *MCHR1* and *MCHR2* genes and their corresponding protein structure. (**a**) The *MCHR1* gene is composed of two exons and one intronic sequence (represented by broken lines, not to scale). *MCHR2* displays considerably more genomic complexity and is formed by six exons and five intronic regions of variable lengths. Exons 1a and 1b correspond to splice variants of this gene, with E1b coding for the active form of the receptor. (**b**) A simplified representation of the three-dimensional structure of these receptors. Portions indicated by "TM" correspond to transmembrane domains of the receptor (the membrane has been omitted for clarity)

genomic databases for homologous sequences. They identified a sequence coding for a putative GPCR that, despite an overall low identity, was more similar to *MCHR1* than to any other gene. This gene, now identified as human *MCHR2* (Gene ID: 84539), is located on chromosome 6q16.2, spans over 23 kb, and has six exons and five intronic sequences, displaying considerably more genomic complexity than *MCHR1*. The ORF of *MCHR2* is 1023 bp long and encodes a 340 aa protein. The first two exons of *MCHR2* are splice variants, with Exon 1b coding for the "extended" version of *MCHR2* and Exon 1a coding for an *MCHR2* truncated in the N-terminal portion. Exon 1b encodes the first 60 aa of *MCHR2*, Exon 2 encodes 70 aa, Exon 3 encodes 66 aa, Exon 4 encodes 40 aa, and Exon 5 encodes the final 104 aa. Exon 1a corresponds to a 24 aa sequence, resulting in a truncated version of MCHR2 that has 304 aa (Fig. 2). In addition to the characteristic 7-TM structure, MCHR2 also displays several other conserved properties of GPCRs, such as two N-linked glycosylation sites, a DRY motif located at the end of TM3, an Asp residue in TM3, an NPXXXY motif in TM7, and a potential palmitoylation site in the C-terminal region (Fig. 2).

As is the case with MCHR1, MCHR2 is highly selective for MCH as its ligand. Transfected cells respond to both mammalian and salmonid MCH in [Ca^{2+}] assays, with an EC_{50} that ranges from 0.085 to 31 nM, depending on the cell line used, the transfection stability, and the method used to measure Ca^{2+} mobilization. These EC_{50} values are similar in magnitude to the values of MCHR1 in similar conditions, if not slightly lower. There is no response from MCHR2-transfected cells to other peptides of the MCH peptidergic system, such as NEI, NGE, and MGOP; to natriuretic peptides, such as rat ANP (1-28 and 3-28), human C-type natriuretic peptide-22, and human brain natriuretic peptide-32; to somatostatin and somatostatin-related peptides, such as somatostatin-14, somatostatin-28, cortistatin-14, and cortistatin-29; to opioid ligands, such as γ-endorphin, dynorphin A, and dynorphin B; or to another member of the melanocortinergic system, α-MSH. No effect of MCH or any other peptide was observed for the truncated version of MCHR2.

A remarkable aspect of MCHR2 is its low similarity to MCHR1, despite their obvious paralogous relationship. The two proteins share a mere 38% identity at the aa level, a remarkably low amount for proteins of the same family that bind to the same ligand. Such a high level of divergence between two paralogues suggests that they appeared early in vertebrate evolution and that there was weak selective pressure on at least one of them. The presence of orthologues of both *MCHR1* (Gene ID: 103189252) and *MCHR2* (Gene ID: 103178350) in the elephant shark (*Callorhinchus milii*) genome supports this early appearance of a family of receptors for MCH. The low selective pressure on MCHR2 is supported by the loss of this receptor in the species of the *Glires* clade. In the rabbit (*Oryctolagus cuniculus*—order Lagomorpha), a base substitution in Exon 2 resulted in the transformation of a CGA codon (Gly^{85} in the human gene) to a TGA stop codon. In the guinea pig (*Cavia porcellus*—order Rodentia), a frameshift mutation occurred by the introduction of an additional A after bp 47, resulting in an early stop codon that truncated the protein. Rats, mice, and hamsters have no indication of an *MCHR2* orthologue in their genomes, suggesting the removal of truncated sequences after the mutations that probably occurred in the last common ancestor of the *Glires* superorder (Tan et al. 2002). The lack of information regarding MCHR2 functions impairs, for now, our ability to speculate about what kind of environmental aspects could have loosened the evolutionary constraint on MCHR2, or even what gains, if any, the animals of the *Glires* clade could have obtained by losing this receptor.

3 Morphological and Neurochemical Aspects of MCH/NEI Neurons

The first complete description of MCH and NEI immunoreactivity in the mammalian brain using a specific antibody was published in 1992, using male animals of albino (Sprague Dawley) and pigmented (Long-Evans) rat strains (Bittencourt et al. 1992). A vast amount of work was invested from then on to expand our knowledge of the anatomical features of the MCH peptidergic system in other mammals. There are now available descriptions of the distribution of MCH-ir neurons in the rat, mouse, Siberian hamster, Syrian hamster, tufted capuchin monkey (a New World primate), human, pig, sheep, and cat (as reviewed in Bittencourt 2011). The description of this peptidergic system in other physiological models, such as females in different reproductive stages (Knollema et al. 1992; Rondini et al. 2007, 2010; Alvisi et al. 2016), also contributed to a greater understanding of this system, but detailed studies regarding its sexual dimorphism are still lacking. In addition to its description in different mammals, several works also contributed to refining the distribution of this system, revealing new neurochemical markers associated with MCH-containing cells, and unveiling the networks that integrate this peptidergic system into functional circuits. It is noteworthy, however, that the lack of systematization regarding anatomical nomenclature has hindered the field, with several different names used to refer to the same structures.

A considerable amount of information was gathered about MCH neurons in the last two decades. For example, it was demonstrated that 80–90% of MCH neurons also contain the mRNA coding for the glutamate decarboxylase (GAD67) enzyme, which suggests GABA synthesis in these cells (Rondini et al. 2007, 2010; Elias et al. 2008). Recent reports, however, suggest that a subset of MCH neurons may express the machinery for glutamate transport and release (Chee et al. 2015). A recent work further complicates our understanding of neurotransmitter use by MCH neurons, suggesting almost all of these neurons contain the machinery for GABA synthesis (*Gad1* and *Gad2*) but lack the vesicular GABA transporter needed for the canonical pathway of GABA release (*Slc32a1*) (Mickelsen et al. 2017). On the other hand, almost all of MCH neurons express the vesicular glutamate transporter 2 (*Slc17a6*), suggesting a glutamatergic phenotype for these neurons (Mickelsen et al. 2017). Neuronal MCH is also extensively colocalized with CART, ranging from 70% to 95% colocalization depending on the region analyzed (Elias et al. 2001). There is also colocalization of MCH with nesfatin-1 (Brailoiu et al. 2007; Fort et al. 2008; Foo et al. 2008), neuronal pentraxin 1 (Reti et al. 2002), acetylcholinesterase in the absence of acetyltransferase (Risold et al. 1989; Chou et al. 2004), monocyte chemoattractant protein-1/CCL2 (Banisadr et al. 2005), secretogranins II and III (Tanabe et al. 2007; Hotta et al. 2009), and α-dystrobrevin (Hazai et al. 2008).

MCH neurons also contain a myriad of receptors, which render them responsive to several inputs originating broadly in the CNS. These neurons receive both GABAergic and glutamatergic inputs, which are processed by the ionotropic $GABA_A$ (containing subunits α2 and α3), NMDA, and AMPA receptors and the metabotropic receptors mGlu1 and mGlu5 present in these cells (Bäckberg et al. 2004; van den Pol et al. 2004; Huang and van den Pol 2007). These neurons are also responsive to the neighboring ORX cells through the $ORXR_1$ and $ORXR_2$ receptors (van den Pol et al. 2004; Parks et al. 2014b). Melanin-concentrating hormone neurons are also responsive to endogenous opioids, probably through kappa-opioid and nociceptin/orphanin receptors (Parsons and Hirasawa 2011; Parks et al. 2014b). Other receptors present in MCH cells are the V1a receptor for vasopressin and OXTR for oxytocin (Yao et al. 2012), NK3 for neurokinin B (Griffond et al. 1997), 5-HT_{1A} for serotonin (Collin et al. 2002), KissR for kisspeptin, SSTR1 and SSTR2 for somatostatin, NTSR1 for neurotensin, NPSR for neuropeptide S, CCKAR for cholecystokinin (Parks et al. 2014b), H3R for histamine (Parks et al.

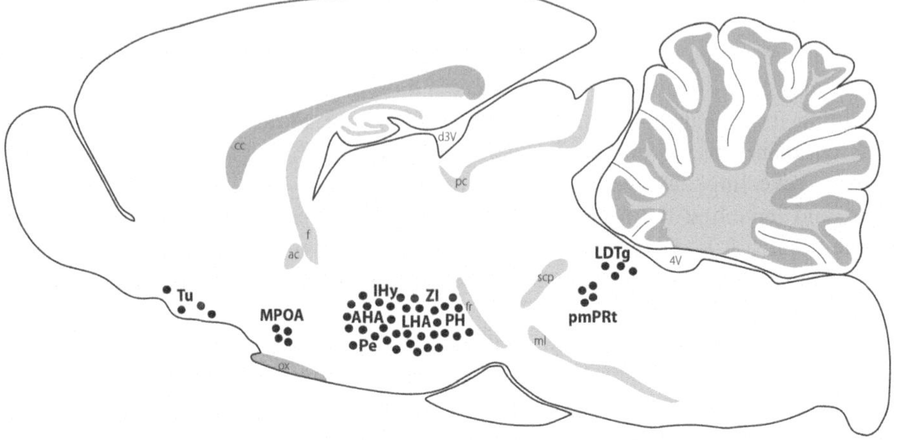

Fig. 3 The distribution of MCH-containing neurons in the rat neuroaxis. Schematic representation of a parasagittal slice of a rat brain. Significant fiber tracts are indicated in gray to provide reference points to the reader. In rostrocaudal order, the primary loci of MCH neurons are as follows: the olfactory tubercle (Tu), the medial preoptic area (MPOA) of lactating females, the incertohypothalamic area (IHy), the periventricular nucleus of the hypothalamus (Pe), the anterior hypothalamic area (AHA), the lateral hypothalamic area (LHA), the *zona incerta* (ZI), the posterior hypothalamic area (PH), the paramedian pontine reticular formation (pmPRt), and the laterodorsal tegmental nucleus (LDTg) only in females. Other abbreviations: *4V* fourth ventricle, *ac* anterior commissure, *cc corpus callosum*, *d3V* dorsal third ventricle, *f* fornix, *fr fasciculus retroflexus*, *ml* medial lemniscus, *ox* optic chiasm, *pc* posterior commissure, *scp* superior cerebellar peduncle

2014a), and the adrenergic receptor α_{2A}, which responds to both noradrenergic and dopaminergic inputs (Modirrousta et al. 2005; Alberto et al. 2011; Conductier et al. 2011). MCH neurons are also responsive to insulin through the insulin receptor (Hausen et al. 2016), as well as to glucose, probably through the GLUT3 glucose transporter (Burdakov et al. 2005; Kong et al. 2010).

In the following sections, the distribution of MCH along the rostrocaudal axis of the rat CNS will be described (Bittencourt et al. 1992), as this is the most detailed description available, and the pertinent parallels to what is known about the human MCH system will be drawn (Elias et al. 1998; Bittencourt 2011; Krolewski et al. 2010). A schematic representation of each area that will be discussed in the section is presented in Fig. 3. Since there is over 96% colocalization between MCH and NEI in the rat brain (Bittencourt et al. 1992), all descriptions in this chapter are valid for both peptides, although we will refer only to "MCH" for the sake of simplicity. The peptide NGE will not be discussed in this section, as the scarcity of data renders impossible any discussion of its immunoreactivity in vivo.

3.1 *Olfactory Tubercle*

The anterior-most group of MCH-ir cells is found in the olfactory tubercle (Tu), in the basal forebrain area, ranging from +0.45 to −0.45 mm from bregma. These small and multipolar immunoreactive cells are found in the polymorph or multiform layer of the Tu (Fig. 4a–a′). A small contingent of cells is found bordering the islands of Calleja, a neighboring structure, while a few cells were found dorsally within the *substantia innominata* and caudally within the anterior amygdaloid area. This population is most easily labeled through immunohistochemistry in colchicine-treated animals or through *in situ* hybridization, suggesting that these cells have a low rate of MCH synthesis or a high rate of degradation. While MCH-ir cells are found in both male and female rat brains, there is no evidence of this population in other species, including humans. It is uncertain, at this time, whether this Tu population is exclusive to rats or poorer immunogenicity in other species generates false negative results.

3.2 *Diencephalon*

3.2.1 Medial Preoptic Area

The medial preoptic area (MPOA) represents the most anterior group of *Pmch* mRNA-expressing and MCH-ir neurons in the diencephalon and is one of the two known areas where MCH is sexually dimorphic. While not observable in the male rat brain, in females during the estrous cycle or in pregnant females, *Pmch* mRNA can be detected in the MPOA of lactating rats as soon as the 5th postpartum day. In addition to the mRNA expression, immunoreactive neurons can be identified in high numbers in the ventromedial aspects of the MPOA (Fig. 4b–b′), with some of them invading the limits of the periventricular preoptic nucleus of the hypothalamus and the rostralmost parts of the paraventricular nucleus of the hypothalamus (PVH) (Knollema et al. 1992) (Fig. 4c–c′). This population of neurons has a very specific temporal development pattern, with increasing synthesis of MCH as lactation progresses, reaching a peak between the 15th and 19th days of lactation and disappearing after weaning (Rondini et al. 2010; Alvisi et al. 2016). Recently, it has been demonstrated that MCH synthesis in this area is directly correlated to the postpartum, but not the prepartum, litter size (Ferreira et al. 2017b).

3.2.2 Incerto-hypothalamic Area

The first major group of MCH-ir and *Pmch* mRNA-expressing cells found in the neuraxis of both male and female animals is located at the level of the caudal PVH and was initially designated the rostromedial *zona incerta* (ZI) group. The designation of

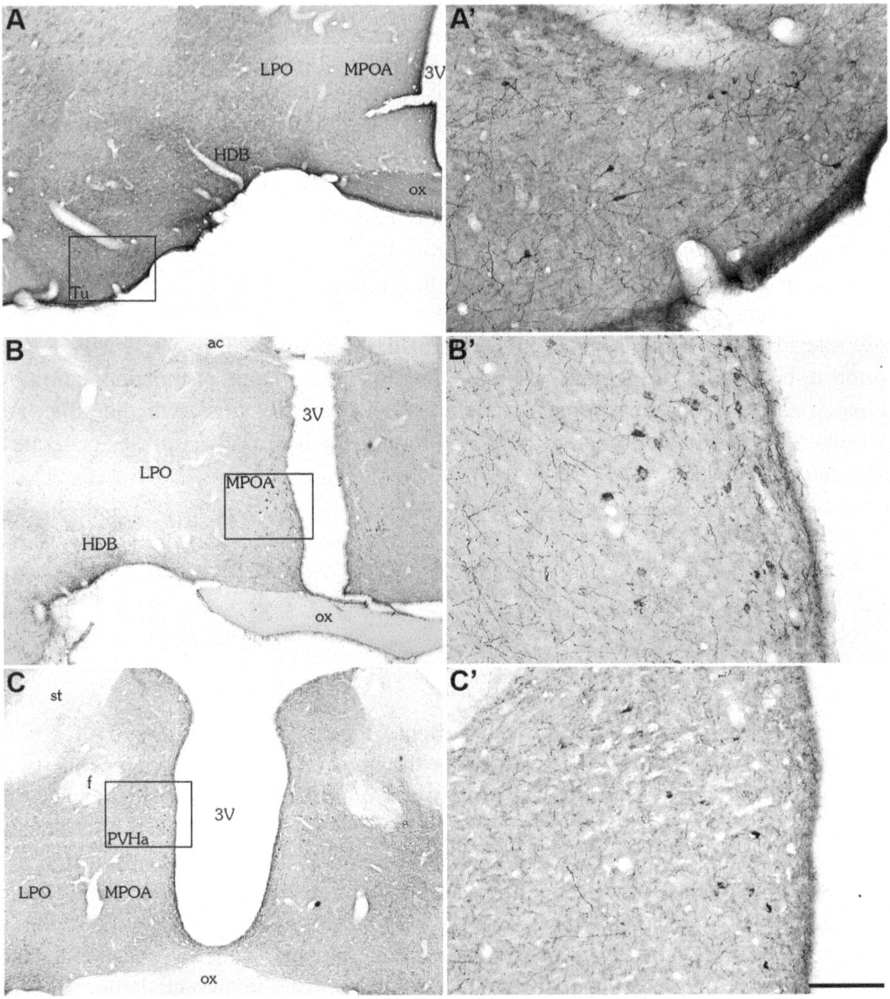

Fig. 4 Areas containing MCH-producing neurons in the rat CNS. Brightfield photomicrographs of frontal rat slices subjected to immunohistochemistry for MCH localization. (**a**) Small immunoreactive neurons are found in the olfactory tubercle (Tu) of rats of both sexes. (**a′**) Higher magnification of the rectangle in **a**. Immunoreactive neurons in the Tu are smaller than hypothalamic neurons. (**b**) Neurons that synthesize MCH are detectable in the medial preoptic area (MPOA) of females exclusively during the lactation period. (**b′**) Higher magnification of the rectangle in **b**. Neurons in the MPOA are concentrated in the ventromedial part of this nucleus and are not as densely stained as neurons in other areas. (**c**) In addition to neurons in the MPOA, the anterior part of the paraventricular nucleus of the hypothalamus (PVHa) also contains MCH-ir neurons in lactating animals. (**c′**) Higher magnification of the rectangle in **c**. Other abbreviations: *3V* third ventricle, *ac* anterior commissure, *f* fornix, *HDB* horizontal limb of the diagonal band of Broca, *LPOA* lateral preoptic area, *ox* optic chiasm, *st stria terminalis*. Scale bar: **a**–**c**, 500 μm; **a′**–**c′**, 100 μm

this area was, perhaps, the most controversial involving MCH neurons, as the same area has been referred to as the rostromedial *zona incerta*, medial *zona incerta*, and rostral *zona incerta* by different authors, owing to the fact that it is not charted in most atlases. It was not until 2003, when an extensive neurochemical and hodological study was performed, that the area was properly renamed the incerto-hypothalamic area (IHy). Although this region can be considered a subthalamic region from a strictly anatomical standpoint, its hodology, neurochemistry, and cytoarchitecture indicate that this area is better described as a hypothalamic transitional region.

Morphologically, the IHy is found at the dorsoventral level of the tip of the third ventricle (3V), and on the lateromedial axis, it is centered around an imaginary line drawn over the mammillothalamic tract (mt), laterally limited by a line drawn over the fornix (f) (Fig. 5a–a′). The IHy can be identified, in cytological staining, as an ovoid-shaped area of five or six layers of cells that display a preferred lateromedial orientation. Neurochemically, this area is characterized by the presence of MCH- and CART-ir neurons intermingled with dopaminergic neurons (immunoreactive for tyrosine hydroxylase—TH-ir). These dopaminergic neurons are part of the A13 group, which partially overlaps with the IHy and the rostral *zona incerta* proper. Although MCH/CART-ir neurons are found in great proximity to TH-ir neurons, there is no colocalization between these markers, and the relationship between these chemically distinct populations is yet to be explored in the literature. The IHy can be found in both rat and human samples of both sexes.

3.2.3 Anterior Periventricular Nucleus

At the caudal level of the PVH, another smaller population of MCH-ir neurons inhabits the periventricular nucleus of the hypothalamus. These cells are preferentially distributed along the dorsal part of this nucleus (Fig. 5b–b′), sometimes constituting a cap-like structure at the top of the third ventricle (3Vdc) (Fig. 5c–c′). This population is typically composed of elongated neurons, some oriented parallel to the 3V wall and some radially oriented with respect to the ventricle, particularly on the 3Vdc. This group of cells is stained in both male and female rat brains, but it cannot be seen in the mouse brain. There is also no evidence of this population in the human brain.

3.2.4 Medial Hypothalamus

Two populations of neurons can be identified in the medial zone of the hypothalamus. The first is a population of neurons that are caudal to the posterior part of the PVH, although never within its borders. This population predominantly occupies the caudal aspect of the anterior hypothalamic area (AHA) but extends beyond its borders to the medial hypothalamus rostral to the dorsomedial nucleus. The rostrocaudal distribution of this population partially overlaps those of the IHy and the anterior periventricular populations, but these neurons can be distinguished easily from the periventricular

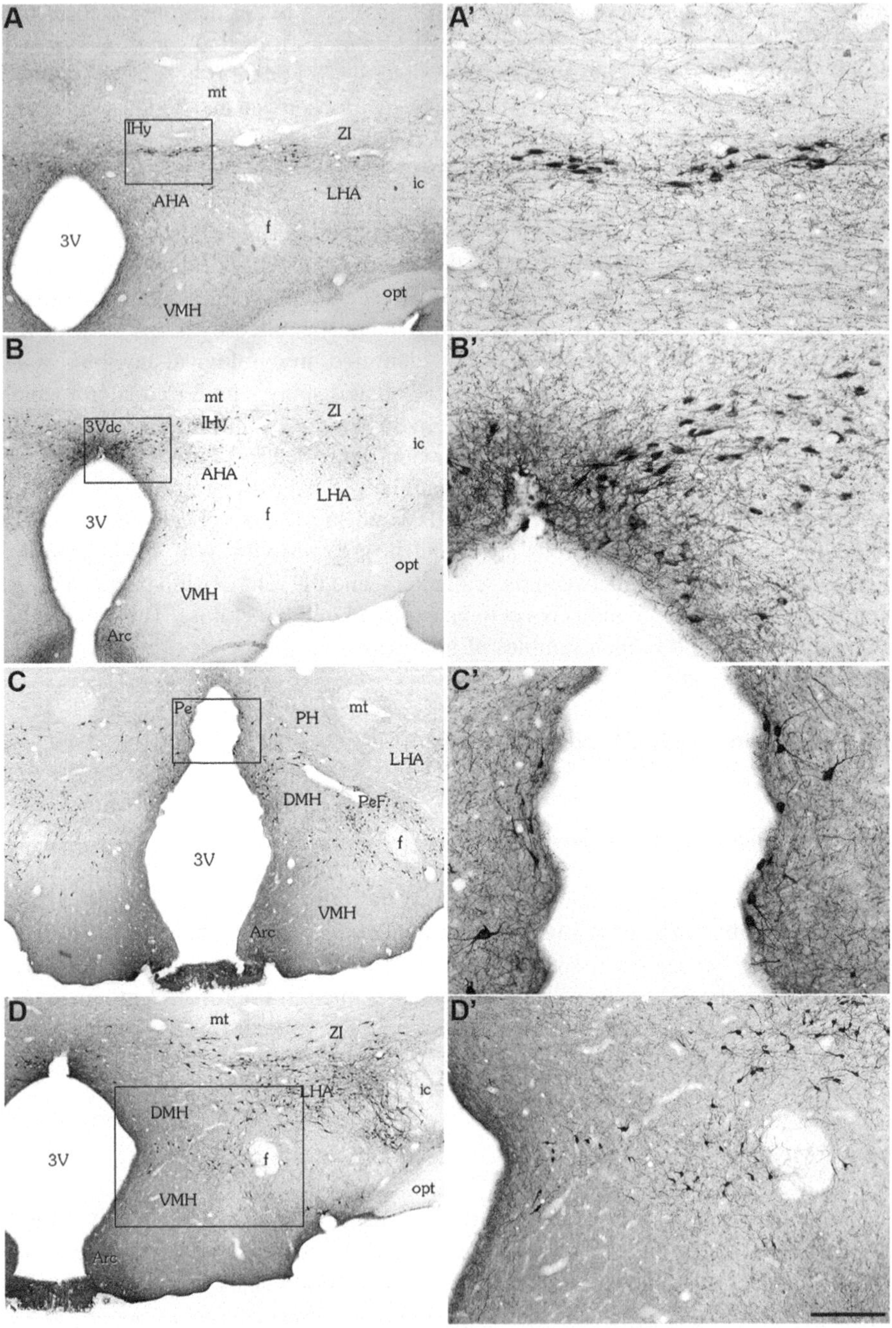

Fig. 5 Areas containing MCH-producing neurons in the rat CNS, continued. Brightfield photomicrographs of frontal rat slices subjected to immunohistochemistry for MCH localization. (**a**) In the diencephalon of rats of both sexes, MCH-ir neurons are found in the incertohypothalamic area

group as there is a gap between these two populations. The separation between the IHy and the medial hypothalamic group is harder to visualize, as these populations appear continuous in the dorsoventral axis. Melanin-concentrating neurons of the AHA, however, can be separated from the IHy by their distinct orientation: while IHy neurons are mostly bipolar and elongated in the lateromedial axis, AHA neurons are multipolar and without clear directionality. Although some neurons are found in the neuropil surrounding the ventral aspect of the PVH or the dorsal margin of the ventromedial nucleus, these neurons are never found within the limit of these nuclei.

A second population of MCH-ir neurons is found more caudally in the medial hypothalamus, at the level of the dorsomedial and ventromedial nuclei. These neurons are found outside the margins of these two nuclei, occupying the area between them (called the internuclear area) and forming a thin layer of cells stretching from the fornix to the periventricular nucleus (Fig. 5d–d′). Both medial hypothalamus populations are found in male and female rats, as well as in human brains.

3.2.5 Lateral Hypothalamic Area

The LHA is the main site in the hypothalamus that harbors MCH-containing cells. Following the classical division of the LHA into three rostrocaudal subdivisions, i.e., anterior, tuberal, and posterior (Saper et al. 1979), MCH-ir cells are mostly found in the tuberal and posterior divisions of this structure. Owing to the size and complexity of this area, three subgroups of MCH-ir neurons can be identified. The first group consists of immunoreactive neurons that occupy the space between the internal capsule (ic) and the fornix (f) in the lateromedial axis and the space between the ventral margin of the ZI and the level of the fornix (Fig. 6a–a′). Neurons in this group are strongly labeled, medium to large and multipolar, giving rise to two to five

Fig. 5 (continued) (IHy). The IHy can be identified as a strip of MCH neurons between the third ventricle (3V) and the mammillothalamic tract (mt), at the dorsoventral level of the *zona incerta* (ZI). (**a′**) Higher magnification of the rectangle in **a**. Neurons in the IHy are bipolar and frequently oriented in the lateromedial axis. (**b**) A distinctive set of neurons can be identified in the area surrounding the dorsal tip of the 3V, forming a cap of neurons (3Vdc). (**b′**) Higher magnification of the rectangle in **b**. Neurons in the 3Vdc are often radially oriented with respect to the 3V. (**c**) Immunoreactive neurons also occupy the periventricular nucleus of the hypothalamus (Pe) along its tuberal extension. (**c′**) Higher magnification of the rectangle in **c**. Labeled cells in the Pe are found parallel or radially oriented with respect to the 3V and extend projections to the subventricular zone. (**d**) Several MCH-ir neurons are found in the medial zone of the hypothalamus. At the level of the dorsal (DMH) and ventromedial (VMH) hypothalamic nuclei, these neurons are found along the margin of those nuclei, in the internuclear area. (**d′**) Higher magnification of the rectangle in **d**. The internuclear group of MCH neurons extends itself from the fornix (f) to the 3V. Other abbreviations: *AHA* anterior hypothalamic area, *Arc* arcuate nucleus, *ic* internal capsule, *LHA* lateral hypothalamic area, *opt* optic tract, *PeF* perifornical nucleus. Scale bar: **a–d**, 500 μm; **d′**, 200 μm; and **a′–c′**, 100 μm

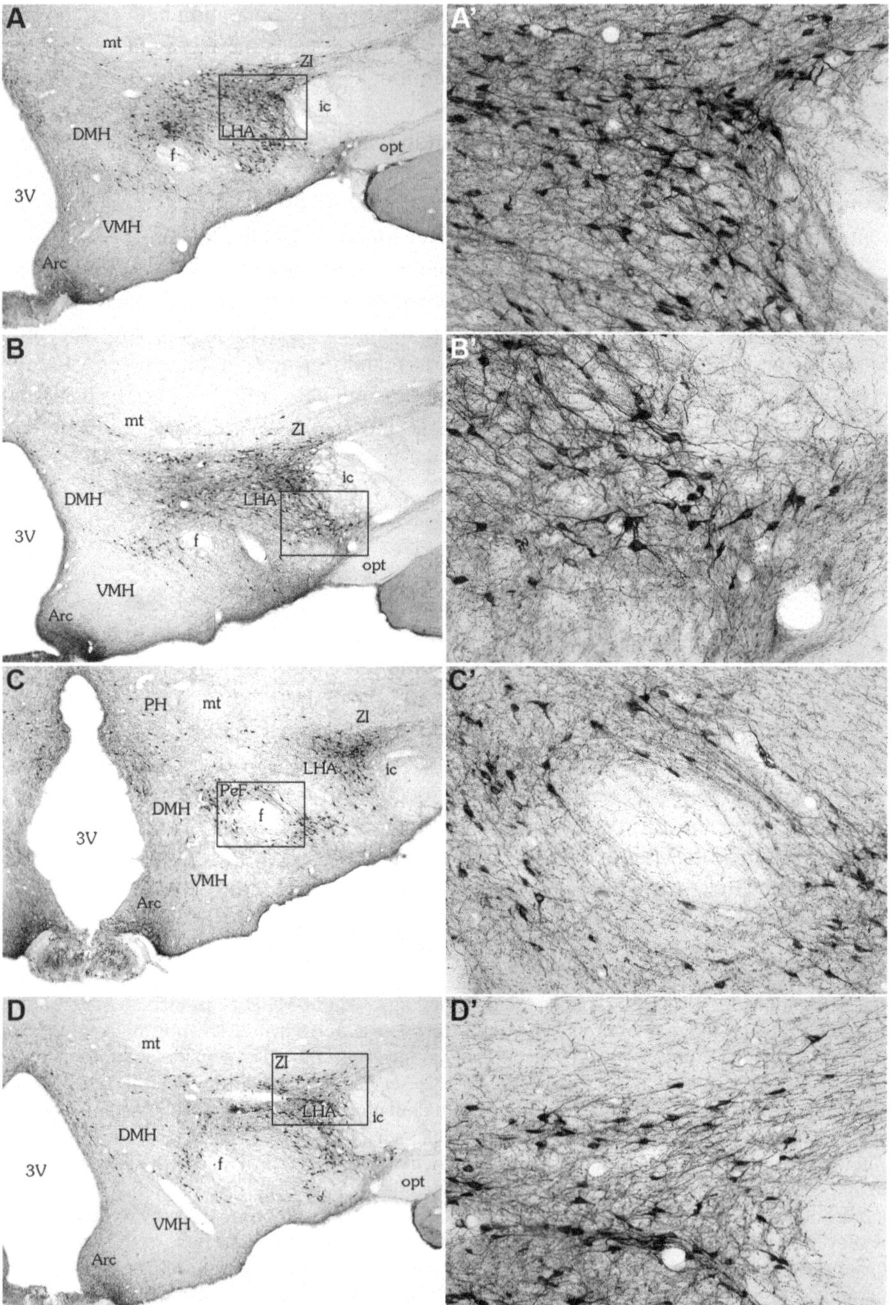

Fig. 6 Areas containing MCH-producing neurons in the rat CNS, continued. Brightfield photomicrographs of frontal rat slices subjected to immunohistochemistry for MCH localization. The most

primary dendrites. Secondary dendritic branching is frequently observed, although tertiary branching is less common. These neurons form the bulk of the MCH-ir neurons in the anterior tuberal hypothalamus, and in certain cases, they appear to be continuous with the stained neurons of the medial hypothalamus. The second group is formed by very large immunoreactive neurons, located in the region dorsal to the optic tract (opt) and ventromedial to the tip of the internal capsule (Fig. 6b–b′). Owing to the unique size of these neurons, this area is called the magnocellular group of the LHA, and it is more evident at the levels at which both ventromedial and dorsomedial nuclei are also present. Third, as the LHA progresses in the rostrocaudal axis, a group of MCH-ir neurons is found surrounding the fornix, in the perifornical nucleus (Fig. 6c–c′). This population is intermingled at more rostral levels with the group that spans the space between the fornix and the internal capsule, which makes it somewhat hard to isolate. At more caudal levels, it becomes better delineated as the number of MCH-ir neurons medial to the internal capsule decreased owing to the increased presence of white matter patches of the *mfb*. It is noteworthy that some authors believe there to be a fourth population, located dorsally to the peduncular part of the lateral hypothalamus and lateral to the subincertal nucleus, called the DL-PLH. In all likelihood, this population is composed of LHA neurons that are pushed to the dorsal part of the internal capsule as the *mfb* expands into the LHA.

In humans, the distribution of MCH in the LHA is similar to that of the rat, but not identical, owing to the smaller lateromedial extent of the hypothalamus in primates as a result of the expansion of the internal capsule. Therefore, a large contingent of immunoreactive neurons is found medial to the fornix, in the space between the fornix and the third ventricle, although some immunoreactive neurons are still found intermingled with fibers of the internal capsule. There are no known sex-related differences in the LHA MCH population.

Fig. 6 (continued) densely populated MCH-ir area in the rat CNS is the lateral hypothalamic area (LHA). Due to its complexity, the MCH neurons in this area can be divided into subgroups. (**a**) The main subgroup of MCH-ir neurons is found in the dorsomedial part of the LHA, ventral to the *zona incerta* (ZI), and medial to the internal capsule (ic), extending approximately to the level of the fornix (f). (**a′**) Higher magnification of the rectangle in **a**. Neurons in the LHA are large, densely stained, multipolar, and without clear orientation. Projections from these neurons can be seen intermingled with the white matter patches of the medial forebrain bundle (*mfb*). (**b**) In the area immediately ventromedial to the ic and dorsal to the optic tract (opt), several very large immunostained neurons can be found, corresponding to the magnocellular group of the LHA. (**b′**) Higher magnification of the rectangle in **b**. (**c**) Neurons immunoreactive to MCH are also found encircling the fornix, in the perifornical nucleus (PeF). This group is more easily identified at caudal levels of the tuberal hypothalamus. (**c′**) Higher magnification of the rectangle in **c**, showing the concentric distribution of MCH-ir neurons around the fornix. (**d**) In close proximity to the LHA, a small number of MCH-ir neurons can be labeled in the ZI. (**d′**) Higher magnification of the rectangle in **d**. Despite the spatial proximity, an apparently "empty" band of neuropil separates the ZI and the LHA neurons. Other abbreviations: *3V* third ventricle, *Arc* arcuate nucleus, *DMH* dorsomedial hypothalamic nucleus, *mt* mammillothalamic tract, *PH* posterior hypothalamic area, *VMH* ventromedial hypothalamic nucleus. Scale bar: **a–d**, 500 μm; **a′–d′**, 100 μm

3.2.6 Zona Incerta

Although MCH-ir neurons do not occupy the rostral ZI, a population of these neurons can indeed be found in the lateral ZI. These neurons occupy the area immediately dorsal to the LHA, forming a three- to four-cell-thick layer in the stratum over the dorsal part of the internal capsule, mostly at the rostrocaudal level of the dorsomedial and ventromedial nuclei (Fig. 6d–d′). Although there is conspicuous proximity between the ZI and the LHA groups, an apparently "empty" band of neuropil separates these two populations. Furthermore, neurons in the ZI are large and multipolar, but their major axis is oriented horizontally, following the overall shape of the area, while LHA neurons have no distinct polarity. Lateromedially, incertal neurons do not go more than halfway between the internal capsule and the fornix. Neurons that synthesize MCH are present in the ZI of rats and humans, regardless of sex.

3.2.7 Tuberomammillary Nucleus, Dorsomedial Part

The dorsomedial part of the tuberomammillary nucleus (TMdm) was one of the unexpected MCH-containing cell groups described during the discovery of MCH in the mammalian brain. This nucleus is localized laterodorsally on a diagonal axis between the brain surface at the level of the ventral premammillary nucleus toward the tip of the third ventricle, where it begins to divide and disappear posteriorly, and its ventral part becomes the mammillary recess (Fig. 7a–a′). This group of MCH-containing cells is intermingled with histaminergic cells, although there is no colocalization between these two neuroactive substances (Casatti et al. 2002). It must be noted that this group of cells is not the tuberal magnocellular nucleus of Bleier (Bleier et al. 1979), as the proper tuberal magnocellular nucleus is localized ventrally, forming a sheet of cells dorsal to the pial surface of the brain at the level of the supramammillary nucleus. This group of cells can be found in both male and female rat brains. No specific mention of this nucleus has been made in studies of the primate brain.

3.2.8 Posterior Hypothalamus

A few scattered immunoreactive neurons are found in the posterior hypothalamic area (PH). These neurons represent the caudal limit of the MCH distribution and have a similar rostrocaudal level to the tuberomammillary nucleus immunoreactive cells (Fig. 7a). In the human brain, likewise, MCH-ir neurons were found extending back into the posterior hypothalamic area. There are no known sex-related differences in this population.

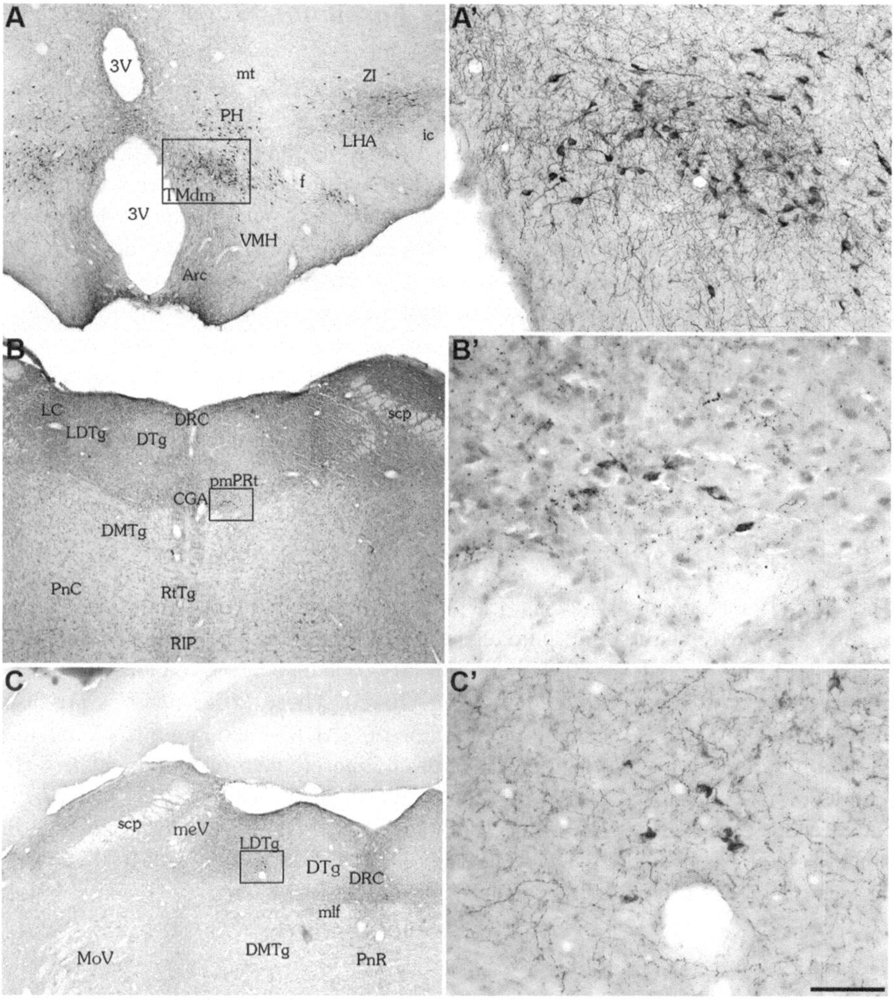

Fig. 7 Areas containing MCH-producing neurons in the rat CNS, continued. Brightfield photomicrographs of frontal rat brain slices subjected to immunohistochemistry for MCH localization. (**a**) At the caudal limit of the hypothalamus, at the transition between the tuberal, and at mammillary subdivisions, the last MCH-ir neurons in the hypothalamus are found. Neurons can be identified in the posterior hypothalamic area (PH) and lateral hypothalamic area (LHA) and in great numbers in the tuberomammillary dorsal nucleus (TMdm). (**a′**) Higher magnification of the rectangle in **a**, highlighting the labeled neurons in the TMdm. (**b**) In the male brainstem, MCH-ir neurons are restricted to the paramedian pontine reticular formation (pmPRt), located adjacent to the α part of the central gray (CGA), and ventromedial to the dorsal tegmental nucleus (DTg). (**b′**) Higher magnification of the rectangle in **b**, highlighting labeled neurons in the pmPRt. (**c**) Labeled neurons are found in the laterodorsal tegmental nucleus (LDTg) exclusively in female rats, regardless of reproductive stage. (**c′**) Higher magnification of the rectangle in **c**. Other abbreviations: *3V* third ventricle, *Arc* arcuate nucleus, *DMTg* dorsomedial tegmental area, *DRC* caudal part of the dorsal raphe nucleus, *f* fornix, *ic* internal capsule, *LC locus coeruleus*, *LHA* lateral hypothalamic area, *meV* mesencephalic V tract, *mlf* medial longitudinal *fasciculus*, *MoV* motor nucleus of the V nerve, *mt*

3.3 Paramedian Pontine Reticular Formation

During the original anatomical characterization of MCH in the male rat, an unexpected group of MCH-containing cells was found in the reticular formation, in an area designated the paramedian pontine reticular formation (pmPRt). These neurons can be seen intermingled with the fibers that constitute the superior cerebellar peduncle at the level of the caudal part of the laterodorsal tegmental nucleus, with some of these neurons located dorsally between the borders of the dorsal raphe nucleus and the pontine central gray matter (Fig. 7b–b′). So far, there has been no further neurochemical characterization of these neurons, with no description regarding other neuropeptides or classical neurotransmitters they may synthesize. This group of cells can be found in both male and female rat brains, but it has never been reported in humans.

3.4 Laterodorsal Tegmental Nucleus

The laterodorsal tegmental nucleus (LDT) was the second nucleus in the rat CNS to be discovered as sexually dimorphic regarding *Pmch* mRNA expression and peptide synthesis (Rondini et al. 2007). Labeling in this nucleus is observed only in female rats, regardless of reproductive status (Fig. 7c–c′). These MCH neurons are also GABAergic, as revealed by their co-expression of GAD67-encoding mRNA, but not cholinergic, despite a significant presence of cholinergic neurons in this nucleus. In addition to the LDT, some neurons can be found in the adjoining *subcoeruleus/* sublaterodorsal nucleus (SDL). Since both LDT and SDL have been implicated in rapid eye movement (REM) sleep (Garcia et al. 2017), this group of MCH neurons may be involved in the particular features of female REM sleep.

4 The Projections of MCH Neurons

The MCH- and NEI-ir neurons project widely to the CNS, to the point that "every commonly recognized cell group and cortical field in the rat brain was found to contain at least some MCH- and NEI-ir fibers" (Bittencourt et al. 1992). As a complete description of every projection would be too exhaustive, we will simply highlight the general organization of these fibers, and a more detailed description of

---

Fig. 7 (continued) mammillothalamic area, *PH* anterior hypothalamic area, *PnC* caudal part of the pontine reticular nucleus, *PnR* pontine raphe nucleus, *RtTg* reticulotegmental nucleus of the pons, *RIP* raphe interpositus nucleus, *scp* superior cerebellar peduncle, *VMH* ventromedial hypothalamic nucleus. Scale bar: **a–c**, 500 μm; **a′**, 100 μm; and **b′–c′**, 50 μm

projections to sleep-related nuclei will be provided in the next chapter. As was the case with the cellular distribution, there is enough similarity between the patterns of MCH and NEI staining that these two can be used interchangeably. The projections of MCH-ir neurons can be organized into four major pathways: ascending, dorsomedial periventricular, pituitary, and descending, in addition to lateral projections local to the hypothalamus. Figure 8 is a schematic representation of all four pathways. Each pathway will be described in this section.

4.1 Ascending Pathway

The major fiber path taken by the MCH-ir fibers is the *mfb*, reinforcing the intimate connection of this complex fiber bundle to the LHA and its neurons. In frontal slices of the LHA, immunoreactive fibers can be seen occupying a large area of the *mfb*, with a special concentration in the ventromedial aspect of compartment *c* of this bundle (Nieuwenhuys et al. 1982). As the *mfb* ascends to the rostral portion of the brain, the first branching of MCH-ir fibers occurs, with these projections exiting laterally through the *ansa peduncularis* to reach the amygdaloid complex, the basal nuclei, and the *substantia innominata*. Some of these fibers may also cross the subcortical structures to reach the external capsule and innervate the lateral portion of the cortical mantle. At the level of the preoptic area, fibers coursing through the *mfb* enter the *stria medullaris*, destined for the habenular nuclei, and others enter the *stria terminalis* to complement innervation of the amygdaloid complex. At the level of the preoptic recess, a major split occurs in the *mfb*-traversing MCH-ir fibers: the majority, occupying the most medial aspect of the nucleus of the diagonal band, bend dorsomedially and spread to reach several cortical areas, the septal region, and the hippocampal formation, the latter in close association with the fornix. A smaller contingent of fibers, on the other hand, exits the *mfb* and travels through the anterior commissure to reach the anterior olfactory area and the olfactory bulbs.

There is ample innervation of the cortical mantle by MCH-ir fibers (Bittencourt et al. 1992; Saper et al. 1986; Cvetkovic et al. 2003; Elias et al. 2008). Among the immunoreactivity-rich fields are the infralimbic, prelimbic, anterior cingulate, and agranular insular cortices, with fewer fibers found in the somatosensory, auditory, and visual cortices. The claustrum is one of the most densely innervated cortical regions. The motor cortex also receives significant input through the MCH-ir fibers (Fig. 9a–a′), most of which originate from neurons in the LHA (Elias et al. 2008). There is a laminar distribution of MCH-ir fibers, with layers I and IV containing more fibers, while a moderate density is observed in layers II and Vb. Layers III, IV, and Va are the most sparsely innervated. A substantial number of immunoreactive fibers are found in olfactory-linked areas, such as the main olfactory bulbs, the anterior olfactory nucleus, the olfactory tubercle, the piriform cortex, and the *taenia tecta*.

Many subcortical areas are also innervated by MCH-ir fibers. The septal area contains some of the highest densities of immunoreactive fibers, including both en

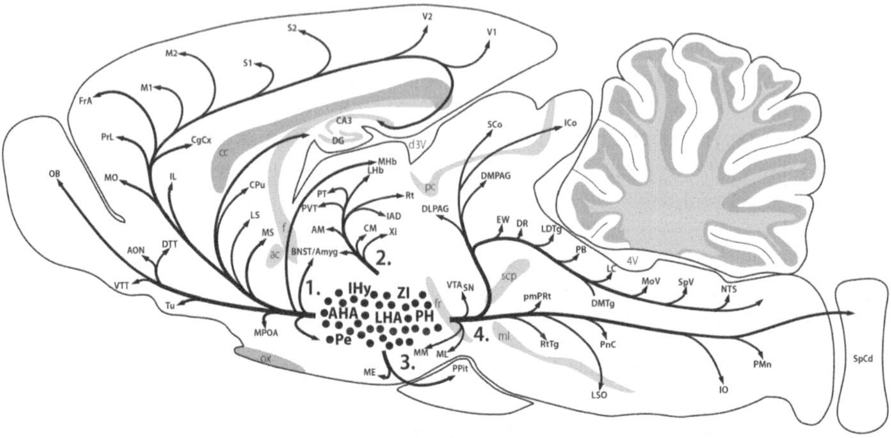

Fig. 8 The main pathways of MCH-ir fibers in the rat neuroaxis. Schematic representation of a parasagittal slice of a rat brain. Major fiber tracts are indicated in gray to provide reference points to the reader. There are four main projection pathways of MCH-ir fibers in the rat brain: (1) The ascending pathway; (2) the periventricular dorsomedial pathway; (3) the pituitary pathway; and (4) the descending pathway. Abbreviations: *4V* fourth ventricle, *ac* anterior commissure, *AM* anteromedial thalamic nucleus, *Amyg* amygdaloid complex, *AON* anterior olfactory nucleus, *BNST* bed nucleus of the *stria terminalis*, *CA3 cornu ammonis* 3 field of the hippocampus proper, *cc corpus callosum*, *CgCx* cingulate cortex, *CM* centromedial thalamic nucleus, *CPu* caudate nucleus and putamen, *d3V* dorsal third ventricle, *DG* dentate gyrus, *DLPAG* dorsolateral periaqueductal gray matter, *DMPAG* dorsomedial periaqueductal gray matter, *DMTg* dorsomedial tegmental area, *DR* dorsal raphe nucleus, *DTT* dorsal *taenia tecta*, *EW* Edinger-Westphal nucleus, *f* fornix, *fr fasciculus retroflexus*, *FrA* frontal association cortex, *IAD* interanterodorsal thalamic nucleus, *ICo* inferior colliculus, *IL* infralimbic, *IO* inferior olive, *LC locus coeruleus*, *LDTg* laterodorsal tegmental nucleus, *LHb* lateral habenular nucleus, *LS* lateral septal nucleus, *LSO* lateral superior olive, *M1* primary motor cortex, *M2* secondary motor cortex, *ME* median eminence, *MHb* medial habenular nucleus, *ML* lateral mammillary nucleus, *ml* medial lemniscus, *MM* medial mammillary nucleus, *MO* medial orbital cortex, *MoV* motor nucleus of the V nerve, *MPOA* medial preoptic area, *MS* medial septal nucleus, *NTS* nucleus of the solitary tract, *OB* olfactory bulb, *ox* optic chiasm, *PB* parabrachial nucleus, *pc* posterior commissure, *PH* posterior hypothalamic area, *PMn* paramedian reticular nucleus, *pmPRt* paramedian pontine reticular formation, *PPit* posterior pituitary, *PrL* prelimbic cortex, *PT* paratenial thalamic nucleus, *PVT* paraventricular thalamic nucleus, *Rt* reticular nucleus, *RtTg* reticulotegmental nucleus of the pons, *S1* primary somatosensory cortex, *S2* secondary somatosensory cortex, *SCo* superior colliculus, *scp* superior cerebellar peduncle, *SN substantia nigra*, *SpCd* spinal cord, *SpV* spinal trigeminal nucleus, *Tu* olfactory tubercle, *V1* primary visual cortex, *V2* secondary visual cortex, *VTA* ventral tegmental area, *VTT* ventral *taenia tecta*, *Xi* xiphoid thalamic nucleus

passant fibers and terminal boutons. Fibers are found at very high density in the medial nucleus and the diagonal band of Broca and at moderate to high density in the lateral nuclei, the bed nucleus of the *stria terminalis*, and the septofimbrial and triangular nuclei (Fig. 9b–c′). The hippocampal formation proper also receives a high density of fibers, including all *cornu ammonis* (CA) subfields, although there is greater fiber density in the dorsal hippocampal formation than in the ventral part, as

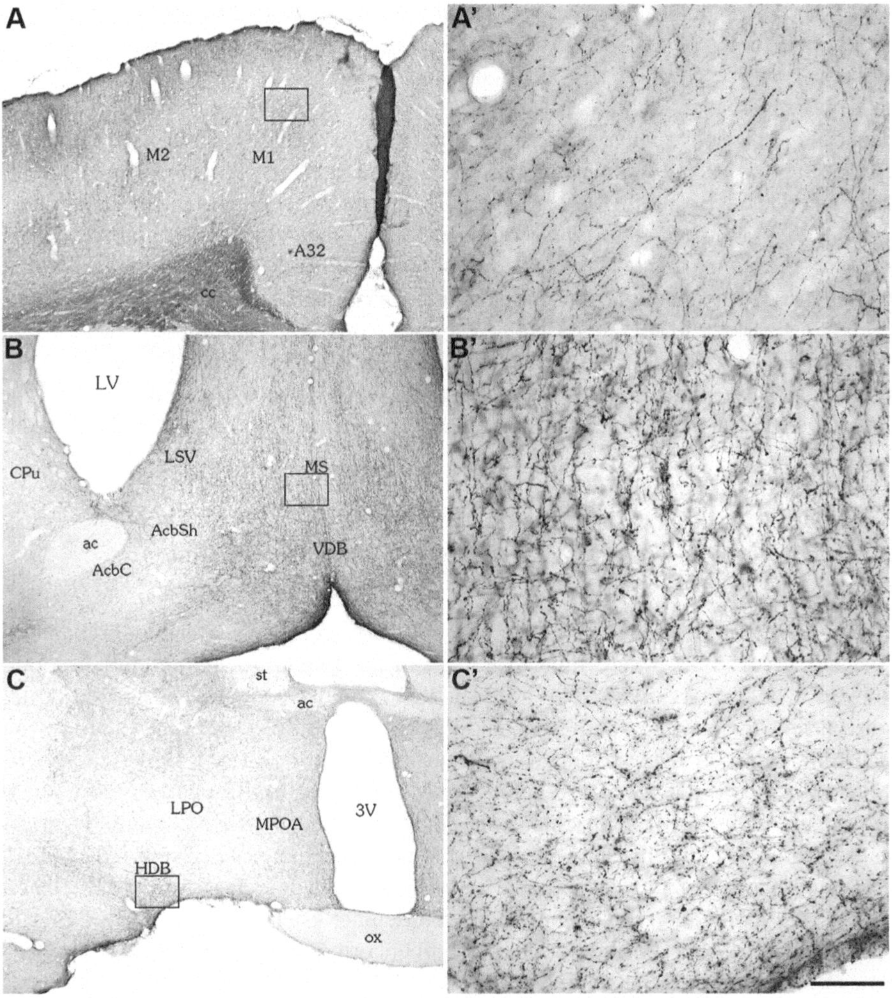

Fig. 9 The distribution of MCH-ir fibers in the rat neuroaxis. Brightfield photomicrographs of frontal rat slices subjected to immunohistochemistry for MCH localization. (**a**) Several cortical fields are innervated by MCH neurons, including the primary motor cortex (M1). (**a′**) Higher magnification of the rectangle in **a**, highlighting a long and varicose MCH-ir fiber. (**b**) The medial septal nucleus (MS) is one of the densest areas of MCH innervation in the rat brain. (**b′**) Higher magnification of **b**. Numerous en passant fibers and terminal boutons can be found in the MS. (**c**) The horizontal limb of the diagonal band of Broca also contains a medium degree of innervation. (**c′**) Higher magnification of the rectangle in **c**. Numerous terminal boutons are found in the HDB. Other abbreviations: *3V* third ventricle, *A32* area 32 of the cingulate cortex, *ac* anterior commissure, *AcbC* core subdivision of the *nucleus accumbens*, *AcbSh* shell subdivision of the *nucleus accumbens*, *cc corpus callosum*, *CPu* caudate nucleus and putamen, *LPO* lateral preoptic area, *LSV* ventral part of the lateral septal nucleus, *M2* secondary motor cortex, *MPOA* medial preoptic area, *ox* optic chiasm, *st stria terminalis*, *VDB* vertical limb of the diagonal band of Broca. Scale bar: **a–c**, 500 μm; **a′–c′**, 50 μm

well as a preferential localization of fibers in CA3 (Lima et al. 2013) (Fig. 10a). These projections to CA3 may be linked to memory retrieval in food-seeking tasks, as suggested by Sita et al. (2016). The dentate gyrus, on the other hand, receives low to moderate innervation, although in some cases immunoreactive fibers can be observed delineating the subgranular zone. The amygdaloid complex, including the medial, cortical, central, lateral, basolateral, basomedial, and intercalated nuclei, receives moderate innervation. Regarding the basal nuclei, there is a moderate to high density of fibers in the *globus pallidus* and the *fundus* of the *striatum*, a moderate to low number of fibers in the caudate-putamen nucleus and a low contingent of fibers in the *nucleus accumbens*. In the latter, fibers are found concentrated in the dorsomedial portion, close to the tip of the third ventricle, in the area identified as the septal pole (Haemmerle et al. 2015) (Fig. 10b–b′). Kampe et al. (2009) demonstrated the existence of MCH neurons in the LHA that project simultaneously to the *nucleus accumbens* and the cingulate and insular cortices, suggesting that MCH may participate in a corticomesolimbic integration circuit.

4.2 Dorsomedial Periventricular and Pituitary Pathways

Immunoreactive fibers are found at moderate density in the hypothalamic periventricular zone, where they extend from the tuberal to the anterior preoptic nucleus. These fibers give rise to a plexus that occupies a significant portion of the subventricular zone, sometimes extending to the ventricle surface. Although the matter remains open to speculation, these fibers may be linked to the control of ciliary beating, as described by Conductier et al. (2013a, b). A cohort of these fibers ascend through the periventricular zone and are joined by fibers apparently originating from the ZI to innervate the midline thalamic nuclei. As periventricular fibers are also found in the subventricular zone of the lateral ventricles, it is possible that a plexus of MCH-ir fibers lines the whole ventricular system of the brain.

A distinctive cohort of immunoreactive fibers originating in the tuberal hypothalamus course through the median eminence (ME) to reach the posterior pituitary. Fibers are reliably found in the adjoining arcuate nucleus (Fig. 11a–a′), especially on the lateral part of this nucleus, and then reach the ME, with a moderate number of fibers present in the internal lamina, while sparser fibers are found in the external lamina (Fig. 11b–b′). It is noteworthy that, in both cases, the fibers are highly varicose, suggesting a putative release of MCH directly into the ME. As other works have demonstrated variations in MCH immunoreactivity in the ME linked to the reproductive status of females (Chiocchio et al. 2001; Gallardo et al. 2004), it is possible that MCH contributes to the modulation of medianosomes in this area. As varicose fibers are found in proximity to blood vessels in the external lamina, a release of MCH or NEI in the portal vasculature cannot be ruled out. A moderate number of fibers are found in the posterior pituitary, preferentially concentrated near the periphery of this gland, and some fibers can be observed in the intermediate lobe. As demonstrated by Cvetkovic et al. (2003), neurons in the 3Vdc and the perifornical

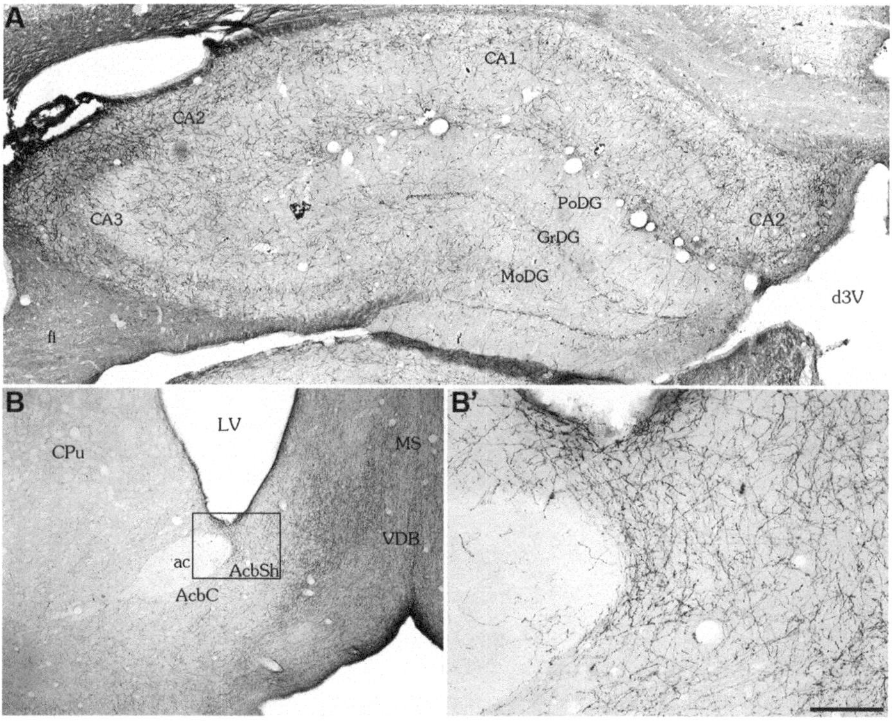

Fig. 10 The distribution of MCH-ir fibers in the rat neuroaxis, continued. Brightfield photomicrographs of frontal rat slices subjected to immunohistochemistry for MCH localization. (**a**) The hippocampal formation receives a high density of MCH-ir fibers along most of its major subdivisions. Although the dentate gyrus (DG) is not innervated as densely as the hippocampus proper, MCH-ir fibers are seen lining the subgranular zone. (**b**) The shell of the *nucleus accumbens* (AcbSh) contains immunoreactive fibers along most of its rostrocaudal extension. (**b′**) Higher magnification of the rectangle in **b**. Most MCH-ir fibers in the AcbSh are found in the area between the anterior commissure (ac) and the ventral tip of the lateral ventricle (LV). Other abbreviations: *AcbC* core subdivision of the *nucleus accumbens*, *CA1 cornu ammonis* 1, *CA2 cornu ammonis* 2, *CA3 cornu ammonis* 3, *CPu* caudate nucleus and putamen, *GrDG* granular layer of the dentate gyrus, *MoDG* molecular layer of the dentate gyrus, *MS* medial septal nucleus, *PoDG* polymorphic layer of the dentate gyrus, *VDB* vertical limb of the diagonal band of Broca. Scale bar: **b**, 500 μm; **a**, 200 μm; and **b′**, 100 μm

nucleus are the main origins of the posterior pituitary projections, and some of these neurons project simultaneously to the cortex.

4.3 Descending Pathway

As is the case with the ascending pathway, the *mfb* constitutes the main descent pathway of fibers to the caudal portion of the CNS. Secondary descending tracts

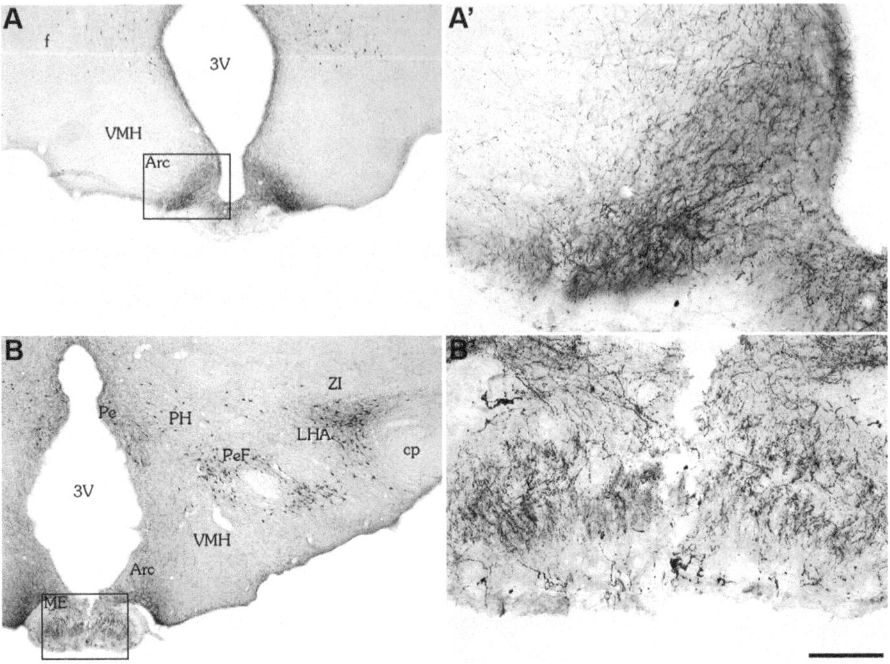

Fig. 11 The distribution of MCH-ir fibers in the rat neuroaxis, continued. Brightfield photomicrographs of frontal rat slices subjected to immunohistochemistry for MCH localization. (**a**) A great number of MCH-ir fibers are found in the lateral part of the arcuate nucleus (Arc). (**a′**) Higher magnification of the rectangle in **a**. (**b**) Labeled fibers can be clearly seen in the median eminence (EM) of the rat. (**b′**) Higher magnification of the rectangle in **b**. There is a preferential distribution of MCH-ir fibers in the internal layer of the ME, although sparser fibers are found in the external layer. Other abbreviations: *3V* third ventricle, *cp* cerebral peduncle, *f* fornix, *LHA* lateral hypothalamic area, *Pe* periventricular nucleus of the hypothalamus, *PeF* perifornical nucleus, *PH* posterior hypothalamic area, *VMH* ventromedial hypothalamic nucleus; *ZI zona incerta*. Scale bar: **a–b**, 500 μm; **a′–b′**, 100 μm

include the fornix, mammillothalamic tract, and *fasciculus retroflexus*, which help supply the mammillary and interpeduncular nuclear complexes (Fig. 12a–c′). At the level of the diencephalic–mesencephalic transition, fibers in the descending pathway are identified as an array of fibers that occupy the space between the *substantia nigra* and the medial lemniscus. This pathway then forks at the level of the red nucleus, with a dorsomedial component innervating the central tegmental fields, the mesencephalic and pontine central gray and other nuclei of the reticular core (Fig. 13a–c′). The mesencephalic periaqueductal gray matter has a topological relationship with diencephalic MCH neurons, with dorsomedial and dorsolateral regions receiving fibers originating in the IHy, while the ventrolateral area receives fibers from the tuberal LHA (Elias and Bittencourt 1997). The ventral component will contribute to the innervation of the ventral tegmental area, the interpeduncular nuclear complex, the raphe nuclei, and the pontine gray. Along its course, nuclei implicated in food

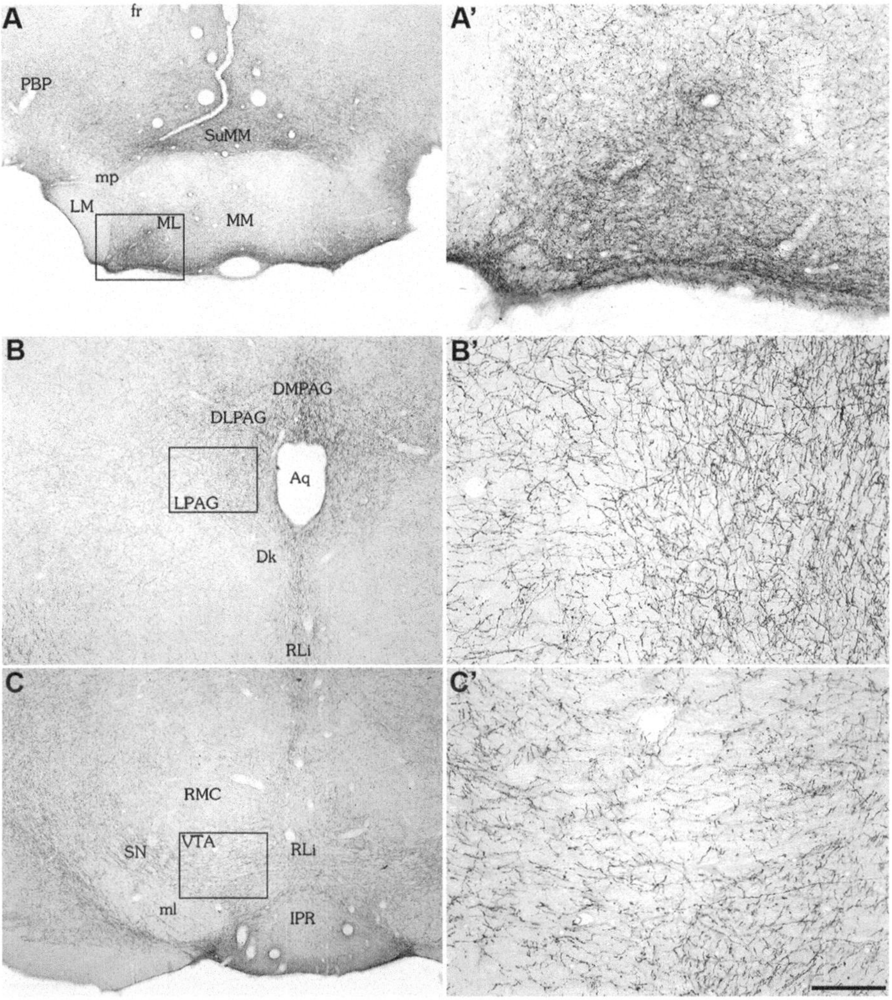

Fig. 12 The distribution of MCH-ir fibers in the rat neuroaxis, continued. Brightfield photomicrographs of frontal rat slices subjected to immunohistochemistry for MCH localization. (**a**) The lateral part of the medial mammillary nucleus (ML) has abundant innervation by MCH-ir fibers, while only a moderate density reaches the adjoining medial part (MM), and there is a virtual absence of fibers in the lateral mammillary nucleus (LM). (**a**′) Higher magnification of the rectangle in **a** highlighting the presence of fibers in the ML. (**b**) The periaqueductal gray matter also receives a high density of labeled fibers, including the very densely labeled dorsomedial (DMPAG) and dorsolateral (DLPAG) subdivisions. (**b**′) Higher magnification of the rectangle in **b**. The lateral PAG (LPAG) contains several radially oriented fibers. (**c**) Midbrain areas containing dopaminergic neurons receive inputs from MCH neurons, including the *substantia nigra* (SN) and the ventral tegmental area (VTA). (**c**′) Higher magnification of the rectangle in **c**, where several MCH-ir terminal fields can be seen in the VTA. Other abbreviations: *Aq* cerebral aqueduct, *Dk* nucleus of Darkschewitsch, *fr fasciculus retroflexus*, *IPR* interpeduncular nucleus, *ml* medial lemniscus, *mp* mammillary peduncle, *PBP* parabrachial pigmented nucleus, *RLi* rostral linear nucleus of raphe, *RMC* magnocellular part of the red nucleus, *SuMM* medial part of the supramammillary nucleus. Scale bar: **a**–**c**, 500 μm; **a**′–**c**′, 100 μm

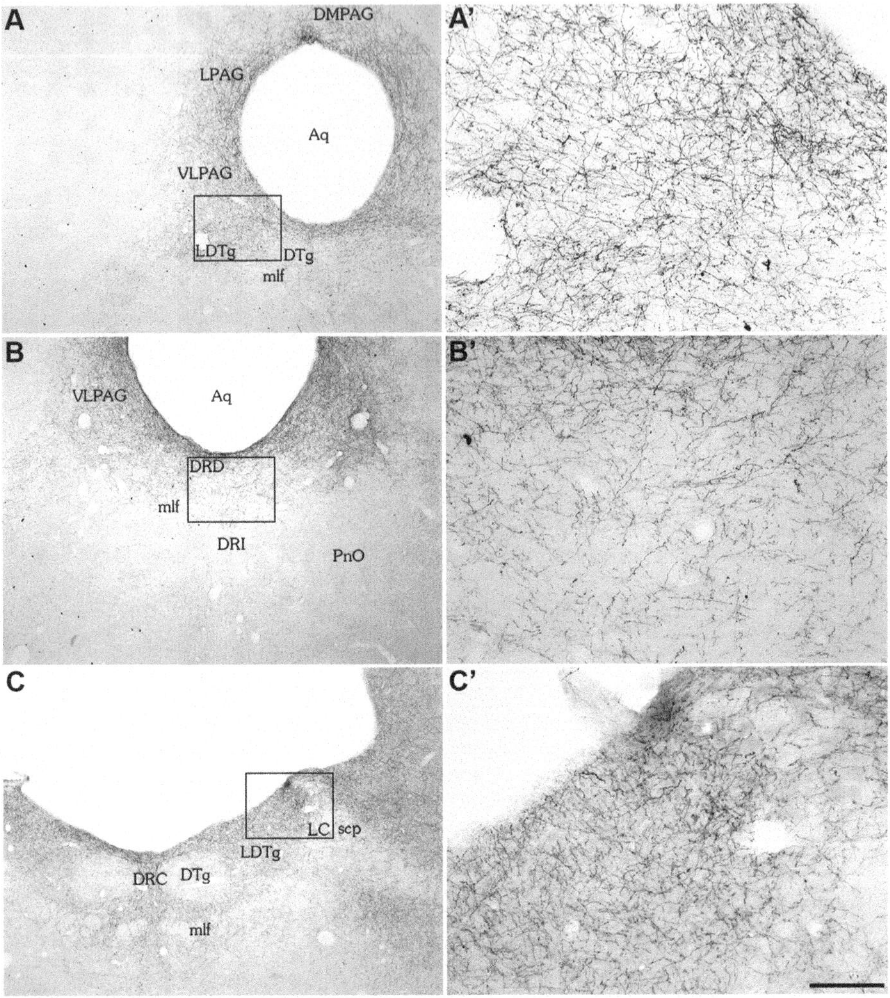

Fig. 13 The distribution of MCH-ir fibers in the rat neuroaxis, continued. Brightfield photomicrographs of frontal rat slices subjected to immunohistochemistry for MCH localization. (**a**) A high contingent of labeled fibers can be found encircling the cerebral aqueduct (Aq), including in the ventrolateral periaqueductal gray matter (VLPAG) and the laterodorsal tegmental nucleus (LDTg). (**a′**) Higher magnification of the rectangle in **a**, showcasing the numerous fibers en passant and terminal boutons found in the LDTg. (**b**) Fibers can also be found in the raphe nuclei, including the dorsal part of the dorsal raphe nucleus (DRD). (**b′**) Higher magnification of the rectangle in **b**. (**c**) The *locus coeruleus* (LC) has inputs originated from MCH neurons. (**c′**) Higher magnification of the rectangle in **c**. Other abbreviations: *DMPAG* dorsomedial part of the periaqueductal gray matter, *DRC* central part of the dorsal raphe nucleus, *DRI* interfascicular part of the dorsal raphe nucleus, *DTg* dorsal tegmental area, *LPAG* lateral periaqueductal gray matter, *mlf* medial longitudinal fasciculus, *PnO* oral part of the pontine reticular nucleus, *scp* superior cerebellar peduncle. Scale bar: **a**–**c**, 500 μm; **a′**–**c′**, 100 μm

consumption, such as the superior salivary nucleus (salivary glands control) and the motor nucleus of the trigeminal nerve (V–masseter innervation) receive fibers from the descending pathway (Pérez et al. 2011). The pedunculopontine tegmental nucleus, an area implicated in motor and arousal function (Saper and Fuller 2017; Garcia-Rill et al. 1987; Rye 1997), also receives a significant input of MCH-ir fibers that originate broadly in GABAergic neurons in the IHy, LHA, and perifornical nucleus. Some of these neurons also project to the motor cortex, suggesting that MCH neurons participate in a cortico mesencephalic integration circuit (Elias et al. 2008). At more caudal levels, the descending pathway is a strip of fibers located at the medial edge of the trigeminal complex, further descending to all levels of the spinal cord through the dorsal funiculi.

4.4 Local Hypothalamic Projections

Virtually the whole rostrocaudal extent of the hypothalamus displays MCH-ir fibers to some degree. In all likelihood, these fibers result from a combination of fibers from the ascending, periventricular, pituitary, and descending pathways, in addition to local extensions of dendrites and axons from MCH-containing neurons. In the anterior hypothalamus, fibers are found in the median preoptic, anteroventral periventricular, ventrolateral preoptic (Fig. 14a–a′), medial preoptic, anterior hypothalamic, suprachiasmatic, supraoptic (Fig. 14b–b′), and paraventricular nuclei, in addition to the medial and lateral preoptic areas. In the tuberal hypothalamus, fibers can be found in the arcuate, dorsomedial, and tuberomammillary nuclei, as well as in the retrochiasmatic and lateral hypothalamic areas. Fibers in the ventromedial nucleus are sparse and often occupy the margins of this nucleus. In the posterior hypothalamus, fibers occupy the premammillary, supramammillary, and lateral and medial mammillary nuclei, with a large contingent of fibers found in the lateral part of the latter, as well as in the posterior hypothalamic area.

5 Anatomical Distribution of MCH Receptor 1

While the distribution of *Pmch* mRNA-expressing neurons has been reported for several mammals, anatomical mappings of *Mchr1* distribution have been restricted to male rats (Hervieu et al. 2000; Saito et al. 2001; Kolakowski et al. 1996) and male mice (Chee et al. 2013), with brief descriptions available for the rhesus monkey, dog, and ferret (Tan et al. 2002). Several other groups provided additional information about additional areas and species through northern blotting, RNA dot blots, PCR, and limited *in situ* hybridization (Chambers et al. 1999; Saito et al. 1999; Hill et al. 2001; Mori et al. 2001; Sailer et al. 2001). In this section, we will briefly explore the areas described to contain *Mchr1* mRNA and highlight regions with conflicting information.

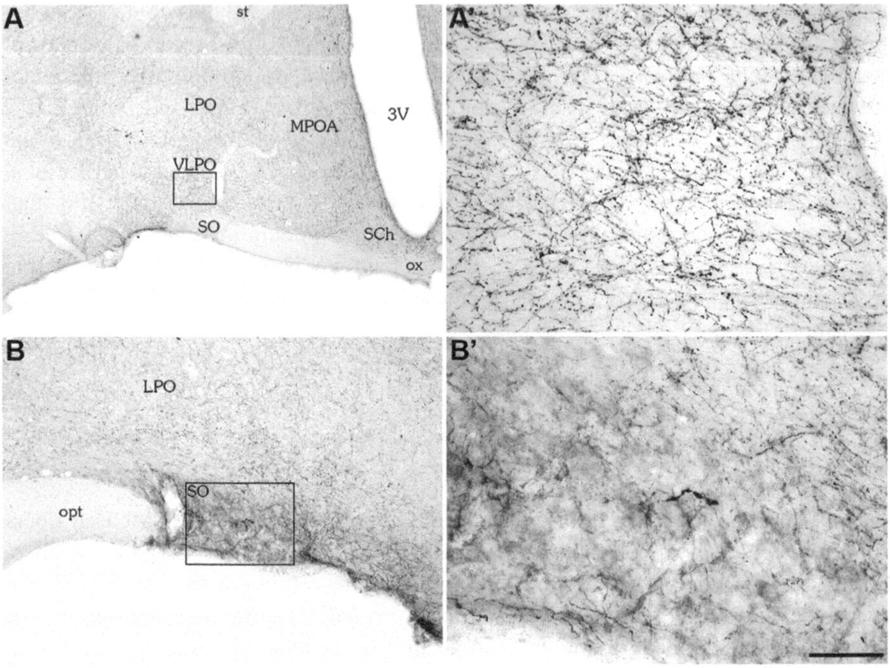

Fig. 14 The distribution of MCH-ir fibers in the rat neuroaxis, continued. Brightfield photomicrographs of frontal rat slices subjected to immunohistochemistry for MCH localization. (**a**) Numerous fibers and terminal boutons are found in the ventrolateral preoptic nucleus (VLPO). (**a**′) Higher magnification of the rectangle in **a**. (**b**) Neuroendocrine nuclei, such as the supraoptic nucleus (SO), receive moderate MCH innervation. (**b**′) Higher magnification of the rectangle in **b**. Fibers in neuroendocrine nuclei are often enlarged and varicose. Other abbreviations: *3V* third ventricle, *LPO* lateral preoptic area, *MPOA* medial preoptic area, *opt* optic tract, *ox* optic chiasm, *SCh* suprachiasmatic nucleus, *st stria terminalis*. Scale bar: **a**, 500 μm; **b**, 200 μm; and **a**′–**b**′, 50 μm

5.1 Cortical Areas

One of the most firmly established features of *Mchr1* mRNA is its high cortical expression. The distribution of *Mchr1* mRNA appears to be preferentially layer-dependent, with the highest levels of expression located in layers II/III and V, although some have reported lower levels in layers IV and VI. Among the cortical fields that contain strong *Mchr1* mRNA signals are the frontal cortex; the primary and secondary motor areas; the primary and secondary somatosensory areas; the parietal, temporal, and orbital cortices; and the agranular and dysgranular insular cortices. Several sensory fields—such as the gustatory, visceral, and primary areas; the ventral and dorsal auditory areas; and the anterolateral, rostrolateral, posterolateral, and primary visual areas—contain receptors. Moderate expression is found in the cingulate cortex, the orbital cortex, and the entorhinal cortex. Low expression is detectable in the retrosplenial agranular and granular cortices. The expression of *Mchr1* mRNA in human cortical areas is supported by PCR data.

In addition to these cortical areas, there are consistent data pointing to high expression of *Mchr1* mRNA in olfactory-linked areas. Among the regions with high expression are the anterior olfactory nucleus, olfactory tubercle, piriform cortex, ventral and dorsal *taenia tecta*, *indusium griseum*, and nucleus of the lateral olfactory tract. Lower levels of expression were detected in the endopiriform nucleus. There are two major points of disagreement in the literature regarding olfactory regions. While Hervieu et al. (2000) described a high expression of *Mchr1* mRNA in the main olfactory bulb and the islands of Calleja, Saito et al. (2001) did not detect signals in these areas. The data of Kolakowski et al. (1996) support the presence of *Mchr1* mRNA in the major islands of Calleja, but further data are necessary to determine whether the MCH receptor is present in these regions. Owing to the diminished olfactory-dedicated areas in the primate brain, information is lacking on this regard.

5.2 Telencephalic Subcortical Areas

There is also ample support for *Mchr1* mRNA in telencephalic subcortical areas, although disagreements exist for some regions. According to Hervieu et al. (2000), the septal area has a high expression of *Mchr1* mRNA, with the highest expression found in the nucleus of the diagonal band of Broca and moderate expression found in the medial septal nucleus, vertical and horizontal limbs of the diagonal band, dorsal and ventral segments of the lateral septum, and bed nucleus of the *stria terminalis* (BNST). On the other hand, Saito et al. (2001) report a distribution that is almost complementary to that of Hervieu et al. (2000), with no labeling detected in the medial septal nucleus and very scarce labeling in the dorsal and ventral lateral septum but heavy labeling in the intermediate division of the lateral septum. However, multiple authors report moderate expression of *Mchr1* mRNA in the septohippocampal nucleus and subdivisions of the BNST. There is no additional information about the septal area, and so the presence of *Mchr1* mRNA in this area in rodents or primates remains debatable.

The hippocampus proper contains some of the highest levels of *Mchr1* mRNA in the whole CNS. According to multiple authors, elevated levels of mRNA are found in all CA subfields (CA1–CA3), especially in the pyramidal layer. Moderate to high expression is also reported for the subiculum. On the other hand, the expression of *Mchr1* mRNA in the dentate gyrus is disputable: while Kolakowski et al. (1996) and Hervieu et al. (2000) describe high expression levels in the granular layer of the dentate gyrus, Saito et al. (2001) report only a few scattered cells labeled in the granular and polymorphic layers. PCR data support the high expression of *Mchr1* mRNA in the human hippocampus proper.

There is also plentiful support for high expression of *Mchr1* mRNA in the amygdaloid complex. Hervieu et al. (2000) report labeling in the basolateral, basomedial, central, medial, cortical, lateral, and posterior amygdaloid nuclei. Saito et al. (2001) describe high expression levels in the posterolateral cortical nucleus, nucleus of the lateral olfactory tract, basolateral amygdala, and anterior amygdaloid nucleus. Moderate expression was found by these authors in the lateral division of the central nucleus; the lateral division of the anterior amygdaloid nucleus; the lateral, medial, and posteromedial cortical nuclei; and the amygdalohippocampal area. Lower levels were found in the anterior cortical and basomedial nuclei and the intercalated masses. Data obtained from PCR experiments report moderate to high concentrations of *Mchr1* mRNA in the human amygdaloid complex.

Finally, there are less coherent reports regarding mRNA expression in the basal nuclei and related structures. Hervieu et al. (2000) report moderate expression in the caudate-putamen nucleus, *nucleus accumbens*, and *substantia innominata*. No labeling was detected in the *globus pallidus*. Saito et al. (2001), on the other hand, describe no expression in the ventral *pallidum*, low expression in the caudate-putamen, and high expression in the *nucleus accumbens* shell, while only scattered signals are found in the core subdivision. This dichotomy between core and shell in the *nucleus accumbens* is further supported by the results of Kolakowski et al. (1996), who also report low expression in the caudate-putamen. Expression in the claustrum is supported by multiple sources, but there is no consensus on the relative amount. In humans, moderate to low expression of *Mchr1* mRNA is supported in the nucleus *accumbens*, caudate-putamen, and *globus pallidus*.

5.3 Thalamus, Subthalamus, and Epithalamus

Hervieu et al. (2000) argue in favor of ample expression of *Mchr1* mRNA in the thalamus. Among the stained areas are the following: high expression (paraventricular, anterodorsal, reticular, ventral posterolateral, and ventral posteromedial thalamic nuclei), moderate (anteromedial, laterodorsal, *reuniens*, ventral anterior–lateral complex, ventral posterolateral, and parafascicular thalamic nuclei), and low (interanterodorsal, anteroventral, centromedial, paratenial, and centrolateral nuclei). In the subthalamic area, moderate staining is found in the subthalamic nucleus and strong labeling in the ZI. In the epithalamus, mRNA is detectable in the medial habenular nucleus but not in the lateral habenular nucleus. Finally, moderate labeling is found in the lateral geniculate nucleus. Saito et al. (2001) described a mostly similar distribution among nuclei but only low to moderate in intensity. Dense labeling in the zona incerta is confirmed by the latter authors, but no labeling was detected in the medial or lateral habenular nuclei. Data from PCR experiments indicate moderate to low levels of expression in the human thalamus.

5.4 Hypothalamus

There are multiple reports of *Mchr1* mRNA expression in the hypothalamus. Most reports indicate expression in the ventromedial and dorsomedial hypothalamic nuclei for both rodents and a nonhuman primate. In addition to those two areas, Hervieu et al. (2000) also describe high expression in the paraventricular, periventricular, and supraoptic hypothalamic nuclei, in addition to the lateral and medial mammillary nuclei. Moderate or low expression was found in the median preoptic nucleus, medial preoptic area, anterior hypothalamic nucleus, arcuate nucleus, and lateral and posterior hypothalamic areas. Saito et al. (2001), on the other hand, describe high expression of *Mchr1* mRNA only in the lateral and medial mammillary nuclei. These authors also found scarce expression in the paraventricular and dorsomedial hypothalamic nuclei; the anterior, lateral, and posterior hypothalamic areas; and the premammillary and supramammillary nuclei. In contrast to the data from rodents, there is a scarcity of data on MCHR1 in the human hypothalamus, with one report describing high expression and another describing low expression. No data are available about individual nuclei for humans.

5.5 Midbrain, Hindbrain, and Spinal Cord

Numerous structures in the reticular formation of the midbrain and hindbrain contain *Mchr1* mRNA expression. The most strongly supported structures are the *substantia nigra* and the ventral tegmental area, structures linked to the function of the basal nuclei. Hervieu et al. (2000) describe expression in several other structures, including the following: *high* (periaqueductal gray matter, interpeduncular nucleus, dorsal tegmental nucleus, olivary complex, and *locus coeruleus*) and *moderate* (anterior pretectal nucleus, raphe nuclei, red nucleus, olivary complex, and pontine reticular formation). Saito et al. (2001) found similar results, reporting high expression in the *locus coeruleus* and red nucleus, along with moderate expression in the pontine and gigantocellular reticular nuclei, dorsal raphe nucleus, and laterodorsal tegmental nucleus. Lower expression is reported by those authors in the parvicellular reticular nucleus, inferior olive, and medullary reticular field. In humans, there are reports of expression in the *locus coeruleus*, pons and medulla oblongata, but there is no consensus regarding the level of expression in the *substantia nigra*.

Several areas linked to sensory and visceral information processing were found to contain *Mchr1* mRNA expression, although reports are somewhat conflicting. Hervieu et al. (2000) report widespread distribution of mRNA, namely, the superficial and intermediate superior colliculus, inferior colliculus, anterior pretectal nucleus, olivary pretectal nucleus, principal sensory nucleus, dorsal and ventral cochlear nuclei, nucleus of the lateral lemniscus, rostral part of the nucleus of the solitary tract, parabrachial nucleus, and Kölliker-Fuse nucleus. Saito et al. (2001), on the other hand, found a very weak signal in the inferior colliculus. Regarding the motor system, labeled areas include the nucleus of the oculomotor nerve (III), spinal trigeminal nucleus, *nucleus ambiguus*, trigeminal motor nucleus, facial nucleus, hypoglossal nucleus, and dorsal motor nucleus of the vagus. The presence of *Mchr1* mRNA in the cerebellum, on the other hand, is highly disputable. While Hervieu et al. (2000) describe strong labeling in the granular layer of the cerebellar cortex and moderate labeling in the interpositus nucleus, Kolakowski et al. (1996) and Saito et al. (2001) found no evidence of expression in the cerebellum. In humans, moderate to low expression has been reported for the cerebellum.

Hervieu et al. (2000) detected *Mchr1* mRNA in the lumbar part of the spinal cord, including all subdivisions of the gray matter, with more pronounced staining in the dorsal horn. Data for humans suggest moderate to low expression of this receptor in the human spinal cord.

6 Anatomical Distribution of MCH Receptor 2

Owing to the phylogenetically restricted expression of *MCHR2* mRNA, there are no data available for this receptor in rodents. Most of the data collected, therefore, come from human tissue that was probed by PCR and northern blotting. Additional *in situ* hybridization data are available for the ferret (*Mustela putorius furo*), dog (*Canis lupus familiaris*), rhesus (*Macaca mulatta*), and African green monkey (*Chlorocebus sabaeus*).

The consensus region of elevated *MCHR2* mRNA expression is the cortex, with several works supporting gene expression in this structure (An et al. 2001; Hill et al. 2001; Mori et al. 2001; Sailer et al. 2001; Wang et al. 2001). Among the cortical regions with demonstrated expression are the frontal, temporal, parietal, and occipital lobes and the cingulate and paracentral gyri. Subcortical areas, such as the hippocampus, amygdaloid complex, caudate-putamen, *globus pallidus*, and nucleus *accumbens*, also express *MCHR2* mRNA (An et al. 2001; Hill et al. 2001; Mori et al. 2001; Sailer et al. 2001; Wang et al. 2001; Rodriguez et al. 2001). Expression in other areas is more contentious. While An et al. (2001) and Sailer et al. (2001) report expression in the diencephalon, Hill et al. (2001), Mori et al. (2001), and Rodriguez et al. (2001) did not detect signals in this structure. The same applies to the midbrain and hindbrain, with An et al. (2001) detecting expression in the pons and *medulla oblongata*, Mori et al. (2001) labeling only the *medulla oblongata*, and Hill et al. (2001) not detecting expression in these structures. A similar situation is encountered for the *substantia nigra*, where Mori et al. (2001) detected labeling, but Hill et al. (2001) did not. The cerebellum and the spinal cord were consistently negative across studies (An et al. 2001; Hill et al. 2001; Mori et al. 2001; Rodriguez et al. 2001). Finally, it is uncertain whether *MCHR2* mRNA expression occurs in fiber bundles, with An et al. (2001) and Mori et al. (2001) detecting labeling in the *corpus callosum* and cerebral peduncles, while Wang et al. (2001) describe a negative result for the *corpus callosum*.

In the African green monkey, there is limited evidence of expression in the dorsomedial hypothalamus. On the other hand, expression is abundant in the anterior and lateral hypothalamic areas, as well as in the ventromedial nucleus (Sailer et al. 2001). In the rhesus monkey, *in situ* hybridization revealed signals in the medial and central amygdala, CA1 of the hippocampus, and entorhinal and temporal cortices, in addition to weaker signals in the hypothalamus, claustrum, and insular cortex. In the dog, *MCHR2* mRNA was detected in the preoptic nuclei of the hypothalamus, the medial septum, and the claustrum. Finally, in the ferret, *MCHR2* mRNA could be detected mainly in the dorsal claustrum (Tan et al. 2002).

7 Conclusion

From the initial discovery of MCH in the salmon pituitary to this day, an impressive amount of data has been collected about the MCH system. An amazing number of functions is associated with it, including modulation of the sleep–wake cycle; sensorimotor integration; ingestive, sexual, and maternal behaviors; energy expenditure; locomotor activity; autonomic modulation; thyroid modulation; reward; hormone secretion; foraging; memory; mood; ventricular homeostasis; and immune response, among others. All these functions are supported by a complex anatomical organization, including two peptides, a putative third peptidergic member of this family, alternative splicing and overprinted sequences, multiple *loci* with MCH-synthesizing neurons, a vast array of projections, and two receptor paralogues that have distinct distributions in the CNS. All this complexity is underlined by a rich phylogenetic history in which, apparently, different levels of conservation pressure were applied to each member of this family. All these factors combine to result in a fascinating biological system that, despite everything that has been described about it, still cannot be considered completely understood.

Acknowledgments The writing of this chapter would not be possible without the funding granted by the Fundação de Amparo à Pesquisa do Estado de São Paulo [São Paulo Research Foundation—FAPESP], grants #2010/52068-0 and #2016/02224-1 [JCB] and #2016/02748-0 [GBD]. JCB is also a recipient of a grant from the Coordenação de Aperfeiçoamento de Pessoal de Ensino Superior [Agency for the Advancement of Higher Education] and the Comité Français d'Evaluation de la Coopération Universitaire avec le Brésil [French Committee for the Evaluation of Academic and Scientific Cooperation with Brazil], CAPES-COFECUB grant #848/15. JCB is an investigator with the Conselho Nacional de Desenvolvimento Científico e Tecnológico [National Council for Scientific and Technological Development—CNPq, Brazil], recipient of grant #426378/2016-4.

References

Alberto CO, Trask RB, Hirasawa M (2011) Dopamine acts as a partial agonist for α2A adrenoceptor in melanin-concentrating hormone neurons. J Neurosci 31(29):10671–10676. https://doi.org/10.1523/JNEUROSCI.6245-10.2011

Allaeys I, Bouyer K, Loudes C, Faivre-Bauman A, Petit F, Ortola C, Cardinaud B, Epelbaum J, Nahon JL (2004) Characterization of MCH-gene-overprinted-polypeptide-immunoreactive material in hypothalamus reveals an inhibitory role of pro-somatostatin1-64 on somatostatin secretion. Eur J Neurosci 19(4):925–936. https://doi.org/10.1111/j.0953-816X.2004.03187.x

Alvisi R, Diniz G, Da-Silva J, Bittencourt J, Felicio L (2016) Suckling-induced Fos activation and melanin-concentrating hormone immunoreactivity during late lactation. Life Sci 148:241–246. https://doi.org/10.1016/j.lfs.2016.02.038

An S, Cutler G, Zhao JJ, Huang SG, Tian H, Li W, Liang L, Rich M, Bakleh A, Du J, Chen JL, Dai K (2001) Identification and characterization of a melanin-concentrating hormone receptor. Proc Natl Acad Sci USA 98(13):7576–7581. https://doi.org/10.1073/pnas.131200698

Bachner D, Kreienkamp H, Weise C, Buck F, Richter D (1999) Identification of melanin concentrating hormone (MCH) as the natural ligand for the orphan somatostatin-like receptor 1 (SLC-1). FEBS Lett 457(3):522–524. https://doi.org/10.1016/S0014-5793(99)01092-3

Bäckberg M, Ultenius C, Fritschy JM, Meister B (2004) Cellular localization of GABAA receptor α subunit immunoreactivity in the rat hypothalamus: relationship with neurones containing orexigenic or anorexigenic peptides. J Neuroendocrinol 16(7):589–604. https://doi.org/10.1111/j.1365-2826.2004.01207.x

Banisadr G, Gosselin RD, Mechighel P, Kitabgi P, Rostène W, Parsadaniantz SM (2005) Highly regionalized neuronal expression of monocyte chemoattractant protein-1 (MCP-1/CCL2) in rat brain: evidence for its colocalization with neurotransmitters and neuropeptides. J Comp Neurol 489(3):275–292. https://doi.org/10.1002/cne.20598

Bittencourt JC (2011) Anatomical organization of the melanin-concentrating hormone peptide family in the mammalian brain. Gen Comp Endocrinol 172(2):185–197. https://doi.org/10.1016/j.ygcen.2011.03.028

Bittencourt J, Celis ME (2008) Anatomy, function and regulation of neuropeptide EI (NEI). Peptides 29(8):1441–1450. https://doi.org/10.1016/j.peptides.2008.03.012

Bittencourt JC, Presse F, Arias C, Peto C, Vaughan J, Nahon JL, Vale W, Sawchenko PE (1992) The melanin-concentrating hormone system of the rat brain: an immuno- and hybridization histochemical characterization. J Comp Neurol 319(2):218–245. https://doi.org/10.1002/cne.903190204

Bleier R, Cohn P, Siggelkow IR (1979) A cytoarchitectonic atlas of the hypothalamus and hypothalamic third ventricle of the rat. In: Morgane PJ, Panksepp J (eds) Anatomy of the hypothalamus, vol l. Marcel Dekker, New York, pp 137–220

Borsu L, Presse F, Nahon JL (2000) The AROM gene, spliced mRNAs encoding new DNA/RNA-binding proteins are transcribed from the opposite strand of the melanin-concentrating hormone gene in mammals. J Biol Chem 275(51):40576–40587. https://doi.org/10.1074/jbc.M006524200

Brailoiu GC, Dun SL, Brailoiu E, Inan S, Yang J, Chang JK, Dun NJ (2007) Nesfatin-1: distribution and interaction with a G protein-coupled receptor in the rat brain. Endocrinology 148(10):5088–5094. https://doi.org/10.1210/en.2007-0701

Breton C, Presse F, Hervieu G, Nahon JL (1993a) Structure and regulation of the mouse melanin-concentrating hormone mRNA and gene. Mol Cell Neurosci 4(3):271–284. https://doi.org/10.1006/mcne.1993.1035

Breton C, Schorpp M, Nahon JL (1993b) Isolation and characterization of the human melanin-concentrating hormone gene and a variant gene. Brain Res Mol Brain Res 18(4):297–310. https://doi.org/10.1016/0169-328X(93)90093-5

Burdakov D, Gerasimenko O, Verkhratsky A (2005) Physiological changes in glucose differentially modulate the excitability of hypothalamic melanin-concentrating hormone and orexin neurons in situ. J Neurosci 25(9):2429–2433. https://doi.org/10.1523/JNEUROSCI.4925-04.2005

Burgaud JL, Poosti R, Fehrentz JA, Martinez J, Nahon JL (1997) Melanin-concentrating hormone binding sites in human SVK14 keratinocytes. Biochem Biophys Res Commun 241(3):622–629. https://doi.org/10.1006/bbrc.1997.7849

Casatti CA, Elias CF, Sita LV, Frigo L, Furlani VC, Bauer JA, Bittencourt JC (2002) Distribution of melaninconcentrating hormone neurons projecting to the medial mammillary nucleus. Neuroscience 115(3):899–915. https://doi.org/10.1016/S0306-4522(02)00508-0

Chambers J, Ames RS, Bergsma D, Muir A, Fitzgerald LR, Hervieu G, Dytko GM, Foley JJ, Martin J, Liu WS, Park J, Ellis C, Ganguly S, Konchar S, Cluderay J, Leslie R, Wilson S, Sarau HM (1999) Melanin-concentrating hormone is the cognate ligand for the orphan G-protein-coupled receptor SLC-1. Nature 400(6741):261–265. https://doi.org/10.1038/22313

Chee MJ, Pissios P, Maratos-Flier E (2013) Neurochemical characterization of neurons expressing melanin-concentrating hormone receptor 1 in the mouse hypothalamus. J Comp Neurol 521 (10):2208–2234. https://doi.org/10.1002/cne.23273

Chee MJ, Arrigoni E, Maratos-Flier E (2015) Melanin-concentrating hormone neurons release glutamate for feedforward inhibition of the lateral septum. J Neurosci 35(8):3644–3651. https://doi.org/10.1523/JNEUROSCI.4187-14.2015

Chiocchio SR, Gallardo MG, Louzan P, Gutnisky V, Tramezzani JH (2001) Melanin-concentrating hormone stimulates the release of luteinizing hormone-releasing hormone and gonadotropins in the female rat acting at both median eminence and pituitary levels. Biol Reprod 64 (5):1466–1472. https://doi.org/10.1095/biolreprod64.5.1466

Chou TC, Rotman SR, Saper CB (2004) Lateral hypothalamic acetylcholinesterase-immunoreactive neurons co-express either orexin or melanin concentrating hormone. Neurosci Lett 370(2–3):123–126. https://doi.org/10.1016/j.neulet.2004.08.070

Collin M, Bäckberg M, Önnestam K, Meister B (2002) 5-HT1A receptor immunoreactivity in hypothalamic neurons involved in body weight control. Neuroreport 13(7):945–951. https://doi.org/10.1097/00001756-200205240-00009

Conductier G, Nahon J-L, Guyon A (2011) Dopamine depresses melanin concentrating hormone neuronal activity through multiple effects on α2-noradrenergic, D1 and D2-like dopaminergic receptors. Neuroscience 178:89–100. https://doi.org/10.1016/j.neuroscience.2011.01.030

Conductier G, Brau F, Viola A, Langlet F, Ramkumar N, Dehouck B, Lemaire T, Chapot R, Lucas L, Rovère C (2013a) Melanin-concentrating hormone regulates beat frequency of ependymal cilia and ventricular volume. Nat Neurosci 16(7):845–847. https://doi.org/10.1038/nn.3401

Conductier G, Martin A, Risold P-Y, Jego S, Lavoie R, Lafont C, Mollard P, Adamantidis A, Nahon J-L (2013b) Control of ventricular ciliary beating by the melanin concentrating hormone-expressing neurons of the lateral hypothalamus: a functional imaging survey. Front Endocrinol 4:182. https://doi.org/10.3389/fendo.2013.00182

Courseaux A, Nahon J-L (2001) Birth of two chimeric genes in the Hominidae lineage. Science 291 (5507):1293–1297. https://doi.org/10.1126/science.1057284

Cvetkovic V, Brischoux F, Griffond B, Bernard G, Jacquemard C, Fellmann D, Risold PY (2003) Evidence of melanin-concentrating hormone-containing neurons supplying both cortical and neuroendocrine projections. Neuroscience 116(1):31–35. https://doi.org/10.1016/S0306-4522 (02)00557-2

de Lecea L, Kilduff TS, Peyron C, Gao X, Foye PE, Danielson PE, Fukuhara C, Battenberg EL, Gautvik VT, Bartlett FS 2nd, Frankel WN, van den Pol AN, Bloom FE, Gautvik KM, Sutcliffe JG (1998) The hypocretins: hypothalamus-specific peptides with neuroexcitatory activity. Proc Natl Acad Sci USA 95(1):322–327

Diniz GB, Bittencourt JC (2017) The melanin-concentrating hormone as an integrative peptide driving motivated behaviors. Front Syst Neurosci 11:32. https://doi.org/10.3389/fnsys.2017.00032

Douglass J, McKinzie AA, Couceyro P (1995) PCR differential display identifies a rat brain mRNA that is transcriptionally regulated by cocaine and amphetamine. J Neurosci 15(3):2471–2481
Drozdz R, Eberle AN (1995) Binding sites for melanin-concentrating hormone (MCH) in brain synaptosomes and membranes from peripheral tissues identified with highly tritiated MCH. J Recept Signal Transduct Res 15(1–4):487–502. https://doi.org/10.3109/10799899509045235
Drozdz R, Baker BI, Zeller A, Eberle AN (1993) Synthesis and biological activity of highly tritiated rat/human melanin-concentrating hormone. Ann N Y Acad Sci 680(1):489–492. https://doi.org/10.1016/0014-5793(95)00043-9
Drozdz R, Siegrist W, Baker BI, Chluba-de Tapia J, Eberle AN (1995) Melanin-concentrating hormone binding to mouse melanoma cells in vitro. FEBS Lett 359(2–3):199–202. https://doi.org/10.1016/0014-5793(95)00043-9
Elias CF, Bittencourt JC (1997) Study of the origins of melanin-concentrating hormone and neuropeptide EI immunoreactive projections to the periaqueductal gray matter. Brain Res 755 (2):255–271. https://doi.org/10.1016/S0006-8993(97)00104-2
Elias CF, Saper CB, Maratos-Flier E, Tritos NA, Lee C, Kelly J, Tatro JB, Hoffman GE, Ollmann MM, Barsh GS, Sakurai T, Yanagisawa M, Elmquist JK (1998) Chemically defined projections linking the mediobasal hypothalamus and the lateral hypothalamic area. J Comp Neurol 402 (4):442–459. https://doi.org/10.1016/S0006-8993(98)00598-8
Elias CF, Lee CE, Kelly JF, Ahima RS, Kuhar M, Saper CB, Elmquist JK (2001) Characterization of CART neurons in the rat and human hypothalamus. J Comp Neurol 432(1):1–19. https://doi.org/10.1002/cne.1085
Elias CF, Sita LV, Zambon BK, Oliveira ER, Vasconcelos LA, Bittencourt JC (2008) Melanin-concentrating hormone projections to areas involved in somatomotor responses. J Chem Neuroanat 35(2):188–201. https://doi.org/10.1016/j.jchemneu.2007.10.002
Ferreira JGP, Bittencourt JC, Adamantidis A (2017a) Melanin-concentrating hormone and sleep. Curr Opin Neurobiol 44:152–158. https://doi.org/10.1016/j.conb.2017.04.008
Ferreira JGP, Duarte JCG, Diniz GB, Bittencourt JC (2017b) Litter size determines the number of melanin-concentrating hormone neurons in the medial preoptic area of Sprague Dawley lactating dams. Physiol Behav 181:75–79. https://doi.org/10.1016/j.physbeh.2017.08.028
Foo KS, Brismar H, Broberger C (2008) Distribution and neuropeptide coexistence of nucleobindin-2 mRNA/nesfatin-like immunoreactivity in the rat CNS. Neuroscience 156 (3):563–579. https://doi.org/10.1016/j.neuroscience.2008.07.054
Fort P, Salvert D, Hanriot L, Jego S, Shimizu H, Hashimoto K, Mori M, Luppi P-H (2008) The satiety molecule nesfatin-1 is co-expressed with melanin concentrating hormone in tuberal hypothalamic neurons of the rat. Neuroscience 155(1):174–181. https://doi.org/10.1016/j.neuroscience.2008.05.035
Gallardo MG, Chiocchio SR, Tramezzani JH (2004) Changes of melanin-concentrating hormone related to LHRH release in the median eminence of rats. Brain Res 1030(1):152–158. https://doi.org/10.1016/j.brainres.2004.10.005
Garcia SV, Libourel P-A, Lazarus M, Grassi D, Luppi P-H, Fort P (2017) Genetic inactivation of glutamate neurons in the rat sublaterodorsal tegmental nucleus recapitulates REM sleep behaviour disorder. Brain 140(2):414–428. https://doi.org/10.1093/brain/aww310
Garcia-Rill E, Houser C, Skinner R, Smith W, Woodward D (1987) Locomotion-inducing sites in the vicinity of the pedunculopontine nucleus. Brain Res Bull 18(6):731–738. https://doi.org/10.1016/0361-9230(87)90208-5
Griffond B, Ciofi P, Bayer L, Jacquemard C, Fellmann D (1997) Immunocytochemical detection of the neurokinin B receptor (NK3) on melanin-concentrating hormone (MCH) neurons in rat brain. J Chem Neuroanat 12(3):183–189. https://doi.org/10.1016/S0891-0618(96)00200-1
Haemmerle CA, Campos AM, Bittencourt JC (2015) Melanin-concentrating hormone inputs to the nucleus accumbens originate from distinct hypothalamic sources and are apposed to GABAergic and cholinergic cells in the Long-Evans rat brain. Neuroscience 289:392–405. https://doi.org/10.1016/j.neuroscience.2015.01.014

Hausen AC, Ruud J, Jiang H, Hess S, Varbanov H, Kloppenburg P, Brüning JC (2016) Insulin-dependent activation of MCH neurons impairs locomotor activity and insulin sensitivity in obesity. Cell Rep 17(10):2512–2521. https://doi.org/10.1016/j.celrep.2016.11.030

Hazai D, Lien C-F, Hajós F, Halasy K, Górecki DC, Jancsik V (2008) Synaptic alpha-dystrobrevin: localization of a short alpha-dystrobrevin isoform in melanin-concentrating hormone neurons of the hypothalamus. Brain Res 1201:52–59. https://doi.org/10.1016/j.brainres.2008.01.046

Hervieu GJ, Cluderay JE, Harrison D, Meakin J, Maycox P, Nasir S, Leslie RA (2000) The distribution of the mRNA and protein products of the melanin-concentrating hormone (MCH) receptor gene, slc-1, in the central nervous system of the rat. Eur J Neurosci 12(4):1194–1216. https://doi.org/10.1046/j.1460-9568.2000.00008.x

Hill J, Duckworth M, Murdock P, Rennie G, Sabido-David C, Ames RS, Szekeres P, Wilson S, Bergsma DJ, Gloger IS, Levy DS, Chambers JK, Muir AI (2001) Molecular cloning and functional characterization of MCH2, a novel human MCH receptor. J Biol Chem 276 (23):20125–20129. https://doi.org/10.1074/jbc.M102068200

Hintermann E, Tanner H, Talke-Messerer C, Schlumberger S, Zumsteg U, Eberle AN (2001) Interaction of melanin-concentrating hormone (MCH), neuropeptide EI (NEI), neuropeptide GE (NGE), and α-MSH with melanocortin and MCH receptors on mouse B16 melanoma cells. J Recept Signal Transduct Res 21(1):93–116. https://doi.org/10.1081/RRS-100107145

Hotta K, Hosaka M, Tanabe A, Takeuchi T (2009) Secretogranin II binds to secretogranin III and forms secretory granules with orexin, neuropeptide Y, and POMC. J Endocrinol 202(1):111–121. https://doi.org/10.1677/JOE-08-0531

Huang H, van den Pol AN (2007) Rapid direct excitation and long-lasting enhancement of NMDA response by group I metabotropic glutamate receptor activation of hypothalamic melanin-concentrating hormone neurons. J Neurosci 27(43):11560–11572. https://doi.org/10.1523/JNEUROSCI.2147-07.2007

Kampe J, Tschop MH, Hollis JH, Oldfield BJ (2009) An anatomic basis for the communication of hypothalamic, cortical and mesolimbic circuitry in the regulation of energy balance. Eur J Neurosci 30(3):415–430. https://doi.org/10.1111/j.1460-9568.2009.06818.x

Kawauchi H, Kawazoe I, Tsubokawa M, Kishida M, Baker BI (1983) Characterization of melanin-concentrating hormone in chum salmon pituitaries. Nature 305(5932):321–323. https://doi.org/10.1038/305321a0

Knollema S, Brown ER, Vale W, Sawchenko PE (1992) Novel hypothalamic and preoptic sites of prepro-melanin-concentrating hormone messenger ribonucleic acid and peptide expression in lactating rats. J Neuroendocrinol 4(6):709–717. https://doi.org/10.1111/j.1365-2826.1992.tb00222.x

Kolakowski LF Jr, Jung BP, Nguyen T, Johnson MP, Lynch KR, Cheng R, Heng HH, George SR, O'Dowd BF (1996) Characterization of a human gene related to genes encoding somatostatin receptors. FEBS Lett 398(2–3):253–258. https://doi.org/10.1016/S0014-5793(96)01160-X

Kong D, Vong L, Parton LE, Ye C, Tong Q, Hu X, Choi B, Brüning JC, Lowell BB (2010) Glucose stimulation of hypothalamic MCH neurons involves K ATP channels, is modulated by UCP2, and regulates peripheral glucose homeostasis. Cell Metab 12(5):545–552. https://doi.org/10.1016/j.cmet.2010.09.013

Krolewski DM, Medina A, Kerman IA, Bernard R, Burke S, Thompson RC, Bunney WE, Schatzberg AF, Myers RM, Akil H (2010) Expression patterns of corticotropin-releasing factor, arginine vasopressin, histidine decarboxylase, melanin-concentrating hormone, and orexin genes in the human hypothalamus. J Comp Neurol 518(22):4591–4611. https://doi.org/10.1002/cne.22480

Lakaye B, Minet A, Zorzi W, Grisar T (1998) Cloning of the rat brain cDNA encoding for the SLC-1 G protein-coupled receptor reveals the presence of an intron in the gene. Biochim Biophys Acta 1401(2):216–220. https://doi.org/10.1016/S0167-4889(97)00135-3

Lembo PM, Grazzini E, Cao J, Hubatsch DA, Pelletier M, Hoffert C, St-Onge S, Pou C, Labrecque J, Groblewski T, O'Donnell D, Payza K, Ahmad S, Walker P (1999) The receptor for the orexigenic peptide melanin-concentrating hormone is a G-protein-coupled receptor. Nat Cell Biol 1(5):267–271. https://doi.org/10.1038/12978

Lima FF, Sita LV, Oliveira AR, Costa HC, da Silva JM, Mortara RA, Haemmerle CA, Xavier GF, Canteras NS, Bittencourt JC (2013) Hypothalamic melanin-concentrating hormone projections to the septo-hippocampal complex in the rat. J Chem Neuroanat 47:1–14. https://doi.org/10.1016/j.jchemneu.2012.10.003

Macdonald D, Murgolo N, Zhang R, Durkin JP, Yao X, Strader CD, Graziano MP (2000) Molecular characterization of the melanin-concentrating hormone/receptor complex: identification of critical residues involved in binding and activation. Mol Pharmacol 58(1):217–225. https://doi.org/10.1124/mol.58.1.217

Mickelsen LE, Kolling FW, Chimileski BR, Fujita A, Norris C, Chen K, Nelson CE, Jackson AC (2017) Neurochemical heterogeneity among lateral hypothalamic hypocretin/orexin and melanin-concentrating hormone neurons identified through single-cell gene expression analysis. eNeuro 4(5). https://doi.org/10.1523/ENEURO.0013-17.2017

Modirrousta M, Mainville L, Jones BE (2005) Orexin and MCH neurons express c-Fos differently after sleep deprivation vs. recovery and bear different adrenergic receptors. Eur J Neurosci 21 (10):2807–2816. https://doi.org/10.1111/j.1460-9568.2005.04104.x

Moldovan G-L, Dejsuphong D, Petalcorin MI, Hofmann K, Takeda S, Boulton SJ, D'Andrea AD (2012) Inhibition of homologous recombination by the PCNA-interacting protein PARI. Mol Cell 45(1):75–86. https://doi.org/10.1016/j.molcel.2011.11.010

Mori M, Harada M, Terao Y, Sugo T, Watanabe T, Shimomura Y, Abe M, Shintani Y, Onda H, Nishimura O, Fujino M (2001) Cloning of a novel G protein-coupled receptor, SLT, a subtype of the melanin-concentrating hormone receptor. Biochem Biophys Res Commun 283 (5):1013–1018. https://doi.org/10.1006/bbrc.2001.4893

Nahon JL, Presse F, Bittencourt JC, Sawchenko PE, Vale W (1989) The rat melanin-concentrating hormone messenger ribonucleic acid encodes multiple putative neuropeptides coexpressed in the dorsolateral hypothalamus. Endocrinology 125(4):2056–2065. https://doi.org/10.1210/endo-125-4-2056

Nahon J-L, Joly C, Levan G, Szpirer J, Szpirer C (1992) Pro-melanin-concentrating hormone gene (PMCH) is localized on human chromosome 12q and rat chromosome 7. Genomics 12 (4):846–848. https://doi.org/10.1016/0888-7543(92)90323-K

Nahon JL, Presse F, Breton C, Hervieu G, Schorpp M (1993) Structure and regulation of the melanin-concentrating hormone gene. Ann N Y Acad Sci 680:111–129. https://doi.org/10.1111/j.1749-6632.1993.tb19678.x

Nieuwenhuys R, Geeraedts LM, Veening JG (1982) The medial forebrain bundle of the rat. I. General introduction. J Comp Neurol 206(1):49–81. https://doi.org/10.1002/cne.902060106

Parkes D, Vale W (1992) Secretion of melanin-concentrating hormone and neuropeptide-EI from cultured rat hypothalamic cells. Endocrinology 131(4):1826–1831. https://doi.org/10.1210/endo.131.4.1327720

Parks GS, Olivas ND, Ikrar T, Sanathara NM, Wang L, Wang Z, Civelli O, Xu X (2014a) Histamine inhibits the melanin-concentrating hormone system: implications for sleep and arousal. J Physiol 592(10):2183–2196. https://doi.org/10.1113/jphysiol.2013.268771

Parks GS, Wang L, Wang Z, Civelli O (2014b) Identification of neuropeptide receptors expressed by melanin-concentrating hormone neurons. J Comp Neurol 522(17):3817–3833. https://doi.org/10.1002/cne.23642

Parsons MP, Hirasawa M (2011) GIRK channel-mediated inhibition of melanin-concentrating hormone neurons by nociceptin/orphanin FQ. J Neurophysiol 105(3):1179–1184. https://doi.org/10.1152/jn.00791.2010

Pedeutour F, Szpirer C, Nahon JL (1994) Assignment of the human pro-melanin-concentrating hormone gene (PMCH) to chromosome 12q23-q24 and two variant genes (PMCHL1 and PMCHL2) to chromosome 5p14 and 5q12-q13. Genomics 19(1):31–37. https://doi.org/10.1006/geno.1994.1008

Pérez CA, Stanley SA, Wysocki RW, Havranova J, Ahrens-Nicklas R, Onyimba F, Friedman JM (2011) Molecular annotation of integrative feeding neural circuits. Cell Metab 13(2):222–232. https://doi.org/10.1016/j.cmet.2010.12.013

Presse F, Nahon JL, Fischer WH, Vale W (1990) Structure of the human melanin concentrating hormone mRNA. Mol Endocrinol 4(4):632–637. https://doi.org/10.1210/mend-4-4-632

Presse F, Conductier G, Rovere C, Nahon J (2014) The melanin-concentrating hormone receptors: neuronal and non-neuronal functions. Int J Obes Suppl 4:S31–S36. https://doi.org/10.1038/ijosup.2014.9

Reti IM, Reddy R, Worley PF, Baraban JM (2002) Selective expression of Narp, a secreted neuronal pentraxin, in orexin neurons. J Neurochem 82(6):1561–1565. https://doi.org/10.1046/j.1471-4159.2002.01141.x

Risold P, Fellmann D, Lenys D, Bugnon C (1989) Coexistence of acetylcholinesterase-, human growth hormone-releasing factor (1–37)-, α-melanotropin-and melanin-concentrating hormone-like immunoreactivities in neurons of the rat hypothalamus: a light and electron microscope study. Neurosci Lett 100(1):23–28. https://doi.org/10.1016/0304-3940(89)90654-X

Rodriguez M, Beauverger P, Naime I, Rique H, Ouvry C, Souchaud S, Dromaint S, Nagel N, Suply T, Audinot V (2001) Cloning and molecular characterization of the novel human melanin-concentrating hormone receptor MCH2. Mol Pharmacol 60(4):632–639. https://doi.org/10.1210/mend-4-4-632

Rondini TA, Rodrigues Bde C, de Oliveira AP, Bittencourt JC, Elias CF (2007) Melanin-concentrating hormone is expressed in the laterodorsal tegmental nucleus only in female rats. Brain Res Bull 74(1–3):21–28. https://doi.org/10.1016/j.brainresbull.2007.04.006

Rondini TA, Donato J Jr, Rodrigues Bde C, Bittencourt JC, Elias CF (2010) Chemical identity and connections of medial preoptic area neurons expressing melanin-concentrating hormone during lactation. J Chem Neuroanat 39(1):51–62. https://doi.org/10.1016/j.jchemneu.2009.10.005

Rye DB (1997) Contributions of the pedunculopontine region to normal and altered REM sleep. Sleep 20(9):757–788. https://doi.org/10.1093/sleep/20.9.757

Sailer AW, Sano H, Zeng Z, McDonald TP, Pan J, Pong SS, Feighner SD, Tan CP, Fukami T, Iwaasa H, Hreniuk DL, Morin NR, Sadowski SJ, Ito M, Bansal A, Ky B, Figueroa DJ, Jiang Q, Austin CP, MacNeil DJ, Ishihara A, Ihara M, Kanatani A, Van der Ploeg LH, Howard AD, Liu Q (2001) Identification and characterization of a second melanin-concentrating hormone receptor, MCH-2R. Proc Natl Acad Sci USA 98(13):7564–7569. https://doi.org/10.1073/pnas.121170598

Saito Y, Nothacker HP, Wang Z, Lin SH, Leslie F, Civelli O (1999) Molecular characterization of the melanin-concentrating-hormone receptor. Nature 400(6741):265–269. https://doi.org/10.1038/22321

Saito Y, Cheng M, Leslie FM, Civelli O (2001) Expression of the melanin-concentrating hormone (MCH) receptor mRNA in the rat brain. J Comp Neurol 435(1):26–40. https://doi.org/10.1002/cne.1191

Sakurai T, Amemiya A, Ishii M, Matsuzaki I, Chemelli RM, Tanaka H, Williams SC, Richardson JA, Kozlowski GP, Wilson S (1998) Orexins and orexin receptors: a family of hypothalamic neuropeptides and G protein-coupled receptors that regulate feeding behavior. Cell 92(4):573–585. https://doi.org/10.1016/S0092-8674(00)80949-6

Saper CB (2013) The neurobiology of sleep. Continuum: Lifelong Learn Neurol 19(1, Sleep Disorders):19–31. https://doi.org/10.1212/01.CON.0000427215.07715.73

Saper CB, Fuller PM (2017) Wake–sleep circuitry: an overview. Curr Opin Neurobiol 44:186–192. https://doi.org/10.1016/j.conb.2017.03.021

Saper CB, Swanson LW, Cowan WM (1979) An autoradiographic study of the efferent connections of the lateral hypothalamic area in the rat. J Comp Neurol 183(4):689–706. https://doi.org/10.1002/cne.901830402

Saper CB, Akil H, Watson SJ (1986) Lateral hypothalamic innervation of the cerebral cortex: immunoreactive staining for a peptide resembling but immunochemically distinct from pituitary/arcuate alpha-melanocyte stimulating hormone. Brain Res Bull 16(1):107–120. https://doi.org/10.1016/0361-9230(86)90018-3

Shimomura Y, Mori M, Sugo T, Ishibashi Y, Abe M, Kurokawa T, Onda H, Nishimura O, Sumino Y, Fujino M (1999) Isolation and identification of melanin-concentrating hormone as the endogenous ligand of the SLC-1 receptor. Biochem Biophys Res Commun 261(3):622–626. https://doi.org/10.1006/bbrc.1999.1104

Simerly R (2015) Organization of the hypothalamus. In: The rat nervous system, 4th edn. Elsevier, San Diego, pp 267–288

Sita LV, Diniz GB, Canteras NS, Xavier GF, Bittencourt JC (2016) Effect of intrahippocampal administration of anti-melanin-concentrating hormone on spatial food-seeking behavior in rats. Peptides 76:130–138. https://doi.org/10.1016/j.peptides.2015.12.007

Tan CP, Sano H, Iwaasa H, Pan J, Sailer AW, Hreniuk DL, Feighner SD, Palyha OC, Pong SS, Figueroa DJ, Austin CP, Jiang MM, Yu H, Ito J, Ito M, Guan XM, MacNeil DJ, Kanatani A, Van der Ploeg LH, Howard AD (2002) Melanin-concentrating hormone receptor subtypes 1 and 2: species-specific gene expression. Genomics 79(6):785–792. https://doi.org/10.1006/geno.2002.6771

Tanabe A, Yanagiya T, Iida A, Saito S, Sekine A, Takahashi A, Nakamura T, Tsunoda T, Kamohara S, Nakata Y (2007) Functional single-nucleotide polymorphisms in the secretogranin III (SCG3) gene that form secretory granules with appetite-related neuropeptides are associated with obesity. J Clin Endocrinol Metabol 92(3):1145–1154. https://doi.org/10.1210/jc.2006-1808

Thompson RC, Watson SJ (1990) Nucleotide sequence and tissue-specific expression of the rat melanin concentrating hormone gene. DNA Cell Biol 9(9):637–645. https://doi.org/10.1089/dna.1990.9.637

Toumaniantz G, Bittencourt JC, Nahon JL (1996) The rat melanin-concentrating hormone gene encodes an additional putative protein in a different reading frame. Endocrinology 137 (10):4518–4521. https://doi.org/10.1210/en.137.10.4518

Toumaniantz G, Ferreira PC, Allaeys I, Bittencourt JC, Nahon JL (2000) Differential neuronal expression and projections of melanin-concentrating hormone (MCH) and MCH-gene-overprinted-polypeptide (MGOP) in the rat brain. Eur J Neurosci 12(12):4367–4380. https://doi.org/10.1046/j.0953-816X.2000.01340.x

van den Pol AN, Acuna-Goycolea C, Clark KR, Ghosh PK (2004) Physiological properties of hypothalamic MCH neurons identified with selective expression of reporter gene after recombinant virus infection. Neuron 42(4):635–652. https://doi.org/10.1016/S0896-6273(04)00251-X

Vaughan JM, Fischer WH, Hoeger C, Rivier J, Vale W (1989) Characterization of melanin-concentrating hormone from rat hypothalamus. Endocrinology 125(3):1660–1665. https://doi.org/10.1210/endo-125-3-1660

Viale A, Zhixing Y, Breton C, Pedeutour F, Coquerel A, Jordan D, Nahon JL (1997) The melanin-concentrating hormone gene in human: flanking region analysis, fine chromosome mapping, and tissue-specific expression. Mol Brain Res 46(1-2):243–255. https://doi.org/10.1016/S0169-328X(97)00018-1

Viale A, Ortola C, Hervieu G, Furuta M, Barbero P, Steiner DF, Seidah NG, Nahon J-L (1999) Cellular localization and role of prohormone convertases in the processing of pro-melanin concentrating hormone in mammals. J Biol Chem 274(10):6536–6545. https://doi.org/10.1074/jbc.274.10.6536

Von Economo C (1930) Sleep as a problem of localization. J Nerv Ment Dis 71(3):249–259

Wang S, Behan J, O'Neill K, Weig B, Fried S, Laz T, Bayne M, Gustafson E, Hawes BE (2001) Identification and pharmacological characterization of a novel human melanin-concentrating hormone receptor, mch-r2. J Biol Chem 276(37):34664–34670. https://doi.org/10.1074/jbc.M102601200

Yao Y, Fu LY, Zhang X, van den Pol AN (2012) Vasopressin and oxytocin excite MCH neurons, but not other lateral hypothalamic GABA neurons. Am J Physiol Regul Integr Comp Physiol 302(7):R815–R824. https://doi.org/10.1152/ajpregu.00452.2011

Projections of the MCH System to Structures Involved in the Regulation of Sleep and Wakefulness

Hyun Sook Lee

Abstract Although glutamatergic sublaterodorsal tegmental nucleus (SLD) constitutes a core of rapid eye movement (REM) sleep circuits, recent reports assert that REM sleep generation is under hypothalamic control of melanin-concentrating hormone (MCH) neurons. GABAergic MCH cells control the onset and maintenance of REM sleep via direct inhibitory projection to REM-off GABAergic neurons in ventrolateral periaqueductal gray matter which exert a potent inhibitory influence on REM-on SLD cells. It is generally accepted that cholinergic neurons in the pedunculopontine tegmental (PPT) and laterodorsal tegmental (LDT) nuclei serve as supplementary REM-on cells, while noradrenergic locus coeruleus (LC) and serotoninergic dorsal raphe (DR) cells contribute to REM-off circuitry. MCH neurons project heavily to the PPT and LDT, where cholinergic neurons might play a role in the strengthening of non-REM (NREM) to REM transitions once initiated. They also project to the LC and DR; the microinjection of MCH into these nuclei increases REM sleep duration. Likewise, MCH neurons send efferent fibers to the tuberomammillary nucleus (TMN); the activation of these fibers prolongs REM sleep episodes. Among monoaminergic nuclei, the TMN is unique in that (1) histaminergic neurons are active in cataplexy, implying their role in the maintenance of arousal state during the period, and (2) it contains a substantial number of MCH somata within its boundary, whose physiological function remains to be established.

Since the discovery of rapid eye movement (REM) sleep in the late 1950s, reciprocal inhibitory interactions between brain stem monoaminergic and cholinergic neurons were regarded as key components in the regulation of REM sleep. Recent studies, however, demonstrated that selective lesions of either cholinergic or monoaminergic nuclei have relatively limited effects on REM sleep.

Recent studies revealed that the rat sublaterodorsal tegmental nucleus (SLD), also called subcoeruleus nucleus in humans and peri-locus coeruleus alpha in cats, plays a

H. S. Lee (✉)
Department of Anatomy, School of Medicine, Konkuk University, Seoul, Republic of Korea
e-mail: hyunsook.lee@kku.ac.kr

S. R. Pandi-Perumal et al. (eds.), *Melanin-Concentrating Hormone and Sleep*,
https://doi.org/10.1007/978-3-319-75765-0_2

key role in the generation and maintenance of REM sleep (Sapin et al. 2009). In fact, the SLD is composed of heterogeneous groups of cells containing glutamate, gamma-aminobutyric acid (GABA), norepinephrine, or acetylcholine (Ford et al. 1995; Boissard et al. 2002; Bruinstroop et al. 2011). Among these, ventral SLD neurons containing glutamate are proven to be the core of REM circuit. They target GABA/glycinergic ventral medullary REM-on neurons (or spinal inhibitory interneurons) to produce skeletal muscle atonia (Lu et al. 2007). Recent studies further demonstrated that hypothalamic MCH neurons play a crucial role in the generation of REM sleep, while glutamatergic SLD cells are more responsible for the generation of skeletal muscle atonia during the REM period (Luppi et al. 2013; Garcia et al. 2017).

Acetylcholine promotes REM sleep by activating glutamatergic SLD cells (Weng et al. 2014). Cholinergic neurons in the pedunculopontine tegmental (PPT) and laterodorsal tegmental (LDT) nuclei serve as supplementary REM-on cells, while serotoninergic dorsal raphe (DR) and noradrenergic locus coeruleus (LC) neurons contribute to REM-off circuitry (Lu et al. 2006). Based on recent findings showing that melanin-concentrating hormone (MCH) is a powerful hypnogenic factor (Verret et al. 2003; Luppi et al. 2013; Konadhode et al. 2015), we will discuss in this chapter the MCH neuronal projections to structures involved in the regulation (or modulation) of the sleep/wake states.

1 MCH Projections to the Ventrolateral Periaqueductal Gray and Sublaterodorsal Tegmental Nucleus

It is well known that GABAergic neurons in the brain stem play a crucial role in the initiation of REM sleep. A recent study revealed that the rat brain stem contains only one population of REM-off GABAergic neurons but multiple REM-on GABAergic populations (Sapin et al. 2009). GABAergic REM-off cells are located in the ventrolateral periaqueductal gray (vlPAG) and the adjacent dorsal part of the deep mesencephalic reticular nucleus (dDpMe); this region also contains GABAergic REM-on cells. Thus, the REM-off and REM-on populations within the vlPAG/dDpMe inhibit each other during waking (W) and REM periods, respectively. Other REM-on GABAergic neurons are located in the dorsal paragigantocellular and lateral paragigantocellular nuclei and inhibit REM-off neurons in the vlPAG/dDpMe, DR, and LC during REM sleep (Sapin et al. 2009).

During W and non-REM (NREM) sleep, REM-on SLD neurons are under a tonic GABAergic control of the vlPAG/dDpMe (Ford et al. 1995; Boissard et al. 2002; Sapin et al. 2009). Subsequent studies revealed that the onset of REM sleep by MCH is enacted through the removal of the tonic GABAergic control exerted by the vlPAG/dDpMe or intrinsic SLD neurons (Clément et al. 2012). The vlPAG/dDpMe is the main ponto-medullary region containing a large number of GABAergic cells activated during REM deprivation (Sapin et al. 2009). On the other hand, a direct anatomical projection from hypothalamic MCH cells to the SLD has been also reported (Bittencourt et al. 1992), which might suggest the possibility of inhibitory hypothalamic influence on GABAergic SLD interneurons during REM sleep. Furthermore, the local microinjection of MCH into the SLD produced inhibitory effects on REM sleep, which might imply the inhibitory role of MCH on glutamatergic REM-on SLD neurons during W (Monti et al. 2016).

2 MCH Projections to the Mesopontine Cholinergic Complex

It has long been considered that acetylcholine plays an important role in the regulation of REM sleep. Recent studies, however, reported that it does not play a major role in REM initiation but strengthens non-REM (NREM) to REM transitions once initiated (Grace et al. 2014). Likewise, selective optogenetic activation of cholinergic neurons in the PPT or LDT during NREM sleep increases the number of REM episodes and does not change REM sleep episode duration (Van Dort et al. 2015). On the other hand, whole-cell patch-clamp recordings using mice brain slices demonstrated that acetylcholine produces synergistic, excitatory pre- and postsynaptic responses on spinally projecting SLD neurons that in turn probably serve to promote REM atonia (Weng et al. 2014).

A number of evidence suggest that MCH+/GABA+ neurons in the LH might be responsible for the genesis of REM sleep (Luppi et al. 2013; Schwartz and Kilduff 2015). MCH neurons project heavily to cholinergic nuclei in the basal forebrain (i.e., globus pallidus and medial septal complex) as well as brain stem cholinergic nuclei (i.e., PPT and LDT) of the rat (Hong et al. 2011; Lima et al. 2013; Chometton et al. 2014). Microinjection of orexin (ORX) induced long-lasting excitation of cholinergic and non-cholinergic neurons in the rat LDT (Takahashi et al. 2002). MCH neurons generally project to the same targets as ORX cells, and their local interactions seem to determine various behavioral states. The MCH prevents the excitation of ORX neurons, while the ORX directly excites MCH neurons (Rao et al. 2008; Barson et al. 2013). Reciprocal synaptic connectivity exists in the lateral hypothalamus (LH), and ORX receptor-1 immunoreactivity is present in MCH neurons (Bäckberg et al. 2002; Guan et al. 2002). In slice preparation, ORX evoked a direct inward current, increasing excitatory synaptic activity (and spike frequency) in normally silent MCH neurons (van den Pol et al. 2004). Overall, the strength in the activity of MCH or ORX neurons seems to determine the sleep/wake states (Rao et al. 2008; Konadhode et al. 2015).

We showed that MCH fibers provide massive projection to the brain stem cholinergic nuclei (i.e., PPT and LDT) as compared to the monoaminergic nuclei (i.e., DR and LC), while the opposite is true for ORXergic fibers (Hong et al. 2011; Yoon and Lee 2013). The PPT and LDT contain distinct populations of cholinergic, GABAergic, and glutamatergic neurons in the rat (Wang and Morales 2009). Furthermore, all identified cholinergic neurons in the PPT/LDT discharge maximally during W and REM sleep; GABAergic and glutamatergic neurons discharge maximally during REM and W, or specifically during REM or W (Boucetta et al. 2014).

Despite massive MCH fiber projections to the brain stem cholinergic nuclei, their roles in the modulation of the sleep/wake states remain to be established.

3 MCH Projections to Brain Stem Monoaminergic Nuclei as well as the Amygdaloid Complex

There are two MCH receptors, but only MCH receptor-1 (i.e., SLC-1) is present in the rat (Tan et al. 2002). The LC and DR regions are positive for both SLC-1 mRNA and its protein product (Hervieu et al. 2000; Saito et al. 2001). Hypothalamic MCH neurons project to the LC and TMN (Yoon and Lee 2013; Lee et al. 2017), where they antagonize the excitatory effect of ORX on noradrenergic LC and histaminergic TMN neurons (Bisetti et al. 2009). MCH varicosities containing vesicular GABA transporter (vGAT) form GABAergic (gephyrin+) synapses on LC neurons (Del Cid-Pellitero and Jones 2012). Microinjection of MCH into the LC (and DR) increases REM sleep duration (Lagos et al. 2009; Torterolo et al. 2011; Monti et al. 2015). LC (and DR) neurons might, in turn, inhibit REM sleep (thus promote W) through a tonic excitation of REM-off vlPAG/dDpMe neurons (Sapin et al. 2009; Luppi et al. 2011).

The amygdala is another area exhibiting intense MCH receptor-1 mRNA and its protein product in the rat (Hervieu et al. 2000; Saito et al. 2001). MCH axon terminals are abundant in lateral, basolateral (BLA), and basomedial amygdala of the pig (Chometton et al. 2014). A recent study reported that MCH and ORX are transcriptionally upregulated in the BLA neurons after repeated stress in mice depression model (Kim et al. 2015). The anterior basomedial as well as anterior cortical nuclei of the amygdala, which are involved in processing olfactory, gustatory, and visceral information, exert excitatory (i.e., glutamatergic) influence on the MCH-containing LH neurons (Niu et al. 2012). In contrast, the central amygdala (CeA) provides inhibitory (i.e., GABAergic) inputs to the MCH and ORX neurons in the LH (Nakamura et al. 2009).

Considering the involvement of amygdala in processing emotional behaviors, its role in cataplexy (as well as in REM sleep) is of great interest. Positron emission tomography study in man revealed that the amygdaloid complex, which plays a role in the acquisition of emotionally influenced memories, is one of the major regions activated during REM sleep (Maquet et al. 1996). Populations of cells in the central and basal amygdala (which are the major output and input regions of the complex, respectively) increase discharges prior to and during cataplexy (Gulyani et al. 2002). Furthermore, lesions in the amygdala reduce cataplexy in orexin knockout mice (Burgess et al. 2013).

During W, ORX neurons maintain muscle tone by activating the basal forebrain, TMN, DR, and LC (Kiyashchenko et al. 2001; Konadhode et al. 2015). Several factors contribute to skeletal muscle atonia during REM sleep (as well as cataplexy). Acetylcholine contributes to REM sleep atonia through excitation of spinally projecting SLD neurons (Sakai et al. 2001; McCarley 2007; Weng et al. 2014). A

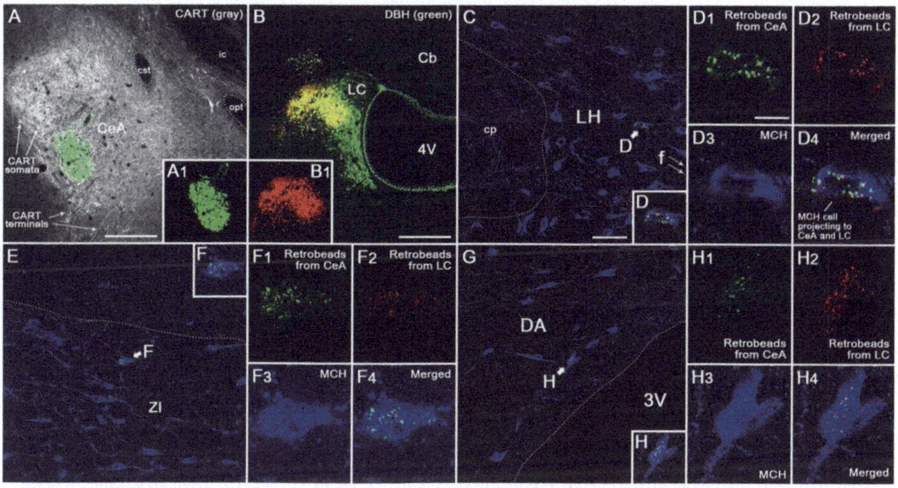

Fig. 1 Following the combined injection of green Retrobeads™ into the central amygdala (CeA) and red one into the locus coeruleus (LC), MCH neurons with dual projections to the two targets were observed in the lateral hypothalamus (LH, **c**–$\mathbf{d}_4$), zona incerta (ZI, **e**–$\mathbf{f}_4$), and dorsal hypothalamic area (DA, **g**–$\mathbf{h}_4$). Images shown are single 3.4 μm (**a** and **b**) and 0.8 μm (**c**–$\mathbf{h}_4$) optical sections. *3(4)V* third (fourth) ventricle, *Cb* cerebellum, *cp* cerebral peduncle, *cst* commissural stria terminalis, *f* fornix, *ic* internal capsule, *opt* optic tract. Scale bars = 150 μm (**a** and **b**), 25 μm (**c**), 5 μm ($\mathbf{d}_1$)

phasic glutamatergic pathway from the CeA to the caudal REM-on SLD neurons might be responsible for inducing muscle atonia during cataplexy (Luppi et al. 2011; Zhang et al. 2012). The cessation of LC neuronal activity is also a major cause of cataplexy-related muscle atonia (Wu et al. 1999; Gulyani et al. 2002). The activation of pontine/medullary motor inhibitory regions reduces LC neuronal discharge, which in turn exerts a disfacilitatory effect on ventral horn cells and produce REM atonia in both decerebrate and intact animals (Mileykovskiy et al. 2000; Lai et al. 2001). Likewise, drugs that increase noradrenaline levels are effective in alleviating cataplexy in humans, dogs, and mice (Nishino and Mignot 2011; Burgess and Peever 2013; Fraigne et al. 2015).

In narcoleptic dogs, REM-off cells in the DR do not cease discharge during cataplexy, which is in great contrast with quiescent LC neurons during the period (Wu et al. 1999, 2004). Interestingly, the CeA provides massive GABAergic input to the LC, while its projection to the DR is minor (Wallace et al. 1992; Lee et al. 2007). We observed extensive cocaine- and amphetamine-regulated transcript (CART) terminals mainly in the central and medial amygdala, which originated from CART/MCH co-containing neurons in the LH, zona incerta (ZI), and dorsal hypothalamic area (DA) (unpublished observations). Since MCH also projects to the LC (Yoon and Lee 2013), we presumed that a portion of hypothalamic MCH neurons might provide axon collaterals to the CeA and LC. The anatomical study revealed that MCH neurons in the LH, ZI, and DA provided divergent axon collaterals to the CeA and LC (Fig. 1). The neurochemical identity of these triple-labeled neurons has not been determined due to the current technical limitations. The possible involvement of these dual-projecting MCH neurons in the regulation of cataplexy (as well as REM sleep) remains elusive.

4 MCH Projections to the Histaminergic Tuberomammillary Nucleus

The TMN is different from brain stem monoaminergic nuclei (including the DR and LC) in that MCH-containing somata exist within adenosine deaminase (ADA, a histamine marker)-immunolabeled TMN boundaries (Casatti et al. 2002; Lee et al. 2017). MCH neurons are located in E_5 (rostral), E_4 (dorsomedial), and E_3 (ventral) subdivisions but not in E_2 (ventrolateral) or E_1 (caudal) regions. Types of cells within the TMN boundaries are MCH+/CART+, MCH+/CART-, or MCH-/CART+ neurons, similar to neuronal populations in the adjacent LH region. A portion of MCH neurons in the E_4 TMN subdivision project to the medial mammillary nucleus, implying their role in spatial memory process (Casatti et al. 2002).

In addition to MCH somata, the TMN contains extensive MCH (and/or CART) terminals of extrinsic origin; their neuronal cell bodies are located in the LH, ZI, and DA (Lee et al. 2017). MCHR1 is expressed in TMN neurons of the rat (Saito et al. 2001). Massive MCH terminals within the TMN might play a role in cataplexy as well as REM sleep. Histaminergic TMN neurons are unique in that they are active in cataplexy at a level similar to or even greater than during quiet waking, implying their possible involvement in the maintenance of arousal state during cataplexy (John et al. 2004). It is in contrast with brain stem monoaminergic nuclei such as the noradrenergic LC neurons which cease discharge during cataplexy of narcoleptic dogs, while serotonergic DR cells exhibit greatly reduced activity (Wu et al. 1999, 2004). Recent optogenetic study revealed that in vitro activation of MCH terminals induces $GABA_A$-mediated inhibitory postsynaptic currents (IPSCs) in histaminergic TMN neurons and in vivo activation of MCH terminals in the TMN prolongs REM sleep episodes (Jego et al. 2013). With regard to the observation of MCH cells within the TMN boundary, additional studies are required to make a functional distinction between extrinsic MCH neurons projecting to the TMN vs. intrinsic MCH cells within the TMN in the modulation of cataplexy and REM sleep.

Acknowledgment The author thanks warmly Dr. Pierre H. Luppi from University Lyon 1, France, for critical reading of the manuscript.

References

Bäckberg M, Hervieu G, Wilson S, Meister B (2002) Orexin receptor-1 (OX-R1) immunoreactivity in chemically identified neurons of the hypothalamus: focus on orexin targets involved in control of food and water intake. Eur J Neurosci 15:315–328

Barson JR, Morganstern I, Leibowitz SF (2013) Complementary roles of orexin and melanin-concentrating hormone in feeding behavior. Int J Endocrinol 2013:1–10

Bisetti A, Cvetkovic V, Bayer L, Jones BE, Serafin M, Muhlethaler M (2009) Melanin concentrating hormone antagonizes the hypocretin/orexin-induced depolarization of neurons in the locus coeruleus and ventral tuberomammillary nuclei. Abstr Soc Neurosci 277.6/EE18

Bittencourt JC, Presse F, Arias C, Peto C, Vaughan J, Nahon JL et al (1992) The melanin-concentrating hormone system of the rat brain: an immuno- and hybridization histochemical characterization. J Comp Neurol 319:218–245

Boissard R, Gervasoni D, Schmidt MH, Barbagli B, Fort P, Luppi PH (2002) The rat ponto-medullary network responsible for paradoxical sleep onset and maintenance: a combined microinjection and functional neuroanatomical study. Eur J Neurosci 16:1959–1973

Boucetta S, Cissé Y, Mainville L, Morales M, Jones BE (2014) Discharge profiles across the sleep–waking cycle of identified cholinergic, GABAergic, and glutamatergic neurons in the pontomesencephalic tegmentum of the rat. J Neurosci 26:4708–4727

Bruinstroop E, Cano G, Vanderhorst VG, Cavalcante JC, Wirth J, Sena-Esteves M, Saper CB (2011) Spinal projections of the A5, A6 (locus coeruleus), and A7 noradrenergic cell groups in rats. J Comp Neurol 520:1985–2001

Burgess CR, Peever JH (2013) A noradrenergic mechanism functions to couple motor behavior with arousal state. Curr Biol 23:1719–1725

Burgess CR, Oishi Y, Mochizuki T, Peever JH, Scammell TE (2013) Amygdala lesions reduce cataplexy in orexin knock-out mice. J Neurosci 33:9734–9742

Casatti CA, Elias CF, Sita LV, Frigo L, Furiani VC, Bauer JA, Bittencourt JC (2002) Distribution of melanin-concentrating hormone neurons projecting to the medial mammillary nucleus. Neuroscience 115:899–915

Chometton S, Franchi G, Houdayer C, Mariot A, Poncet F, Fellmann D, Tillet Y, Risold PY (2014) Different distributions of preproMCH and hypocretin/orexin in the forebrain of the pig (Sus scrofa domesticus). J Chem Neuroanat 61:72–82

Clément O, Sapin E, Libourel PA, Arthaud S, Brischoux F, Fort P, Luppi PH (2012) The lateral hypothalamic area controls paradoxical (REM) sleep by means of descending projections to brainstem GABAergic neurons. J Neurosci 32:16763–16774

Del Cid-Pellitero E, Jones BE (2012) Immunohistochemical evidence for synaptic release of GABA from melanin-concentrating hormone containing varicosities in the locus coeruleus. Neuroscience 223:269–276

Ford B, Holmes CJ, Mainville L, Jones BE (1995) GABAergic neurones in the rat pontomesencephalic tegmentum: codistribution with cholinergic and other tegmental neurones projecting to the posterior lateral hypothalamus. J Comp Neurol 363:177–196

Fraigne JJ, Torontali ZA, Snow MB, Peever JH (2015) REM sleep at its core – circuits, neurotransmitters, and pathophysiology. Front Neurol 6(123):1–9

Garcia SV, Libourel PA, Lazarus M, Grassi D, Luppi PH, Fort P (2017) Genetic inactivation of glutamate neurons in the rat sublaterodorsal tegmental nucleus recapitulates REM sleep behaviour disorder. Brain 140:414–428

Grace KP, Vanstone LE, Horner RL (2014) Endogenous cholinergic input to the pontine REM sleep generator is not required for REM sleep to occur. J Neurosci 34:14198–14209

Guan JL, Uehara K, Lu S, Wang QP, Funahashi H, Sakurai T, Yanagizawa M, Shioda S (2002) Reciprocal synaptic relationships between orexin- and melanin-concentrating hormone-containing neurons in the rat lateral hypothalamus: a novel circuit implicated in feeding regulation. Int J Obes Relat Metab Disord 26:1523–1532

Gulyani S, Wu MF, Nienhuis R, John J, Siegel JM (2002) Cataplexy-related neurons in the amygdala of the narcoleptic dog. Neuroscience 112:355–365
Hervieu GJ, Cluderay JE, Harrison D, Meakin J, Maycox P, Nasir S, Leslie RA (2000) The distribution of the mRNA and protein products of the melanin-concentrating hormone (MCH) receptor gene, slc-1, in the central nervous system of the rat. Eur J Neurosci 12:1194–1216
Hong EY, Yoon YS, Lee HS (2011) Differential distribution of melanin-concentrating hormone (MCH)- and hypocretin (Hcrt)-immunoreactive neurons projecting to the mesopontine cholinergic complex in the rat. Brain Res 1424:20–31
Jego S, Glasgow SD, Herrera CG, Ekstrand M, Reed SJ, Boyce R, Friedman J, Burdakov D, Adamantidis AR (2013) Optogenetic identification of a rapid eye movement sleep modulatory circuit in the hypothalamus. Nat Neurosci 16:1637–1643
John J, Wu MF, Boehmer LN, Siegel JM (2004) Cataplexy-active neurons in the hypothalamus: implications for the role of histamine in sleep and waking behavior. Neuron 42:619–634
Kim TK, Kim JE, Park JY, Lee JE, Choi J, Kim H, Lee EH, Kim SW, Lee JK, Kang HS, Han PL (2015) Antidepressant effects of exercise are produced via suppression of hypocretin/orexin and melanin-concentrating hormone in the basolateral amygdala. Neurobiol Dis 79:59–69
Kiyashchenko LI, Mileykovskiy BY, Lai YY, Siegel JM (2001) Increased and decreased muscle tone with orexin (hypocretin) microinjections in the locus coeruleus and Pontine Inhibitory Area. J Neurophysiol 85(5):2008–2016
Konadhode RR, Pelluru D, Shiromani PJ (2015) Neurons containing orexin or melanin concentrating hormone reciprocally regulate wake and sleep. Front Syst Neurosci 8(244):1–9
Lagos P, Torterolo P, Jantos H, Chase MH, Monti JM (2009) Effects on sleep of melanin-concentrating hormone (MCH) microinjections into the dorsal raphe nucleus. Brain Res 1265:103–110
Lai YY, Kodama T, Siegel JM (2001) Changes in monoamine release in the ventral horn and hypoglossal nucleus linked to pontine inhibition of muscle tone: an *in vivo* microdialysis study. J Neurosci 15:7384–7391
Lee HS, Eum YJ, Jo SM, Waterhouse BD (2007) Projection patterns from the amygdaloid nuclear complex to subdivisions of the dorsal raphe nucleus in the rat. Brain Res 1143:116–125
Lee EY, Hwang YG, Lee HS (2017) Hypothalamic neuronal origin of neuropeptide Y (NPY) or cocaine- and amphetamine-regulated transcript (CART) fibers projecting to the tuberomammillary nucleus of the rat. Brain Res 1657:16–28
Lima FF, Sita LV, Oliveira AR, Costa HC, da Silva JM, Mortara RA, Haemmerle CA, Xavier GF, Canteras NS, Bittencourt JC (2013) Hypothalamic melanin-concentrating hormone projections to the septo-hippocampal complex in the rat. J Chem Neuroanat 47:1–14
Lu J, Sherman D, Devor M, Saper CB (2006) A putative flip-flop switch for control of REM sleep. Nature 441(7093):589–594
Lu JW, Fenik VB, Branconi JL, Mann GL, Rukhadze I, Kubin L (2007) Disinhibition of perifornical hypothalamic neurones activates noradrenergic neurones and blocks pontine carbachol-induced REM sleep-like episodes in rats. J Physiol 582:553–567
Luppi PH, Clément O, Sapin E, Gervasoni D, Peyron C, Léger L, Salvert D, Fort P (2011) The neuronal network responsible for paradoxical sleep and its dysfunctions causing narcolepsy and rapid eye movement (REM) behavior disorder. Sleep Med Rev 15:153–163
Luppi PH, Clément O, Fort P (2013) Paradoxical (REM) sleep genesis by the brainstem is under hypothalmic control. Curr Opin Neurobiol 23:786–792
Maquet P, Péters J, Aerts J, Delfiore G, Dequeldre C, Luxen A, Franck G (1996) Functional neuroanatomy of human rapid-eye-movement sleep and dreaming. Nature 383:163–166
McCarley RW (2007) Neurobiology of REM and NREM sleep. Sleep Med 8:302–330
Mileykovskiy BY, Kiyashchenko L, Kodama T, Lai YY, Siegel JM (2000) Activation of pontine and medullary motor inhibitory regions reduce discharge in neurons located in locus coeruleus and the anatomical equivalent of the midbrain locomotor region. J Neurosci 20:8551–8558
Monti JM, Lagos P, Jantos H, Torterolo P (2015) Increased REM sleep after intra-locus coeruleus nucleus microinjection of melanin-concentrating hormone (MCH) in the rat. Prog Neuropsychopharmacol Biol Psychiatry 56:185–188

Monti JM, Torterolo P, Jantos H, Lagos P (2016) Microinjection of the melanin-concentrating hormone into the sublaterodorsal tegmental nucleus inhibits REM sleep in the rat. Neurosci Lett 630:66–69

Nakamura S, Tsumori T, Yokota S, Oka T, Yasui Y (2009) Amygdaloid axons innervate melanin-concentrating hormone- and orexin-containing neurons in the mouse lateral hypothalamus. Brain Res 1278:66–74

Nishino S, Mignot E (2011) Narcolepsy and cataplexy. Handb Clin Neurol 99:783–814

Niu JG, Yokota S, Tsumori T, Oka T, Yasui Y (2012) Projections from the anterior basomedial and anterior cortical amygdaloid nuclei to melanin-concentrating hormone-containing neurons in the lateral hypothalamus of the rat. Brain Res 1479:31–43

Rao Y, Lu M, Ge F, Marsh DJ, Qian S, Wang AH, Picciotto MR, Gao XB (2008) Regulation of synaptic efficacy in hypocretin/orexin-containing neurons by melanin concentrating hormone in the lateral hypothalamus. J Neurosci 26:9101–9110

Saito Y, Cheng M, Leslie FM, Civelli O (2001) Expression of the melanin-concentrating hormone (MCH) receptor mRNA in the rat brain. J Comp Neurol 435:26–40

Sakai K, Crochet S, Onoe H (2001) Pontine structures and mechanisms involved in the generation of paradoxical (REM) sleep. Arch Ital Biol 139:93–107

Sapin E, Lapray D, Berod A, Goutagny R, Leger L, Ravassard P, Clément O, Hanriot L, Fort P, Luppi PH (2009) Localization of the brainstem GABAergic neurons controlling paradoxical (REM) sleep. PLoS One 4:e4272

Schwartz MD, Kilduff TS (2015) The neurobiology of sleep and wakefulness. Psychiatr Clin North Am 38:615–644

Takahashi K, Koyama Y, Kayama Y, Yamamoto M (2002) Effects of orexin on the laterodorsal tegmental neurones. Psychiatry Clin Neurosci 56:335–336

Tan C, Sano H, Iwaasa H, Pan J, Sailer AW, Hreniuk DL et al (2002) Melanin-concentrating hormone receptor subtypes 1 and 2: species-specific gene expression. Genomics 79:785–792

Torterolo P, Lagos P, Monti JM (2011) Melanin-concentrating hormone: a new sleep factor? Front Neurol 2(14):1–12

van den Pol AN, Acuna-Goycolea C, Clark KR, Ghosh PK (2004) Physiological properties of hypothalamic MCh neurons identified with selective expression of reporter gene after recombinant virus infection. Neuron 42:635–652

Van Dort CJ, Zachs DP, Kenny JD, Zheng S, Goldblum RR, Gelwan NA et al (2015) Optogenetic activation of cholinergic neurons in the PPT or LDT induces REM sleep. Proc Natl Acad Sci USA 112:584–589

Verret L, Goutagny R, Fort P, Cagnon L, Salvert D, Léger L, Boissard R, Salin P, Peyron C, Luppi PH (2003) BioMed Central Neurosci 4:1–10

Wallace DM, Magnuson DJ, Gray TS (1992) Organization of amygdaloid projections to brainstem dopaminergic, noradrenergic, and adrenergic cell groups in the rat. Brain Res Bull 28:447–454

Wang HL, Morales M (2009) Pedunculopontine and laterodorsal tegmental nuclei contain distinct populations of cholinergic, glutamatergic and GABAergic neurons in the rat. Eur J Neurosci 29:340–358

Weng FI, Williams RH, Hawryluk JM, Lu J, Scammell TE, Saper CB, Arrigoni E (2014) Carbachol excites sublaterodorsal nucleus neurons projecting to the spinal cord. J Physiol 592:1601–1617

Wu MF, Gulyani SA, Yau E, Mignot E, Phan B, Siegel JM (1999) Locus coeruleus neurons: cessation of activity during cataplexy. Neuroscience 91:1389–1399

Wu MF, John J, Bochmer LN, Yau D, Nguyen GB, Siegel JM (2004) Activity of dorsal raphe cells across the sleep-waking cycle and during cataplexy in narcoleptic dogs. J Physiol 554:202–215

Yoon YS, Lee HS (2013) Projections from melanin-concentrating hormone (MCH) neurons to the dorsal raphe or the nuclear core of the locus coeruleus in the rat. Brain Res 1490:72–82

Zhang J, Xi M, Fung SJ, Sampogna S, Chase MH (2012) Projections from the central nucleus of the amygdala to the nucleus pontis oralis in the rat: an anterograde labelling study. Neurosci Lett 525:157–162

The Role of Melanin-Concentrating Hormone in the Regulation of the Sleep/Wake Cycle: Sleep Promoter or Arousal Modulator?

Xiao-Bing Gao

Abstract The neuronal system that resides in the perifornical/lateral hypothalamus (Pf/LH) and synthesizes the neuropeptide melanin-concentrating hormone (MCH) participates in critical brain functions across species from fish to human. The MCH system was originally reported to regulate neural activity responsible for the change of skin color in lower vertebrates (such as fish), homeostatic functions (e.g., feeding and energy balance), and complex behaviors (e.g., reward seeking/addiction, depression/anxiety) in higher animals. The most recent evidence indicates that MCH participates in the regulation of sleep in animals as well. In this chapter, we summarize the evidence supporting the role of MCH in the regulation of sleep homeostasis in animals and discuss its physiological implications in the context of MCH as a central node of the overall brain circuitry controlling physiological functions and complex behaviors in animals.

1 Introduction

The periodic rotation of the earth on its axis in relation to the sun induces the light/dark cycle on the planet, based upon which animals (higher animals in particular) exhibit a periodic activity-rest pattern. Although its exact significance is still under debate, the activity-rest pattern (sleep/wake cycle) is critical to the survival of individual animals and their species from *C. elegans* to *Homo sapiens*. The brain of higher animals has been established as the organ that controls the sleep/wake cycle and regulates physiological and behavioral functions based upon this cycle. Among many brain regions, the hypothalamus emerged as a critical area that regulates the sleep/wake state as early as the first half of the twentieth century, when von Economo described that the disruption of the posterior hypothalamus led to sleepiness and coma in human patients (von Economo 1930).

X.-B. Gao (✉)
Department of Comparative Medicine, Program on Integrative Cell Signaling and Neurobiology of Metabolism (ICSNM), Yale University School of Medicine, New Haven, CT, USA
e-mail: xiao-bing.gao@yale.edu

S. R. Pandi-Perumal et al. (eds.), *Melanin-Concentrating Hormone and Sleep*,
https://doi.org/10.1007/978-3-319-75765-0_3

In the past two decades, investigations on the role played by the hypothalamus in the regulation of the sleep/wake cycle intensively centered on the neuropeptide hypocretin/orexin (Hcrt) and its receptors based on the discovery of Hcrt as a potent wake/arousal promoter in animals. Hcrt was originally discovered by two groups of researchers independently (de Lecea et al. 1998; Sakurai et al. 1998), which is selectively synthesized in neurons in the perifornical and lateral hypothalamus (Pf/LH). Two distinct peptidergic hormones named hypocretin 1 and 2 (orexin-A and B) were determined to be the products of proteolytic cleavage of a pre-prohormone (de Lecea et al. 1998). Two G protein-coupled receptors, orexin receptor 1 (hypocretin receptor 1) and orexin receptor 2 (hypocretin receptor 2), were identified as mediators of the effects of both peptides (Sakurai et al. 1998).

The fact that Hcrt exhibits a strong arousal/wake-promoting effect on animals is supported by current data. First, a deficiency in Hcrt peptide or its receptors leads to narcolepsy in dogs, mice, and human patients (Chemelli et al. 1999; Lin et al. 1999; Nishino et al. 2000; Thannickal et al. 2000; Ripley et al. 2001). Second, the concentration of Hcrt fluctuates according to the behavioral states of animals. Hcrt-1 levels are high in the cerebrospinal fluid during the active (wake) phase and low during the inactive (sleep) period (Fujiki et al. 2001; Yoshida et al. 2001; Zeitzer et al. 2003). These findings are consistent with those seen when activity levels of Hcrt cells across various behavioral states were assessed by c-Fos expression and detection of spike firing in these cells during the normal sleep/wake cycle and during sleep deprivation (Estabrooke et al. 2001; Lee et al. 2005; Mileykovskiy et al. 2005). Correspondingly, a direct and selective stimulation of Hcrt neurons by the optogenetic approach increases the transition from either SW sleep or REM sleep to wakefulness in rats (Adamantidis et al. 2007). Lastly, a dual Hcrt receptor antagonist has recently been developed as an effective treatment for insomnia (Roecker et al. 2016).

Although the evidence briefly summarized above seems sufficient to address the original observations of impaired wakefulness and arousal in patients suffering from the disruption of the lateral hypothalamus (von Economo 1930), nerve cells that are specifically activated during the sleep phase have been identified in the LH as well (Steininger et al. 1999; Alam et al. 2002). A growing body of data reveals that neurons in the LH may promote sleep in animals. Nerve cells synthesizing the neuropeptide melanin-concentrating hormone (MCH), which are also distributed in the LH and intermingle with Hcrt cells, exert this exact function in animals.

2 An Overview of MCH and Its Receptors in Animals

2.1 Discovery of MCH and MCH Receptors in Animals

MCH was originally isolated from the pituitary gland and hypothalamus in teleost fishes and found to be a regulator of skin color in these species (Rance and Baker 1979). The release of MCH in teleost fishes was regulated by the background color of the environment (Rance and Baker 1979). Later, MCH-like immunoreactivity was identified predominantly in the posterior hypothalamic area (including the medial forebrain bundle-lateral hypothalamic area, subzona incerta, and perifornical area) in rats and humans (Skofitsch et al. 1985; Bresson et al. 1987). It is now clear that MCH is a cyclic 17-amino-acid peptide in fishes and 19-amino-acid peptide in mammals. MCH and two other peptides (NGE and NEI) are cleavage products of the same prohormone in mammals (Nahon et al. 1989).

The receptors for MCH were not identified until a decade after the discovery of this peptide. An orphan G protein-coupled receptor, GPCR SLC-1 (also called GPR24), was eventually identified as the receptor for MCH in rodents and humans by several groups independently (Saito et al. 1999; Chambers et al. 1999; Lembo et al. 1999). Another orphan G protein-coupled receptor, GPCR SLT, was later identified as an MCH receptor in humans only (Mori et al. 2001). These two receptors were then named MCH receptor 1 (MCHR1) and 2 (MCHR2), respectively.

2.2 *Connectomics of MCH Neurons in Animals*

It is now clear that MCH is synthesized in the central nervous system (CNS) mainly by neurons in the lateral hypothalamus (LH) in mammals (Skofitsch et al. 1985; Bresson et al. 1987; Nahon et al. 1989). The expression of MCH begins as early as the late embryo and early postnatal period (Bresson et al. 1987; Brischoux et al. 2001). MCH-containing efferent axons innervate a wide variety of regions in the CNS from the cortex to the spinal cord (Skofitsch et al. 1985; Nahon et al. 1989; Bittencourt et al. 1992). Consistent with this wide distribution pattern of MCH-containing fibers, MCHR1 is also widely distributed throughout the CNS (Lembo et al. 1999; Saito et al. 2001; Kilduff and de Lecea 2001). Among all the brain structures innervated by MCH-containing fibers, the highest density of MCH axons and boutons is found in the LH (Nahon et al. 1989; Bittencourt et al. 1992). By using an immunohistochemical approach with an antiserum against MCH and antisense against MCH mRNA, Bittencourt et al. (1992) reported that the most intensely immunoreactive nerve fibers were in the lateral hypothalamus, medial nucleus of the septum, and certain areas of the reticular formation. The MCH-containing fibers were also identified extensively in brain areas regulating the sleep/wake cycle, such as dorsal raphe (DR), locus coeruleus (LC), and tuberomammillary nucleus (TMN), which suggests a critical role of MCH in the regulation of wake/arousal in animals. Conversely, afferent pathways onto MCH-containing neurons from various parts of the brain are as abundant as projections from the MCH system to other regions all over the brain (González et al. 2016). This piece of evidence indicates that MCH cells are under the modulation of cues that encode physiological, environmental, and behavioral factors.

Within the perifornical and LH areas, current data indicate that MCH cells reside in a microcircuitry comprising several neuronal populations (Fig. 1). There is evidence that MCH neurons and neighboring Hcrt neurons reciprocally innervate each other (Guan et al. 2002; van den Pol et al. 2004). However, since MCH does not

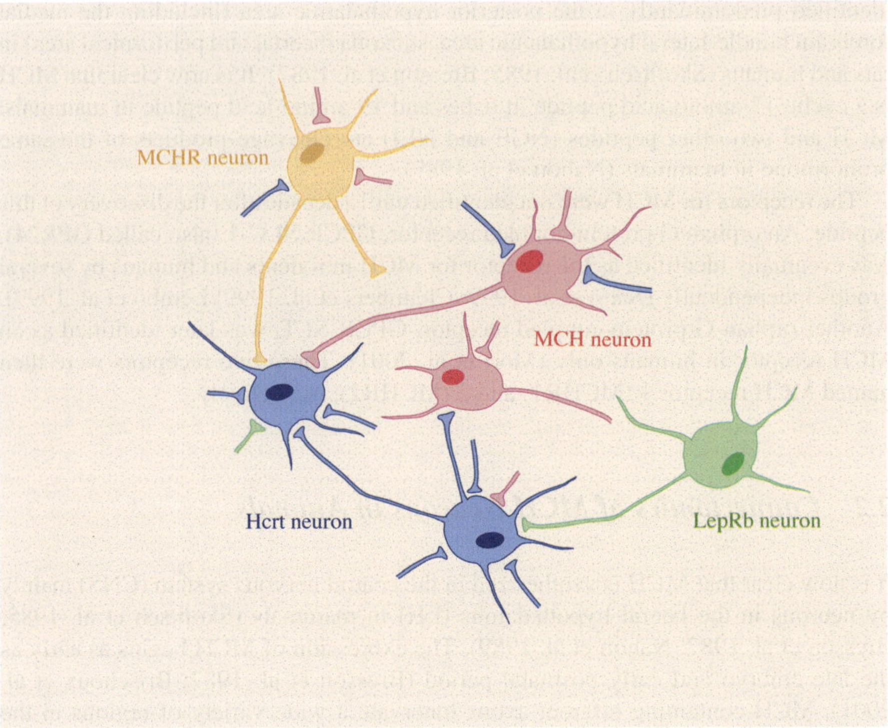

Fig. 1 A diagram of a simplified local circuitry containing MCH neurons in the LH. MCH cells may reciprocally innervate nearby Hcrt neurons directly and modulate Hcrt neurons indirectly through the innervation of MCH receptor-expressing (MCHR) interneurons

have direct but rather has indirect effects on Hcrt neurons (Rao et al. 2008) and the latter does not belong to the neuronal population expressing MCHR1 (Chee et al. 2013), it is possible that the MCH system may also regulate wake-promoting Hcrt neurons with its peptidergic content through MCHR1-expressing interneurons (Fig. 1).

2.3 A Diverse Role of MCH in the Regulation of Physiological Functions in Animals

The original function of the peptide MCH was identified to modulate pigmentation in lower vertebrates (such as fish, amphibians, and reptiles) (Kawauchi et al. 1983; Eberle 1988), for which the peptide was named. In fish, in addition to a strong neuroendocrine role in the inhibition of corticotropin-releasing hormone (Baker et al. 1985; Eberle 1988), MCH also exhibits appetite-inhibiting actions in some species

(Shimakura et al. 2006; Matsuda et al. 2007) and appetite-stimulating effects in others (Tuziak and Volkoff 2012, 2013).

In mammals, however, MCH plays a profound role in the regulation of food intake and energy balance in the brain. Firstly, the acute administration of MCH directly into the arcuate, paraventricular, or dorsomedial nucleus evokes feeding or elevates food intake in rodents (Qu et al. 1996; Rossi et al. 1999; Della-Zuana et al. 2002; Abbott et al. 2003; Clegg et al. 2003), while a chronic infusion of MCH or activation of MCHR1 significantly increases food intake, body weight, white adipose tissue (WAT) mass, and liver mass and elevates levels of insulin and leptin in rodents (Ito et al. 2003; Shearman et al. 2003). Overexpression of MCH in the LH leads to obesity and resistance to insulin in mice (Ludwig et al. 2001). Conversely, a deficiency in MCH peptide causes reduced body weight and leanness due to a lessened food intake and an enhanced metabolism in mice (Shimada et al. 1998). The inhibition of MCH neurons by physiological factors such as estrogen or antagonism of MCHR1 leads to a sustained decrease in food intake, weight loss, and reductions in body weight/fat gain (Mystkowski et al. 2000; Shearman et al. 2003; Morton et al. 2004). In addition, MCH has been reported to be involved in the regulation of learning and memory, depression-like behaviors, anxiety, and addiction (Borowsky et al. 2002; Adamantidis et al. 2005; Georgescu et al. 2005; David et al. 2007; Chung et al. 2009; Lagos et al. 2011b; García-Fuster et al. 2012; Domingos et al. 2013). The latest evidence indicates that MCH participates in the regulation of sleep, which will be discussed in the latter part of this chapter.

3 Cellular and Molecular Effects of MCH on Nerve Cells in Animals

3.1 *MCH Acts as an Inhibitory Neuropeptide in Animals*

When MCH receptors were initially identified, it was clear that G protein-coupled signaling pathways might mediate biological effects of MCH. In in vitro experiments with a non-neuronal expression system (*Xenopus* oocyte) and cell lines (HEK and CHO), the activation of MCH receptor-1 led to an increase in intracellular calcium concentrations, a decrease in cyclic AMP levels, and induction of G protein-gated inwardly rectifying potassium currents (GIRKs) (Bächner et al. 1999; Chambers et al. 1999; Saito et al. 1999; Shimomura et al. 1999), suggesting diverse effects of MCH depending on the signaling pathways (such as G_i, G_o, or G_q pathways) downstream to the activation of MCH receptors.

Conversely, in the primary culture of neurons from the hypothalamus, MCH was demonstrated to act as an inhibitory peptide (Gao and van den pol 2001, 2002) in terms of direct effects on individual neurons. MCH inhibited the release of glutamate from presynaptic terminals and glutamate receptors at the postsynaptic membrane in cultured neurons from the LH in rats (Gao and van den pol 2001). In contrast to the

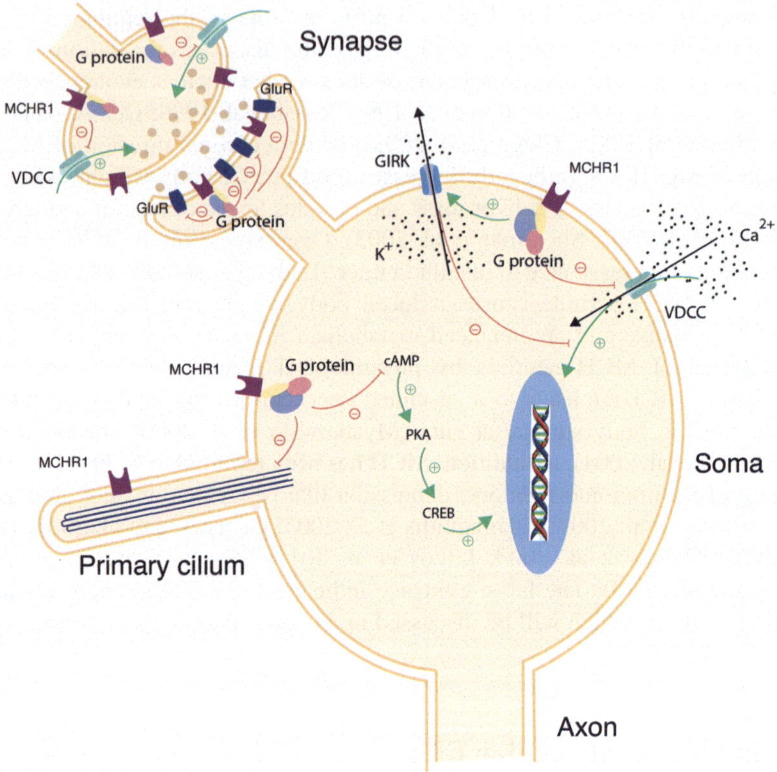

Fig. 2 A diagram depicting signaling pathways underlying the inhibitory effects of MCH on nerve cells in the brain. MCH may directly modulate synaptic transmission, neurotransmitter receptor responses, voltage-gated ion channels (such as Ca^{2+} and K^{+} channels), secondary messenger systems, and primary cilia in neurons

activation of GIRKs in *Xenopus* oocytes, MCH inhibited voltage-dependent calcium channels (VDCCs) in hypothalamic neurons in primary culture (Gao and van den pol 2002). The G_i/G_o pathway was suggested to mediate these inhibitory effects (Gao and van den pol 2001, 2002). In other regions of the brain in rodents, the majority of results demonstrated that MCH exerted inhibitory effects on central neurons as well. For instance, MCH inhibited spontaneous and evoked excitatory postsynaptic currents in NTS neurons in medulla slices from rats (Zheng et al. 2005); it induced a direct inhibition through an activation of barium-sensitive potassium currents in a subpopulation of GnRH neurons in the medial septum/diagonal band of Broca (MSDB) (Wu et al. 2009); and it depressed action potential firing in medium spiny neurons (MSNs) in the shell of nucleus accumbens (AcbSh) in rat brain slices and freely moving animals (Sears et al. 2010) (Fig. 2).

In addition to the modulation of cellular functions through direct and indirect effects on the excitability of neurons, MCH may also modify morphological properties of nerve cells and lead to fine-tuning of neuronal responses to afferent signals. Primary cilia are microtubule-based cellular organelles projecting from surfaces of many cell types including central neurons. Emerging evidence has showed that neuronal cilia could serve as an important modulator of neuronal functions by monitoring the ambient environment of nerve cells (Guemez-Gamboa et al. 2014; Oh et al. 2015). The disruption of neuronal cilia has been implicated in many neurological and mental diseases, particularly in obesity (Guemez-Gamboa et al. 2014; Oh et al. 2015). Recent data indicate that MCHR1 is exclusively expressed in the primary cilia and is responsible for the reduction of cilium length (Berbari et al. 2008; Hamamoto et al. 2016; Tomoshige et al. 2017), which may be critical to the brain functions regulated by MCH, such as energy balance (Oh et al. 2015) (Fig. 2).

Earlier studies suggested that neurons synthesizing MCH were GABAergic in nature, which has been repeatedly demonstrated by many groups using immunocytochemical and in situ hybridization methods (Sapin et al. 2010; Del Cid-Pellitero and Jones 2012). However, more recent data raised the possibility that at least a subpopulation of MCH neurons may be glutamatergic (Chee et al. 2015; Mickelsen et al. 2017). The heterogeneity in MCH neurons in terms of neurotransmitters released from these cells adds another layer of complexity in the regulation of the targets of the MCH system at the circuit level.

3.2 *Signaling Pathways Underlying the Effects Mediated by MCH*

As presented above, the expression of exogenous MCHR1 in non-neuronal expressing systems led to coupling of the receptor with various G protein-mediated pathways (such as G_i, G_o and G_q pathways) that exist in those cells (Bächner et al. 1999; Chambers et al. 1999; Saito et al. 1999; Shimomura et al. 1999). In neuronal primary cultures and acute brain slices from the central nervous system of rodents, current evidence suggests the coupling of MCHR1 with inhibitory G protein pathways, for example, the G_i/G_o pathway (Gao and van den Pol 2001), through which MCH may directly regulate excitability via ion channels (VDCCs and K^+ channels) and neurotransmitter release in nerve cells (Gao and van den Pol 2001, 2002; Wu et al. 2009; Sears et al. 2010). MCH-induced shortening of neuronal cilia may be mediated by Akt phosphorylation downstream to the G_i/G_o pathway activation (Hamamoto et al. 2016; Tomoshige et al. 2017).

4 Regulation of Sleep by MCH in Animals

4.1 Regulation of the Sleep/Wake Cycle by MCH in Animals

As a potent promoter of food intake in higher animals, MCH's action in the regulation of sleep homeostasis was revealed rather late, especially when compared to the discovery of the neighboring Hcrt cells as wake/arousal promoters (Chemelli et al. 1999; Lin et al. 1999). The first clue of a possible role for the PF/LH area in the promotion of sleep was the finding that certain neuronal populations were activated during the paradoxical sleep (PS) phase in rats (Steininger et al. 1999; Alam et al. 2002). To identify these PS-active cells in the PF/LH area, Verret and colleagues reported that MCH neurons, but not Hcrt cells, expressed c-Fos immunoreactivity in rats undergoing recovery sleep after PS deprivation (Verret et al. 2003). Furthermore, they reported that i.c.v. application of MCH through the lateral ventricles increased the amount of PS and slow-wave sleep (SWS) during the dark phase in rats in a dose-dependent manner (Verret et al. 2003). This is the first piece of evidence that MCH promoted sleep in rodents, which was later confirmed by other investigators (Modirrousta et al. 2005; Hanriot et al. 2007). It is interesting to note, however, that MCH neurons appeared not to express c-Fos (activation) during quiet sleep or active sleep induced by carbachol microinjected directly into the nucleus pontis oralis (NPO) in cats (Torterolo et al. 2006). A later report clarified that a direct infusion of MCH into the NPO could increase active sleep and decrease the latency to active sleep in felines (Torterolo et al. 2009). These data suggest that the role of MCH in sleep regulation may be state dependent.

Consistent with the co-relationship between activation of MCH neurons and various stages of sleep in rodents, the application of MCHR1 antagonist revealed more details regarding the role of MCH in sleep regulation. Subcutaneous administration of two selective MCHR1 antagonists significantly decreased NREM, REM, and intermediate stage sleep duration and enhanced wake time during the first 4 h after administrations in rats (Ahnaou et al. 2008). The loss-of-function studies with genetically modified animals revealed similar results. Mice with a deficiency in MCH peptide exhibited a reduced sleep in both dark and light phases, suppressed REM sleep, and hyperactivity during food deprivation, which indicate an enhanced vigilance/arousal state in these animals (Willie et al. 2008). In a strain of MCHR1-deficient mice with a behavioral phenotype of hyperactivity during the dark phase of the light-dark cycle (Marsh et al. 2002), an elevated arousal state as demonstrated by a sensitization to the wake-promoting effect of the psychostimulant, modafinil, through the Hcrt pathway was observed (Rao et al. 2008), despite an altered vigilant state and response to modafinil in another strain of MCHR1 mice (Adamantidis et al. 2005, 2008). Nevertheless, most of the available evidence supports the concept that MCHR1 deficiency leads to enhanced wake/arousal and compromised sleep (Marsh et al. 2002; Rao et al. 2008; Ahnaou et al. 2011; but not Adamantidis et al. 2008).

The use of in vivo pharmacological and transgenic approaches has certainly led to the discovery of regulation of sleep by the MCH system in animals. However, the lack of specificity and possible developmental and compensatory effects resulting from genetic mutations will surely compromise the interpretations of results obtained from studies summarized above. New techniques such as optogenetic and chemogenetic methods and in vivo imaging with calcium-sensitive reporter proteins have shed new light on the regulation of sleep by the MCH system. By selectively stimulating MCH neurons with light-sensitive channgelrhodopsin-2 (ChR2) for a 24-h period, Konadhode and colleagues showed that stimulation of MCH cells during the dark phase of the light-dark cycle significantly reduced the latency of sleep onset, decreased the length of wake bouts, and increased the total time in both REM and NREM sleep in mice (Konadhode et al. 2013). Activation of MCH cells during the light phase enhanced sleep intensity in these animals (Konadhode et al. 2013). At the same time, Jego and colleagues reported that an acute stimulation of MCH neurons in the LH led to an extended REM but not NREM sleep at the onset of REM sleep (Jego et al. 2013), which could also be achieved by stimulating MCH-containing nerve fibers in the tuberomammillary nucleus (TMN) or medial septum (Jego et al. 2013). Interestingly, Tsunematsu and colleagues showed that acute stimulation of MCH neurons led to a transition from NREM sleep to REM sleep with an increased REM sleep time and decreased NREM sleep time in mice (Tsunematsu et al. 2014), suggesting that the activation of MCH neurons can induce REM sleep even if mice are at the stage of NREM sleep (Tsunematsu et al. 2014). In rats, optogenetic stimulation of MCH neurons induced similar changes in sleep profiles as reported by Konadhode and colleagues in mice (Blanco-Centurion et al. 2016). Moreover, a selective stimulation of MCH neurons with chemogenetic methods significantly increased REM sleep without changes in NREM sleep and wakefulness in mice (Vetrivelan et al. 2016).

4.2 *Neuronal Circuits Participating in MCH-Mediated Sleep Promoting Effects*

Since MCH-containing nerve fibers innervate structures all over the brain including many arousal-promoting centers (Bittencourt et al. 1992), it is not surprising that MCH may promote sleep through multiple loci in the brain. Since MCH-containing fibers are extensively distributed in the LH and MCH-containing nerve terminals are identified on wake-promoting Hcrt neurons (Guan et al. 2002; van den Pol et al. 2004), it is intriguing to understand whether the neighboring neuronal systems are targets of MCH cells in the regulation of the sleep/wake cycle. By using in vivo single-unit recording and juxtacellular labeling techniques in head-fixed rats, Hassani and colleagues reported that MCH neurons fired action potentials selectively during sleep with maximal firing during the PS phase but not during wakefulness (Hassani et al. 2009). During the transition between wake and sleep, wake off/sleep on MCH cells generated action potentials in a reciprocal way as compared to wake on/sleep off Hcrt neurons (Hassani et al. 2009). This is the first evidence that shows a possible interaction between wake-promoting and sleep-promoting neuronal systems in the LH. Microinjection of MCH directly into the dorsal raphe nucleus (DR), a region extensively innervated by MCH-containing nerve fibers, increased both REM and slow-wave sleep (Lagos et al. 2009), while an administration of anti-MCH antibody into the DR induced an opposite effect (Lagos et al. 2011a). A direct infusion of MCH into the lateral basal forebrain decreased wake time and increased REM sleep in rats (Lagos et al. 2012). A direct infusion of MCH into the NPO could increase active sleep and decrease the latency to active sleep in cats (Torterolo et al. 2009). The bilateral infusion of MCH into the ventrolateral preoptic area (VLPO) and adjacent dorsal lateral preoptic region increased light sleep and total non-REM sleep (Benedetto et al. 2013). MCH neurons were identified to participate in the inhibition of PS-off GABAergic neurons in the ventrolateral part of the periaqueductal gray (VLPAG) and the adjacent dorsal deep mesencephalic nucleus (dDpMe), which promotes PS in animals (Clément et al. 2012). The infusion of MCH into the locus coeruleus (LC) increased REM sleep (Monti et al. 2015). In summary, MCH acts at multiple brain sites to generate its effects on sleep in animals.

4.3 Physiological Significance of MCH-Mediated Promotion of Sleep Homeostasis

Although the role of MCH as a sleep regulator is emerging, the exact physiological significance of this function by the MCH system in higher animals is not yet clear. Based on information presented here, it is reasonable to propose that the MCH system might be an arousal regulator rather than a sleep promoter in animals.

First, compared with other systems such as GABAergic neurons in the medullary brainstem parafacial zone (PZ^{GABA}) (Anaclet et al. 2014; Anaclet and Fuller 2017), MCH-mediated effects on sleep are not robust and are rather selective. In fact, current data as summarized above indicate that MCH promotes REM or paradoxical sleep in most cases, although total wake time and NREM sleep are compromised in some instances (Steininger et al. 1999; Alam et al. 2002; Verret et al. 2003; Modirrousta et al. 2005; Hanriot et al. 2007; Torterolo et al. 2009; Konadhode et al. 2013; Jego et al. 2013; Tsunematsu et al. 2014; Blanco-Centurion et al. 2016; Vetrivelan et al. 2016). The modulation of the MCH system alone is not sufficient to alter the sleep/wake cycle leading to phenotypes that are similar to sleep disorders in animals (Konadhode et al. 2013; Jego et al. 2013; Tsunematsu et al. 2014; Blanco-Centurion et al. 2016; Vetrivelan et al. 2016).

Second, compared with other physiological and behavioral functions governed or regulated by the MCH system, such as feeding and energy balance, changes in phenotypes resulting from modifications of the output of MCH cells by pharmacological, genetic, and optogenetic/chemogenetic methods are far less profound in the case of sleep regulation. Loss of function studies show that a deficiency in the MCH system leads to profound changes in gross metabolism, body weight, and depression-like behaviors but not in the sleep/wake cycle (Borowsky et al. 2002; Marsh et al. 2002).

Third, the interactions between the MCH system and other arousal-promoting systems, in particular, Hcrt, are emerging and exemplify how MCH may modulate arousal levels in higher animals. It has been shown that MCH peptide or receptor deficiency leads to enhanced locomotor activity, response to psychostimulants, metabolic rate, response to stress, and impaired learning and memory (Marsh et al. 2002; Adamantidis et al. 2005; Rao et al. 2008). These changes in behavioral phenotypes may be due to enhanced arousal levels in these mutated mice, as shown in MCHR1 KO mice (Rao et al. 2008). Elevated activity and excitatory synaptic inputs onto Hcrt neurons have been reported in MCHR1 KO mice, which facilitates the activation of the Hcrt cells (Rao et al. 2008). It will be both intriguing and essential to examine whether MCH has similar effects on neurons in other arousal-promoting systems.

Therefore, we propose here that the regulation of sleep by MCH is a physiologically significant part of the overall brain circuitry regulating general arousal levels in animals. The role of MCH has evolved from a peptide that regulates skin color to blend with the environment as a protection from predators in lower vertebrates to a player in basic homeostatic regulations and complex behaviors including emotion control, learning and memory in higher animals. Having the control to respond at an optimal arousal level during each physiological process or behavioral task is required for the successful performance of that specific process or task. For instance, it may be required to maintain a high arousal level to forage for food, while not so high of a level is necessary during food consumption (Borbély 1977). A lowered arousal level might be beneficial for conservation and storage of energy after a meal, which is essential to animals in their natural habitats where food supply is not always readily available. The role of MCH in the promotion of positive energy balance and restriction of arousal levels may be a hallmark of the basic physiological function of this system in higher animals, in which the regulation of REM sleep may be a by-product of the MCH system. In addition, since MCH is also required for the maintenance of cognitive functions (Adamantidis et al. 2005) and REM sleep is critical in the consolidation of spatial and contextual memory (Boyce et al. 2017), it is intriguing to identify whether the regulation of sleep by MCH is an essential part of the role of MCH in the promotion of higher brain functions in animals.

5 Conclusions

In summary, the role of MCH in the regulation of sleep is still an emerging picture. MCH promotes REM sleep in most cases, while it enhances NREM sleep and decreases wake time in other instances in animals. Consistent with its wide projection field in the brain, the MCH system executes its sleep regulatory effects through several well-defined wake-promoting and sleep-promoting centers in the brain including the LH, DR, LC, VLPO, etc. Since MCH exhibits inhibitory effects through the activation of G_i/G_o pathways on central neurons in the brain, it is expected that MCH is able to modulate wake-promoting and sleep-promoting neurons directly or indirectly depending on the local circuits where MCH receptors are expressed.

The physiological significance of MCH-mediated effects on sleep is not yet clearly defined. More questions than answers have been raised on how exactly the MCH system participates in sleep regulation. From a broader point of view, is the sleep-regulatory effect of MCH only the tip of the iceberg? One future direction may be trying to understand how the regulatory effects of MCH on sleep interact with its other functions as a regulator of energy balance, anxiety, and depression-like behaviors, which are all closely associated with sleep. Another avenue for future study is a complete investigation of neuronal mechanisms underlying the role of MCH in the regulation of different stages of sleep (REM, NREM, quiet wake, and activity wake), which would lead to a comprehensive understanding of the role of MCH in sleep regulation and arousal modulation. Lastly, it is also essential to identify whether MCH-mediated sleep regulation contributes to sleep disorders resulting from altered metabolism (such as obesity), behaviors (such as drug addiction), and other diseases in humans. Nevertheless, the understanding of sleep regulation by MCH under various physiological and pathological conditions would greatly advance our knowledge on the control of sleep/wake cycle in animals and sleep disorders resulting from a dysregulated MCH system in humans.

Acknowledgments The author's work has been supported by NIH/NIDA grant DA040782-01A1. The author would like to thank Ms. Marya Shanabrough for assistance in preparing the manuscript.

References

Abbott CR, Kennedy AR, Wren AM, Rossi M, Murphy KG, Seal LJ, Todd JF, Ghatei MA, Small CJ, Bloom SR (2003) Identification of hypothalamic nuclei involved in the orexigenic effect of melanin-concentrating hormone. Eur J Pharmacol 475:37–47

Adamantidis A, Thomas E, Foidart A, Tyhon A, Coumans B, Minet A, Tirelli E, Seutin V, Grisar T, Lakaye B (2005) Disrupting the melanin-concentrating hormone receptor 1 in mice leads to cognitive deficits and alterations of NMDA receptor function. Eur J Neurosci 21:2837–2844

Adamantidis AR, Zhang F, Aravanis AM, Deisseroth K, de Lecea L (2007) Neural substrates of awakening probed with optogenetic control of hypocretin neurons. Nature 450:420–424

Adamantidis A, Salvert D, Goutagny R, Lakaye B, Gervasoni D, Grisar T, Luppi PH, Fort P (2008) Sleep architecture of the melanin-concentrating hormone receptor 1-knockout mice. Eur J Neurosci 27:1793–1800

Ahnaou A, Drinkenburg WH, Bouwknecht JA, Alcazar J, Steckler T, Dautzenberg FM (2008) Blocking melanin-concentrating hormone MCH1 receptor affects rat sleep-wake architecture. Eur J Pharmacol 579:177–188

Ahnaou A, Dautzenberg FM, Huysmans H, Steckler T, Drinkenburg WH (2011) Contribution of melanin-concentrating hormone (MCH1) receptor to thermoregulation and sleep stabilization: evidence from MCH1 (−/−) mice. Behav Brain Res 218:42–50

Alam MN, Gong H, Alam T, Jaganath R, McGinty D, Szymusiak R (2002) Sleep-waking discharge patterns of neurons recorded in the rat perifornical lateral hypothalamic area. J Physiol 538:619–631

Anaclet C, Fuller PM (2017) Brainstem regulation of slow-wave-sleep. Curr Opin Neurobiol 44:139–143

Anaclet C, Ferrari L, Arrigoni E, Bass CE, Saper CB, Lu J, Fuller PM (2014) The GABAergic parafacial zone is a medullary slow wave sleep-promoting center. Nat Neurosci 17:1217–1224

Baker BI, Eberle AN, Baumann JB, Siegrist W, Girard J (1985) Effect of melanin concentrating hormone on pigment and adrenal cells in vitro. Peptides 6:1125–1130

Bächner D, Kreienkamp H, Weise C, Buck F, Richter D (1999) Identification of melanin concentrating hormone (MCH) as the natural ligand for the orphan somatostatin-like receptor 1 (SLC-1). FEBS Lett 457:522–524

Benedetto L, Rodriguez-Servetti Z, Lagos P, D'Almeida V, Monti JM, Torterolo P (2013) Microinjection of melanin concentrating hormone into the lateral preoptic area promotes non-REM sleep in the rat. Peptides 39:11–15

Berbari NF, Lewis JS, Bishop GA, Askwith CC, Mykytyn K (2008) Bardet-Biedl syndrome proteins are required for the localization of G protein-coupled receptors to primary cilia. Proc Natl Acad Sci USA 105:4242–4246

Bittencourt JC, Presse F, Arias C, Peto C, Vaughan J, Nahon J-L, Vale W, Sawchenko PE (1992) The melanin-concentrating hormone system of the rat brain: an immuno- and hybridization histochemical characterization. J Comp Neurol 319:218–245

Blanco-Centurion C, Liu M, Konadhode RP, Zhang X, Pelluru D, van den Pol AN, Shiromani PJ (2016) Optogenetic activation of melanin-concentrating hormone neurons increases non-rapid eye movement and rapid eye movement sleep during the night in rats. Eur J Neurosci 44:2846–2857

Borbély AA (1977) Sleep in the rat during food deprivation and subsequent restitution of food. Brain Res 124:457–471

Borowsky B, Durkin MM, Ogozalek K, Marzabadi MR, DeLeon J, Lagu B, Heurich R, Lichtblau H, Shaposhnik Z, Daniewska I, Blackburn TP, Branchek TA, Gerald C, Vaysse PJ, Forray C (2002) Antidepressant, anxiolytic and anorectic effects of a melanin-concentrating hormone-1 receptor antagonist. Nat Med 8:825–830

Boyce R, Williams S, Adamantidis A (2017) REM sleep and memory. Curr Opin Neurobiol 44:167–177

Bresson JL, Clavequin MC, Fellmann D, Bugnon C (1987) Ontogenetic data on the peptidergic interneuronal population in immunoreactivity of the GRF 37 type serum of the human posterolateral hypothalamus. Immunocytochemical studies using anti-GRF 37 and anti-MCH (melanin-concentrating hormone) immune sera. C R Seances Soc Biol Fil 181:376–382

Brischoux F, Fellmann D, Risold PY (2001) Ontogenetic development of the diencephalic MCH neurons: a hypothalamic 'MCH area' hypothesis. Eur J Neurosci 13:1733–1744

Chambers J, Ames RS, Bergsma D, Muir A, Fitzgerald LR, Hervieu G, Dytko GM, Foley JJ, Martin J, Liu WS, Park J, Ellis C, Ganguly S, Konchar S, Cluderay J, Leslie R, Wilson S, Sarau HM (1999) Melanin-concentrating hormone is the cognate ligand for the orphan G-protein-coupled receptor SLC-1. Nature 400:261–265

Chee MJ, Pissios P, Maratos-Flier E (2013) Neurochemical characterization of neurons expressing melanin-concentrating hormone receptor 1 in the mouse hypothalamus. J Comp Neurol 521:2208–2234

Chee MJ, Arrigoni E, Maratos-Flier E (2015) Melanin-concentrating hormone neurons release glutamate for feedforward inhibition of the lateral septum. J Neurosci 35:3644–3651

Chemelli RM, Willie JT, Sinton CM, Elmquist JK, Scammell T, Lee C, Richardson JA, Williams SC, Xiong Y, Kisanuki Y, Fitch TE, Nakazato M, Hammer RE, Saper CB, Yanagisawa M (1999) Narcolepsy in orexin knockout mice: molecular genetics of sleep regulation. Cell 98:437–451

Chung S, Hopf FW, Nagasaki H, Li CY, Belluzzi JD, Bonci A, Civelli O (2009) The melanin-concentrating hormone system modulates cocaine reward. Proc Natl Acad Sci USA 106:6772–6777

Clegg DJ, Air EL, Benoit SC, Sakai RS, Seeley RJ, Woods SC (2003) Intraventricular melanin-concentrating hormone stimulates water intake independent of food intake. Am J Physiol Regul Integr Comp Physiol 284:R494–R499

Clément O, Sapin E, Libourel PA, Arthaud S, Brischoux F, Fort P, Luppi PH (2012) The lateral hypothalamic area controls paradoxical (REM) sleep by means of descending projections to brainstem GABAergic neurons. J Neurosci 32:16763–16774

David DJ, Klemenhagen KC, Holick KA, Saxe MD, Mendez I, Santarelli L, Craig DA, Zhong H, Swanson CJ, Hegde LG, Ping XI, Dong D, Marzabadi MR, Gerald CP, Hen R (2007) Efficacy of the MCHR1 antagonist N-[3-(1-{[4-(3,4-difluorophenoxy) phenyl]methyl}(4-piperidyl))-4-methylphenyl]-2-m ethylpropanamide (SNAP 94847) in mouse models of anxiety and depression following acute and chronic administration is independent of hippocampal neurogenesis. J Pharmacol Exp Ther 321:237–248

de Lecea L, Kilduff TS, Peyron C, Gao X, Foye PE, Danielson PE, Fukuhara C, Battenberg EL, Gautvik VT, Bartlett FS 2nd, Frankel WN, van den Pol AN, Bloom FE, Gautvik KM, Sutcliffe JG (1998) The hypocretins: hypothalamus-specific peptides with neuroexcitatory activity. Proc Natl Acad Sci USA 95:322–327

Del Cid-Pellitero E, Jones BE (2012) Immunohistochemical evidence for synaptic release of GABA from melanin-concentrating hormone containing varicosities in the locus coeruleus. Neuroscience 223:269–276

Della-Zuana O, Presse F, Ortola C, Duhault J, Nahon JL, Levens N (2002) Acute and chronic administration of melanin-concentrating hormone enhances food intake and body weight in Wistar and Sprague-Dawley rats. Int J Obes Relat Metab Disord 26:1289–1295

Domingos AI, Sordillo A, Dietrich MO, Liu ZW, Tellez LA, Vaynshteyn J, Ferreira JG, Ekstrand MI, Horvath TL, de Araujo IE, Friedman JM (2013) Hypothalamic melanin concentrating hormone neurons communicate the nutrient value of sugar. Elife 2:e01462

Eberle AN (1988) Melanin-concentrating hormone. In: Eberle AN (ed) The melanotropins. Karger, Basel, pp 321–332

Estabrooke IV, McCarthy MT, Ko E, Chou TC, Chemelli RM, Yanagisawa M, Saper CB, Scammell TE (2001) Fos expression in orexin neurons varies with behavioral state. J Neurosci 21:1656–1662

Fujiki N, Yoshida Y, Ripley B, Honda K, Mignot E, Nishino S (2001) Changes in CSF hypocretin-1 (orexin A) levels in rats across 24 hours and in response to food deprivation. Neuroreport 12:993–997

García-Fuster MJ, Parks GS, Clinton SM, Watson SJ, Akil H, Civelli O (2012) The melanin-concentrating hormone (MCH) system in an animal model of depression-like behavior. Eur Neuropsychopharmacol 22:607–613

Gao XB, van den Pol AN (2001) Melanin concentrating hormone depresses synaptic activity of glutamate and GABA neurons from rat lateral hypothalamus. J Physiol 533:237–252

Gao XB, van den Pol AN (2002) Melanin-concentrating hormone depresses L-, N-, and P/Q-type voltagedependent calcium channels in rat lateral hypothalamic neurons. J Physiol 542:273–286

Georgescu D, Sears RM, Hommel JD, Barrot M, Bolanos CA, Marsh DJ, Bednarek MA, Bibb JA, Maratos-Flier E, Nestler EJ, DiLeone RJ (2005) The hypothalamic neuropeptide melanin-concentrating hormone acts in the nucleus accumbens to modulate feeding behavior and forced-swim performance. J Neurosci 25:2933–2940

González JA, Iordanidou P, Strom M, Adamantidis A, Burdakov D (2016) Awake dynamics and brain-wide direct inputs of hypothalamic MCH and orexin networks. Nat Commun 7:11395

Guan JL, Uehara K, Lu S, Wang QP, Funahashi H, Sakurai T, Yanagizawa M, Shioda S (2002) Reciprocal synaptic relationships between orexin- and melanin-concentrating hormone-containing neurons in the rat lateral hypothalamus: a novel circuit implicated in feeding regulation. Int J Obes Relat Metab Disord 26:1523–1532

Guemez-Gamboa A, Coufal NG, Gleeson JG (2014) Primary cilia in the developing and mature brain. Neuron 82:511–521

Hamamoto A, Yamato S, Katoh Y, Nakayama K, Yoshimura K, Takeda S, Kobayashi Y, Saito Y (2016) Modulation of primary cilia length by melanin-concentrating hormone receptor 1. Cell Signal 28:572–584

Hanriot L, Camargo N, Courau AC, Leger L, Luppi PH, Peyron C (2007) Characterization of the melanin-concentrating hormone neurons activated during paradoxical sleep hypersomnia in rats. J Comp Neurol 505:147–157

Hassani OK, Lee MG, Jones BE (2009) Melanin-concentrating hormone neurons discharge in a reciprocal manner to orexin neurons across the sleep-wake cycle. Proc Natl Acad Sci USA 106:2418–2422

Ito M, Gomori A, Ishihara A, Oda Z, Mashiko S, Matsushita H, Yumoto M, Ito M, Sano H, Tokita S, Moriya M, Iwaasa H, Kanatani A (2003) Characterization of MCH-mediated obesity in mice. Am J Physiol Endocrinol Metab 284(5):E940

Jego S, Glasgow SD, Herrera CG, Ekstrand M, Reed SJ, Boyce R, Friedman J, Burdakov D, Adamantidis AR (2013) Optogenetic identification of a rapid eye movement sleep modulatory circuit in the hypothalamus. Nat Neurosci 16:1637–1643

Kawauchi H, Kawazoe I, Tsubokawa M, Kishida M, Baker BI (1983) Characterization of melanin-concentrating hormone in chum salmon pituitaries. Nature 305:321–323

Kilduff TS, de Lecea L (2001) Mapping of the mRNAs for the hypocretin/orexin and melanin-concentrating hormone receptors: networks of overlapping peptide systems. J Comp Neurol 435:1–5

Konadhode RR, Pelluru D, Blanco-Centurion C, Zayachkivsky A, Liu M, Uhde T, Glen WB Jr, van den Pol AN, Mulholland PJ, Shiromani PJ (2013) Optogenetic stimulation of MCH neurons increases sleep. J Neurosci 33:10257–10263

Lagos P, Torterolo P, Jantos H, Chase MH, Monti JM (2009) Effects on sleep of melanin-concentrating hormone (MCH) microinjections into the dorsal raphe nucleus. Brain Res 1265:103–110

Lagos P, Torterolo P, Jantos H, Monti JM (2011a) Immunoneutralization of melanin-concentrating hormone (MCH) in the dorsal raphe nucleus: effects on sleep and wakefulness. Brain Res 1369:112–118

Lagos P, Urbanavicius J, Scorza MC, Miraballes R, Torterolo P (2011b) Depressive-like profile induced by MCH microinjections into the dorsal raphe nucleus evaluated in the forced swim test. Behav Brain Res 218:259–266

Lagos P, Monti JM, Jantos H, Torterolo P (2012) Microinjection of the melanin-concentrating hormone into the lateral basal forebrain increases REM sleep and reduces wakefulness in the rat. Life Sci 90:895–899

Lee MG, Hassani OK, Jones BE (2005) Discharge of identified orexin/hypocretin neurons across the sleep-waking cycle. J Neurosci 25:6716–6720

Lembo PMC, Grazzini E, Cao J, Hubatsch DA, Pelletier M, Hoffert C, St-Onge S, Pou C, Labrecque J, Groblewski T, O'Donnell D, Payza K, Ahmad S, Walker P (1999) The receptor for the orexigenic peptide melanin-concentrating hormone is a G-protein-coupled receptor. Nat Cell Biol 1:267–271

Lin L, Faraco J, Li R, Kadotani H, Rogers W, Lin X, Qiu X, de Jong PJ, Nishino S, Mignot E (1999) The sleep disorder canine narcolepsy is caused by a mutation in the hypocretin (orexin) receptor 2 gene. Cell 98:365–376

Ludwig DS, Tritos NA, Mastaitis JW, Kulkarni R, Kokkotou E, Elmquist J, Lowell B, Flier JS, Maratos-Flier E (2001) Melanin-concentrating hormone overexpression in transgenic mice leads to obesity and insulin resistance. J Clin Invest 107:379–386

Marsh DJ, Weingarth DT, Novi DE, Chen HY, Trumbauer ME, Chen AS, Guan XM, Jiang MM, Feng Y, Camacho RE, Shen Z, Frazier EG, Yu H, Metzger JM, Kuca SJ, Shearman LP, Gopal-Truter S, MacNeil DJ, Strack AM, MacIntyre DE, Van der Ploeg LH, Qian S (2002) Melanin-concentrating hormone 1 receptor-deficient mice are lean, hyperactive, and hyperphagic and have altered metabolism. Proc Natl Acad Sci USA 99(5):3240

Matsuda K, Shimakura SI, Miura T, Maruyama K, Uchiyama M, Kawauchi H et al (2007) Feeding-induced changes of melanin-concentrating hormone (MCH)-like immunoreactivity in goldfish brain. Cell Tissue Res 328:375–382

Mickelsen LE, Kolling FW 4th, Chimileski BR, Fujita A, Norris C, Chen K, Nelson CE, Jackson AC (2017) Neurochemical heterogeneity among lateral hypothalamic hypocretin/orexin and melanin-concentrating hormone neurons identified through single-cell gene expression analysis. eNeuro 4(5)

Mileykovskiy BY, Kiyashchenko LI, Siegel JM (2005) Behavioral correlates of activity in identified hypocretin/orexin neurons. Neuron 46:787–798

Modirrousta M, Mainville L, Jones BE (2005) Orexin and MCH neurons express c-Fos differently after sleep deprivation vs. recovery and bear different adrenergic receptors. Eur J Neurosci 21:2807–2816

Monti JM, Lagos P, Jantos H, Torterolo P (2015) Increased REM sleep after intra-locus coeruleus nucleus microinjection of melanin-concentrating hormone (MCH) in the rat. Prog Neuro-Psychopharmacol Biol Psychiatry 56:185–188

Mori M, Harada M, Terao Y, Sugo T, Watanabe T, Shimomura Y, Abe M, Shintani Y, Onda H, Nishimura O, Fujino M (2001) Cloning of a novel G protein-coupled receptor, SLT, a subtype of the melanin-concentrating hormone receptor. Biochem Biophys Res Commun 283:1013–1018

Morton GJ, Mystkowski P, Matsumoto AM, Schwartz MW (2004) Increased hypothalamic melanin concentrating hormone gene expression during energy restriction involves a melanocortin-independent, estrogen-sensitive mechanism. Peptides 25:667–674

Mystkowski P, Seeley RJ, Hahn TM, Baskin DG, Havel PJ, Matsumoto AM, Wilkinson CW, Peacock-Kinzig K, Blake KA, Schwartz MW (2000) Hypothalamic melanin-concentrating hormone and estrogen-induced weight loss. J Neurosci 20:8637–8642

Nahon JL, Presse F, Bittencourt JC, Sawchenko PE, Vale W (1989) The rat melanin-concentrating hormone messenger ribonucleic acid encodes multiple putative neuropeptides coexpressed in the dorsolateral hypothalamus. Endocrinology 125:2056–2065

Nishino S, Ripley B, Overeem S, Lammers GJ, Mignot E (2000) Hypocretin (orexin) deficiency in human narcolepsy. Lancet 355:39–40

Oh EC, Vasanth S, Katsanis N (2015) Metabolic regulation and energy homeostasis through the primary cilium. Cell Metab 21:21–31

Qu D, Ludwig DS, Gammeltoft S, Piper M, Pelleymounter MA, Cullen MJ, Mathes WF, Przypek R, Kanarek R, Maratos-Flier E (1996) A role for melanin-concentrating hormone in the central regulation of feeding behaviour. Nature 380:243–247

Rance T, Baker BI (1979) The teleost melanin-concentrating hormone—a pituitary hormone of hypothalamic origin. Gen Comp Endocrinol 37:64–73

Rao Y, Lu M, Ge F, Marsh DJ, Qian S, Wang AH, Picciotto MR, Gao XB (2008) Regulation of synaptic efficacy in hypocretin/orexin-containing neurons by melanin concentrating hormone in the lateral hypothalamus. J Neurosci 28:9101–9110

Ripley B, Overeem S, Fujiki N, Nevsimalova S, Uchino M, Yesavage J, Di Monte D, Dohi K, Melberg A, Lammers GJ, Nishida Y, Roelandse FW, Hungs M, Mignot E, Nishino S (2001) CSF hypocretin/orexin levels in narcolepsy and other neurological conditions. Neurology 57:2253–2258

Roecker AJ, Cox CD, Coleman PJ (2016) Orexin receptor antagonists: new therapeutic agents for the treatment of insomnia. J Med Chem 59:504–530

Rossi M, Beak SA, Choi SJ, Small CJ, Morgan DGA, Ghatei MA, Smith DM, Bloom SR (1999) Investigation of the feeding effects of melanin concentrating hormone on food intake – action independent of galanin and the melancortin receptors. Brain Res 846:164–170

Saito Y, Nothacker HP, Wang ZW, Lin SHS, Leslie F, Civelli O (1999) Molecular characterization of the melanin-concentrating-hormone receptor. Nature 400:265–269

Saito Y, Cheng M, Leslie FM, Civelli O (2001) Expression of the melanin-concentrating hormone (MCH) receptor mRNA in the rat brain. J Comp Neurol 435:26–40

Sakurai T, Amemiya A, Ishii M, Matsuzaki I, Chemelli RM, Tanaka H, Williams SC, Richardson JA, Kozlowski GP, Wilson S, Arch JR, Buckingham RE, Haynes AC, Carr SA, Annan RS, McNulty DE, Liu WS, Terrett JA, Elshourbagy NA, Bergsma DJ, Yanagisawa M (1998) Orexins and orexin receptors: a family of hypothalamic neuropeptides and G protein-coupled receptors that regulate feeding behavior. Cell 92:573–585

Sapin E, Bérod A, Léger L, Herman PA, Luppi PH, Peyron C (2010) A very large number of GABAergic neurons are activated in the tuberal hypothalamus during paradoxical (REM) sleep hypersomnia. PLoS One 5:e11766

Sears RM, Liu RJ, Narayanan NS, Sharf R, Yeckel MF, Laubach M, Aghajanian GK, DiLeone RJ (2010) Regulation of nucleus accumbens activity by the hypothalamic neuropeptide melanin-concentrating hormone. J Neurosci 30:8263–8273

Shearman LP, Camacho RE, Sloan Stribling D, Zhou D, Bednarek MA, Hreniuk DL, Feighner SD, Tan CP, Howard AD, Van der Ploeg LH, MacIntyre DE, Hickey GJ, Strack AM (2003) Chronic MCH-1 receptor modulation alters appetite, body weight and adiposity in rats. Eur J Pharmacol 475:37–47

Shimada M, Tritos NA, Lowell BB, Flier LS, Maratos-Flier E (1998) Mice lacking melanin-concentrating hormone are hypophagic and lean. Nature 396:670–674

Shimomura Y, Mori M, Sugo T, Ishibashi Y, Abe M, Kurokawa T, Onda H, Nishimura O, Sumino Y, Fujino M (1999) Isolation and identification of melanin-concentrating hormone as the endogenous ligand of the SLC-1 receptor. Biochem Biophys Res Commun 261:622–626

Shimakura S, Marayama K, Miura T, Uchiyama M, Kawauchi H, Takahashi A et al (2006) Effect of melanin-concentrating hormone on feeding behavior and locomotor activity in the goldfish, Carassius auratus. Regul Pept 135:156

Skofitsch G, Jacobowitz DM, Zamir N (1985) Immunohistochemical localization of a melanin concentrating hormone-like peptide in the rat brain. Brain Res Bull 15:635–649

Steininger TL, Alam MN, Gong H, Szymusiak R, McGinty D (1999) Sleep-waking discharge of neurons in the posterior lateral hypothalamus of the albino rat. Brain Res 840:138–147

Thannickal TC, Moore RY, Nienhuis R, Ramanathan L, Gulyani S, Aldrich M, Cornford M, Siegel JM (2000) Reduced number of hypocretin neurons in human narcolepsy. Neuron 27:469–474

Tomoshige S, Kobayashi Y, Hosoba K, Hamamoto A, Miyamoto T, Saito Y (2017) Cytoskeleton-related regulation of primary cilia shortening mediated by melanin-concentrating hormone receptor 1. Gen Comp Endocrinol 253:44–52

Torterolo P, Sampogna S, Morales FR, Chase MH (2006) MCH-containing neurons in the hypothalamus of the cat: searching for a role in the control of sleep and wakefulness. Brain Res 1119:101–114

Torterolo P, Sampogna S, Chase MH (2009) MCHergic projections to the nucleus pontis oralis participate in the control of active (REM) sleep. Brain Res 1268:76–87

Tsunematsu T, Ueno T, Tabuchi S, Inutsuka A, Tanaka KF, Hasuwa H, Kilduff TS, Terao A, Yamanaka A (2014) Optogenetic manipulation of activity and temporally controlled cell-specific ablation reveal a role for MCH neurons in sleep/wake regulation. J Neurosci 34:6896–6909

Tuziak SM, Volkoff H (2012) A preliminary investigation of the role of melanin-concentrating hormone (MCH) and its receptors in appetite regulation of winter flounder (*Pseudopleuronectes americanus*). Mol Cell Endocrinol 348:281–296

Tuziak SM, Volkoff H (2013) Melanin-concentrating hormone (MCH) and gonadotropin-releasing hormones (GnRH) in Atlantic cod, *Gadus morhua*: tissue distributions, early ontogeny and effects of fasting. Peptides 50:109–118

van den Pol AN, Acuna-Goycolea C, Clark KR, Ghosh PK (2004) Physiological properties of hypothalamic MCH neurons identified with selective expression of reporter gene after recombinant virus infection. Neuron 42:635–652

Verret L, Goutagny R, Fort P, Cagnon L, Salvert D, Léger L, Boissard R, Salin P, Peyron C, Luppi PH (2003) A role of melanin-concentrating hormone producing neurons in the central regulation of paradoxical sleep. BMC Neurosci 4:19

Vetrivelan R, Kong D, Ferrari LL, Arrigoni E, Madara JC, Bandaru SS, Lowell BB, Lu J, Saper CB (2016) Melanin-concentrating hormone neurons specifically promote rapid eye movement sleep in mice. Neuroscience 336:102–113

von Economo C (1930) Sleep as a problem of localization. J Nerv Ment Dis 71:249–259

Willie JT, Sinton CM, Maratos-Flier E, Yanagisawa M (2008) Abnormal response of melanin-concentrating hormone deficient mice to fasting: hyperactivity and rapid eye movement sleep suppression. Neuroscience 156:819–829

Wu M, Dumalska I, Morozova E, van den Pol A, Alreja M (2009) Melanin-concentrating hormone directly inhibits GnRH neurons and blocks kisspeptin activation, linking energy balance to reproduction. Proc Natl Acad Sci U S A 106:17217–17222

Yoshida Y, Fujiki N, Nakajima T, Ripley B, Matsumura H, Yoneda H, Mignot E, Nishino S (2001) Fluctuation of extracellular hypocretin-1 (orexin A) levels in the rat in relation to the light-dark cycle and sleep-wake activities. Eur J Neurosci 14:1075–1081

Zeitzer JM, Buckmaster CL, Parker KJ, Hauck CM, Lyons DM, Mignot E (2003) Circadian and homeostatic regulation of hypocretin in a primate model: implications for the consolidation of wakefulness. J Neurosci 23:3555–3560

Zheng H, Patterson LM, Morrison C, Banfield BW, Randall JA, Browning KN, Travagli RA, Berthoud HR (2005) Melanin concentrating hormone innervation of caudal brainstem areas involved in gastrointestinal functions and energy balance. Neuroscience 135:611–625

Optogenetic Control of the Melanin-Concentrating Hormone Expressing Neurons

Carlos Blanco-Centurion, Meng Liu, and Priyattam Shiromani

Abstract Since the discovery of the neuropeptide melanin-concentrating hormone (MCH) more than half a century ago, MCH neurons have spawned hundreds of scientific studies. MCH neurons are a phylogenetically well-conserved group of hypothalamic neurons. The MCH neurons project diffusely throughout the CNS but only reside within a restricted area encompassing the incerto-lateral and perifornical hypothalamus. Here we review optogenetic studies focused on understanding the functions of the MCH neurons, particularly their role as sleep modulators. We attempted to put optogenetic findings in context with other studies focused on neuronal/behavioral modulation by MCH or its receptors. We also laid a theoretical framework to understand better the data on MCH neuronal activity in relation to the activity of the orexin neurons. For readers not familiar with optogenetics, we also went over major developments in this new field.

1 Introduction

The implementation of optogenetic technology to the study of brain functions has made possible a sophisticated control over one of its chief biological units; the neurons. This leap forward in technology has been applied to numerous studies of the brain functions including the understanding of one of its most basic functions, the wake–sleep behavior. Here we are reviewing optogenetic findings from our own laboratory as well as from others that have used it to understand the activity of one important phenotype among hypothalamic neurons, neurons expressing the melanin-

C. Blanco-Centurion (✉) · M. Liu
Department of Psychiatry and Behavioral Sciences, Medical University of South Carolina, Charleston, SC, USA
e-mail: blancoce@musc.edu

P. Shiromani
Ralph Johnson Medical Center, US Department of Veterans Affairs, Charleston, SC, USA

Department of Psychiatry and Behavioral Sciences, Medical University of South Carolina, Charleston, SC, USA

S. R. Pandi-Perumal et al. (eds.), *Melanin-Concentrating Hormone and Sleep*,
https://doi.org/10.1007/978-3-319-75765-0_4

concentrating hormone (MCH). Ahead of the main topic, we introduced the reader not acquainted with optogenetics with a chronological account of its major developments. Readers already familiar with optogenetics may certainly skip this section. We divide the review optogenetic studies in those experiments that stimulated and those that inhibited the activity of MCH neurons. At the end, we put forward a model for interpretation of all sort of data, i.e., sleep and other behaviors. Our model proposes a counterbalance between MCH and orexin neuronal activity.

2 Development of the Optogenetic Tools

Before reviewing the studies that used optogenetic tools to control the excitability of MCH neurons, it is helpful to follow its major developments. Optogenetics only has been around for 12 years. Yet its development has been breathtaking. This biotechnology could not have fully developed without the parallel development of molecular biology tools and photonics (optics + electronics). Laser invention, and particularly laser application to imaging, opened the door for using laser pulses as surrogate of the electrical stimulator. In early years medical laser applications like tumor laser heating gave incentive to study the optical properties of intact brain tissue, i.e., absorbance, diffraction, etc. Optical measurements indicated that, for instance, blue light from lasers penetrates 0.5 mm into white matter and 1.84 mm into the gray matter making a light-neuron interface theoretically possible (Yaroslavsky et al. 2002). Two decades after laser was invented, Farber and Grinvald "photostimulated" neurons in the mollusk *Aplysia* by synthetizing a photoexcitable dye (Farber and Grinvald 1983). Alas, Grinvald's Lab later focused heavily on the development of fluorescent voltage sensors and stopped pursuing the development of photostimulation tools. This application had to wait another decade until Callaway and Katz used laser to generate the photochemical release of the excitatory neurotransmitter glutamate, i.e., "uncaging." Using either one- or two-photon laser scanning microscopy, it became possible to activate individual neuronal synapsis in vitro by photolysis of caged neurotransmitters (Callaway and Katz 1993; Denk 1994). Photon-driven uncaging of neurotransmitters requires the caged compounds to be loaded into all neurons, hence lacked cell specificity. Also, caged neurotransmitters are loaded by bath applications making this approach only amenable to in vitro preparations.

Advances toward activity control of specific neurons were first made by transfecting nematode-derived genes of ligand-gated chloride ion channels into mammalian neurons, i.e., ivermectin inhibition. This strategy of genetic inhibition of neurons swiftly moved up from in vitro constitutive (Johns et al. 1999) to in vitro reversible (Lechner et al. 2002; Slimko et al. 2002), and ultimately it was applied to freely behaving mice (Lerchner et al. 2007). Still this approach is dependent on the slow pharmacokinetics of the drug binding to the ion channel. This problem was partially solved by coupling genetic control of excitability with photo-uncaging.

Ectopic ionotropic excitatory receptors were transfected into specific neurons, while its ligands were caged and bath-loaded. Thus successfully transfected neurons were forced to fire when light pulses shone over them (Zemelman et al. 2003). A subsequent refinement of photolysis uncaging was the creation of a mutant potassium channel that could be chemically gated depending on light wavelength (Banghart et al. 2004). This technology is the forbearer of what few years later will be the "step function bidirectional control of excitability." Time resolution of neuronal response by photolysis uncaging is in the order of seconds, better than minutes-hours taken by drugs but still far slower than traditional electrical stimulation, i.e., milliseconds.

Gero Miesenböck's research group (currently in Oxford University) provided the first proof of principle behind the modern optogenetic approach. A chimeric protein made up of arrestin-2, rhodopsin and the subunit of a G-coupled excitatory protein "chARGe" was transfected into hippocampal neurons. White light pulses triggered action potentials (AP) only in the chARGe transfected neurons (Zemelman et al. 2002). Same research group took this approach one step further evoking behavioral responses in fruit flies that had chARGe receptors expressed in specific neurons. Caged ligands for the ectopic ion channel receptors were microinjected into the fruit flies. Then 150 ms duration white light pulses uncaged the ligands stimulating the transgenic neurons (Zemelman et al. 2003). The photochemical strategy had nonetheless two limitations. First, either the cofactor retinol or the uncaged ligands were still required to be applied prior to the experiment. Also the kinetic of the neuronal response lagged behind the precise time resolution of traditional electrical stimulation. An ectopic photo-driven cation channel with ultra-rapid response and no requirement of cofactors loads was needed to advance the field.

The discovery of microbial ion channels responsive to light (opsins) kindled the field of optogenetics. The proton pump channelrhodopsin-1 (crChR1) and the mono/divalent cation channel channelrhodopsin-2 (crChR2) were the first microbial opsins cloned from the unicellular freshwater green algae *Chlamydomonas reinhardtii* (Nagel et al. 2002, 2003). A couple of years after its cloning, Karl Deisseroth's Lab (Stanford University) used for the first time crChR2 to activate neurons by giving millisecond (ms)-duration pulses of blue light (Boyden et al. 2005). Cultured hippocampal neurons transfected with lentivirus containing the mammalian codon-optimized crChR2 gene showed fast and phase-locked spiking in response to 5–30 Hz frequency light pulses. Light-evoked spikes were indistinguishable from those produced by intracellular depolarizing current. Light pulses at low intensity were also able to elicit subthreshold responses. Electrophysiological and cell viability markers were minimally affected by crChR2 transfection or its photoactivation. Despite crChR2 requires retinal, no cofactor was needed to add to the culture media. It seems that mammalian stores of retinal are large enough to fuel the chromophore. Two independent labs corroborated the stimulatory effects of crChR2 (Li et al. 2005; Ishizuka et al. 2006). These labs reported that in mice brain slices, photostimulation mediated by crChR2 elicited depolarization/repolarization events occurring at a faster rate than the neuron membrane time constant. The fast response of crChR2 made the neuronal spiking get phase-locked to the light pulses up to a frequency of

20 Hz. The turning on phase of crChR2 photocurrent was light intensity dependent with a time constant of only 2.2 ms. By contrast, the turning off phase was independent of light intensity having a slightly slower time constant than the chChR2 opening (Ishizuka et al. 2006). Soon after cloning of ChR2, the first gain-of-function ChR2 mutant was engineered, i.e., $ChR2_{H134R}$. Photostimulation of $ChR2_{H134R}$ evoked larger and faster photocurrents (Nagel et al. 2005). $ChR2_{H134R}$ was then ectopically expressed in muscle and mechanosensory neurons of the round worm *C. elegans* eliciting appropriate behavioral responses. Unlike transfection of ChR2 into rat hippocampal neurons, positive photoactivation of ChR2 in chick embryonic neurons or in the round worm required the presence of retinal. These seminal experiments were also the first to use fast optical switching coupled to a laser/arc lamps/LED as light stimulator.

A crucial milestone in optogenetics came with the development of the first light-neuron interface capable of eliciting a predictable behavioral response in freely behaving rodents (Aravanis et al. 2007). The αCaMKII promoter was selected for targeting expression of crChR2 in excitatory cortical motoneurons controlling whisker movements. Using an optic fiber (200 microns O.D.) and blue light laser (473 nm), photostimulation pulses significantly elicited whisker deflections with a latency of few seconds. Also for the first time, it was described the optical properties of the rat/mouse cerebral cortex. It was estimated that with a laser power of roughly 38 mW/mm^2, a brain volume of roughly 0.5 mm^3 below the tip of the optic fiber would receive enough energy (>1 mW/mm^2) to activate all neurons transfected with crChR2. Considering the mouse neuronal cortical density is 920,000/mm^3 (Schuz and Palm 1989), this energy would activate more than 400,000 neurons. Thus 137 years after Hitzig and Fritsch (Thomas and Young 1993) elicited a predicted muscular contraction by electrical stimulation of the motor cortex, selective optogenetic stimulation of cortical motoneurons exquisitely reproduced this paradigm.

Since 2005, channelrhodopsin-based technology has gone through a vigorous development; for a recent review of this topic, see Wietek and Prigge (2016). Thus other channelrhodopsins from different species of freshwater algae have been both characterized and tested as optogenetic tools. The cation channel opsin from *Volvox carteri* (vChR1) also evokes depolarizing photocurrents in response to light. vChR1 shows red-shifted (520 nm) photosensitivity as compared to crChR2 or $ChR2_{H134R}$ (470 nm) (Zhang et al. 2008). Channelrhodopsin from *Mesostigma viride* (mChR) is another red-shifted (534 nm) opsin with better channel kinetic properties, including broader pH range, that undergoes minimal inactivation upon sustained illumination (Govorunova et al. 2011). Other two channelrhodopsins from two different species of the *Chlamydomonas* genus show similar red-shifted (520 nm) spectral photosensitivity as well as lack of fast desensitization (Hou et al. 2012). These other channelrhodopsins are named CaChR1 (*C. augustae*) and CyChR1 (*C. yellowstonensis*). Red-shifted optimal photosensitivity has two main advantages over blue photosensitivity. First it is well established that longer-wavelength photons penetrate deeper into intact mammalian brain tissue (Yaroslavsky et al. 2002; Al-Juboori et al. 2013); henceforth red-shifted ChRs can excite more neurons with the same intensity of illumination. Likewise, neuronal photostimulation with red-shifted ChRs is compatible with blue light excitable calcium indicators like GCAMP6.

In addition to find more WT ChRs, other research groups used protein chimeric technology (or site-directed mutagenesis) to engineer improved channelrhodopsins. Swapping the six-helix domains between ChR1 and ChR2, it yielded the red-shifted opsin ChRGR (Lin et al. 2009; Wen et al. 2010). ChRGR transfected cells show light-induced photocurrents featuring faster channel kinetics as well as less desensitization upon constant illumination. Site-directed mutagenesis of crChR2 significantly improved the channel kinetics measured during fast frequency illumination (Gunaydin et al. 2010). The resultant mutant called $ChETA_A$ showed ultrafast on response (1 ms), faster of time constant, lack of doublets or missing spikes, and no current plateau when stimulated at frequencies as high as 200 Hz. $ChETA_A$ can be activated with blue-greenish light pulses as brief as 2 ms and still evoke 100% of neuronal spiking. Another mutant $ChR2_{TC}$ can swiftly excite neurons at illumination levels as low as 1.9 mW/mm^2 (Berndt et al. 2011). Readers wishing to know more about the particular performance for each ChRs variant, please consult Mattis et al. (2011).

One of the most interesting developments in ChR biotechnology was the bioengineering of "step function opsins" (SFO). A step function opsin is a photo-induced cation channel with slow closing kinetics allowing long (several seconds) and steady depolarization. SFO undergo rapid channel closing with a brief pulse of yellow light (Gunaydin et al. 2010). Since the original SFO ($ChR2_{E123T}$) was slow for bidirectional control of fast-spiking neurons, a faster double mutant SFO called $ChR2_{E123T/T159C}$ was engineered to evoke step function bidirectional control up to 60 Hz (Berndt et al. 2011). At the same time, vChR1 was fused with crChR1 and subjected to point mutations in order to come up with a fast bistable SFO (Yizhar et al. 2011). The resulting $ChR2_{C128S/D156A}$ double mutant was tested in vivo eliciting significant behavioral changes in mice (Yizhar et al. 2011). Same year Boyden's research group (MIT) also engineered bistable optogenetic expressing vectors fusing different varieties of ChRs (Kleinlogel et al. 2011).

As important as the application of ChR2 to photostimulate neurons, it was the successful application of the photosensitive inhibitory channel halorhodopsin (HR). HR evolved in archaea halophiles, i.e., salt-thriving primitive bacteria. In nature, HR uses the energy from photons to drive an electrogenic chloride pump to keep osmotic balance amidst an hyperosmotic environment (Lanyi 1990). The photosensitive pump cycle of HR only takes a few milliseconds (Tittor et al. 1987). Furthermore, HR shows red-shifted (560–590 nm, yellow) optimal photosensitivity (Bamberg et al. 1993) and hence can be used coupled with ChR2. All these features make HR a good candidate as a photo-gated inhibitory chloride channel. The extracellular concentration of chloride ions is many times higher than the intracellular, and the chloride equilibrium potential is close to the neuron's membrane potential. Thus, chloride ion influx follows the opening of a chloride-selective channel (i.e., $GABA_A$ receptors) bringing the membrane potential toward the resting level counteracting depolarization currents.

Karl Deisseroth's Lab in collaboration with German scientists tested a couple of HRs derived from two species of archaea: *Halobacterium salinarum* (HsHR) and *Natronomonas pharaonis* (NpHR) (Zhang et al. 2007). In vertebrate cells, NpHR showed higher affinity for chloride and was therefore chosen for transfecting mammalian neurons. Mice neurons transfected with mammalian codon-optimized NpHR-YFP showed yellow light-elicited hyperpolarizing currents (−40 mV) with rise/decay time constants of roughly 6 ms. Yellow light pulses silenced neurons being activated either by depolarization currents or by optogenetic activation of ChR2. NpHR-YFP transfected neurons did not show electrophysiological/vital alterations either at baseline or after light-induced inhibition. NpHR inhibited neuronal firing in a light intensity manner (threshold = 7 mW/mm^2; ceiling = 22 mW/mm^2), over a continuous illumination (1–10 min) or at a single spike level over frequencies ranging up to 30 Hz. NpHR inhibition occurred without interfering with other inhibitory currents mediated by $GABA_A$ receptors. In mice Selective expression of NpHR caused complete inhibition of APs and intracellular calcium transients. Similar to ChR2, the mammalian neurons transfected with NpHR did not require the cofactor retinol to evoke photocurrents. Yet retinol was necessary to observe behavioral responses in *C. elegans*. In these invertebrates, NpHR produced reliable photoinhibition of muscles in charge of swimming and cholinergic motoneurons as well. NpHR was also effective in inhibiting swimming behavior elicited by ChR2 activation (Zhang et al. 2007). Boyden's Lab produced similar results as Zhang et al. transfecting cortical projecting neurons both with NpHR and ChR2 (Han and Boyden 2007). Remarkably a Gaussian mode of alternating yellow/blue light pulses was able to alter neuronal spike timing without altering the neuron firing rate. In a further refinement, NpHR was engineered to show increased mammalian neuronal expression and reduced toxicity by removing the motif signals that caused its buildup inside the endoplasmic reticulum. The modified NpHR, called eNpHR2.0, displayed increased peak photocurrent in the absence of aggregations or toxicity and potent optical inhibition both in vitro and in vivo (Gradinaru et al. 2008). Two years later an improved version of NpHR2.0, dubbed NpHR3.0, was engineered (Gradinaru et al. 2010). NpHR3.0 was made modifying the C-terminus of NpHR2.0 with a trafficking signal motif borrowed from the rectifying potassium channel gene Kir2.1. Thanks to this addition, NpHR3.0 mostly expressed along the plasma membrane. As a result, the photocurrents measured in neurons transfected with NpHR3.0 showed a threefold increase, and the ensuing hyperpolarization increased by 50% (−100 mV). The potent hyperpolarization of NpHR3.0 was also observed under orange/reddish light (630 nm) and even at nearly infrared wavelengths (680 nm) (Gradinaru et al. 2010). The large red-shifted photosensitivity of NpHR3.0 allows to be used simultaneously with ChR in the same neuron. This application was made possible creating a bicistronic lentivirus vector containing both ChR2-EYFP and NpHR-mCherry genes, i.e., eNPAC. eNPAC enabled bidirectional (excitation/inhibition) control with blue (excitation) and orange (inhibition) light pulses of the same transfected neuron (Gradinaru et al. 2010).

Around the same time NpHR was being developed, Boyden's group screened several other photosensitive HRs derived from both fungi and archaea. In mice archaerhodopsin-3 (Arch) from *Halorubrum sodomense* showed stronger neuronal inhibition as compared to NpHR (Chow et al. 2010). Data strongly suggested that Arch hyperpolarizes neurons by a protonic outward current. Unlike NpHR which shows long-lasting deactivation after continuous yellow light illumination, Arch recovered its photosensitivity spontaneously in matter of seconds. Inhibitory currents elicited by Arch were larger (hundreds of pA) than those caused by NpHR (100 pA) but with similar on/off time scales. Arch was also more photosensitive (0.3 mW/mm^2) and showed lack of saturation at high intensity illumination (36 mW/mm^2). Inhibitory yellow photocurrents mediated by Arch occurred regardless of the presence of potassium or chloride ions. Arch has shown to be a better inhibitory optogenetic tool than chloride channel eNpHR3.0 in GABAergic synapsis where the photoactivation of chloride pumps changed the reversal potential eliciting after light spiking (Raimondo et al. 2012). Lentivirus having the mammalian codon-optimized Arch gene produced robust plasma membrane neuronal expression. In awake mice, Arch transfected cortical neurons were completely silenced for several seconds. One month after Arch transfection, neurons expressing EYFP had similar passive and active electrophysiological/vital properties as the non-expressing neurons. Boyden's group also developed a blue light-shifted photosensitive proton pump derived from the fungus *Leptosphaeria maculans* (Mac). Mac allowed simultaneous inhibition of two different neuronal clusters using blue (470 nm; −20 mV) and red (630 nm; −30 mV) light pulses. They also developed a more photosensitive inhibitory opsin from the archaea *Halorubrum* strain TP009 (ArchT) (Han et al. 2011). Similar to Arch, ArchT expresses along plasma membrane at somata and terminals of mice and in non-human primates. ArchT showed the same level of photoinhibitory currents as the WT Arch, but it had a threefold increase in photosensitivity (1–10 mW/mm^2) resulting in doubling of the volume of tissue being inhibited. Recently a fusion protein made between Arch and ChR has yielded another SFO (Kleinlogel et al. 2011).

A significant development of the optogenetic field is the creation of transgenic (Tg) lines of mice constitutively expressing ChR2-YFP in different phenotypes of neurons. The first Tg ChR2 mice line was developed to photostimulate projecting neurons (Arenkiel et al. 2007; Wang et al. 2007). On these neurons blue light pulses evoked phase-locked spiking similar to those reported using viral transfection of ChR2. The in vitro studies indicated light-elicited spiking was both dependent on light intensity (threshold $= 0.2$ mW/mm^2; ceiling ≈ 9 mW/mm^2) and frequency of stimulation (5–30 Hz). Despite ChR2-YFP was expressed in both perikarya and processes, APs were only evoked by the illumination of the perikarya. Illumination focused on proximal dendrites only produced subthreshold photocurrents. The optogenetic stimulation of neurons also evoked postsynaptic currents in downstream neurons confirming that light-induced APs cause release of neurotransmitters. The capacity of optogenetics to stimulate neurons whose projections are simultaneously recorded opens the door for the creation of high-resolution circuit brain maps.

The repertoire of Tg mice line expressing ChR2 has grown ever since to include VGAT-ChR2$_{H134R}$-EYFP, ChAT-ChR2$_{H134R}$-EYFP, Tph2-ChR2$_{H134R}$-EYFP, and Pvalb$_{H134R}$-ChR2-EYFP (Zhao et al. 2011). Slice recordings from these Tg mice lines confirmed that blue light pulses reliably stimulated GABAergic, cholinergic, serotonergic, or parvalbumin-expressing neurons. Transgenic lines of mice selectively expressing ChR2-YFP in glutamatergic (Hagglund et al. 2010), sensory olfactory (Dhawale et al. 2010), and striatal medium spiny neurons (Chuhma et al. 2011) have been also engineered.

Another major development in optogenetics has been the bioengineering of multiple Tg lines of mice expressing the Cre-Lox system in different neuronal phenotypes. Currently there are dozens of commercially available Tg Cre mice specific for genes relevant for neurotransmission like TH-Cre, vGABAT-Cre, MCH-Cre, etc. Cre-Lox is a conditional gene expression system requiring the presence of two DNA components. One is the Cre gene inserted downstream of a specific promoter sequence: TH promoter, MCH promoter, etc. The other component is the gene of interest, in this case ChR2, NpHR3.0, or ArchT. Located upstream of the opsin genes, there is a transcription STOP codon flanked by a pair of specific DNA sequences called Lox sites. Without Cre the STOP signal will prevent the opsin to be transcribed. Yet once both sets of genes are present within the same cell, Cre will bind to the Lox sites, and depending on its orientation, it will either cut it (same) or invert it (opposite). Researchers at the Allan Institute for Brain Science bioengineered four Tg lines of mice harboring the Lox flanked STOP codon upstream of ChR2$_{H134R}$-dTomato (Ai27), ChR2$_{H134R}$-EYFP (Ai32), Arch-EGFP (Ai35), or eNpHR3.0-EYFP (Ai39) (Madisen et al. 2012). The progeny of the crossing between a Cre Tg line mouse and a mouse harboring the floxed-stop/opsins will show selective expression of the opsin inside Cre-expressing neurons. As a proof of principle, the Tg opsin mice were first crossed with parvalbumin-Cre mice corroborating that the Cre-dependent opsins performed as expected (Madisen et al. 2012). Alternatively, the Cre-lox system can be used to virally transfect the microbial opsins into a specific subpopulation of Cre-expressing neurons. This system is called double-floxed inverse ORF or DIO for short. DIO vectors contain the gene of interest (ChR2, Arch, NpHR) oriented as antisense and flanked by two different lox sites also in reverse orientation. Once the DIO virus infects the Cre-expressing neuron, Cre will cut both pairs and place the gene in the sense orientation for transcription. After recombination a mismatch Lox pair will remain preventing further Cre binding (Sohal et al. 2009). DIO viruses containing the genes for all optogenetic opsins are commercially available from virus vector cores at Stanford University, University of Pennsylvania, and the University of North Carolina at Chapel Hill.

Now that the reader had acquired a firm foot onto optogenetics, it follows focus on melanin-concentrating hormone (MCH) neurons.

3 Brief Evolutionary Perspective of the MCH Neurons

It is necessary to mention that the name given to MCH does not suit well its various roles in the brain of many vertebrates expressing it. The name of this neuropeptide derived from the discovery that, in bony fishes, MCH functions as depigmentation hormone, i.e., "melanin-concentrating." It can be said this endocrine role evolved as a secondary specialization in jawed fishes, i.e., gnathostomes. Comparative anatomy studies reveal that the original small periventricular neurons expressing MCH in agnathans (e.g., lampreys, hagfish) later evolved in jawed fishes into large neuroendocrine-type cells whose axons mainly project to the neurohypophysis (Baker and Bird 2002). The skin light tone effect of endocrine MCH has been replaced later on in some tetrapod (e.g., frogs, lizards) by the most potent action of the pituitary gland-secreted hormone melatonin (Filadelfi and Castrucci 1994). In most reptiles, and all birds and mammals, the magnocellular incerto-latero hypothalamic MCH neurons send its axons throughout the neuroaxis staying away of the neurohypophysis (Cardot et al. 1994; Vallarino et al. 2009). Thus in more recent vertebrates, the MCH neurons evolved away from endocrine regulation into paracrine (cell to cell) regulation of a wide variety of intra- and extra-hypothalamic neurons. Noteworthy in mammals MCH still plays a role regulating the beat frequency of ependymal cilia and ventricular volume (Conductier et al. 2013) perhaps in the same way the parvocellular periventricular MCH neurons did it. In any case in all mammals hitherto studied, MCH-expressing neurons only reside within a region covering the zona incerta (ZI), the lateral hypothalamus (LH), and the perifornical area (PeF) (Bittencourt 2011). In mammals, the total number of MCH neurons has significantly increased along with brain size and brain complexity. It is estimated that the mouse has roughly 5000 (Toossi et al. 2016; McGregor et al. 2017), whereas the rat has 10,000 (Mikrouli et al. 2011), and humans may have from 90,000 up to 140,000 MCH neurons (Thannickal et al. 2007; Aziz et al. 2008). The number of cognate receptors for MCH has also increased. Rodents only have one MCH receptor (Chambers et al. 1999; Saito et al. 1999), while primates have two types of MCH receptors (Sailer et al. 2001). Nonetheless the primary amino acid sequence of MCH has not evolved, and there is also a high degree of conservation (90%) to the level of its base pair gene sequence (Nahon 1994). In hominids a second MCH precursor gene has evolved (Courseaux and Nahon 2001). In summary, the evolutionary trends in mammals highlight the increasing importance for the brain of having more synapsis involving MCH neurons.

4 Electrical Stimulation of Brain Areas Containing MCH Neurons: Findings and Caveats

Long before optogenetic tools were used to activate hypothalamic neurons, clues of its functions emerged out of electrical stimulation studies; for a review see Berthoud and Munzberg (2011). More than a century ago, it was discovered that discrete electrical currents applied to the brain can evoke discrete and specific movements/behaviors (Thomas and Young 1993). Ever since electrical stimulation has produced direct evidence that activation of specific brain regions is responsible for all sort of brain functions. Nowadays elegant microsimulation experiments still produce meaningful data; for review see Clark et al. (2011). By the middle of the twenty-first century, Nobel Prize awardee Weiss R. Hess discovered that electrical stimulation of the LH elicited biologically crucial emotional responses such as flight or fight, voracious eating, or mating (Hess and Akert 1955). In particular the LH was recognized as the main feeding center (Anand and Brobeck 1951), and later on it was discovered feeding and reward (self-stimulation) were both integrated there (Hoebel and Teitelbaum 1962). Electrical stimulation of the LH also produced changes in motor output (Sinnamon 1993) and emotionally triggered changes in cardiovascular function (Smith et al. 1990). Electrical stimulation of the ZI also has an effect on motor outputs like locomotion/stepping (Sinnamon 1984), cardiovascular function (van der Plas et al. 1995), and thermoregulation (Kelly and Bielajew 1996). Recent studies, still using electrical stimulation, confirm that activation of the ZI is crucial to modulate skeletal muscle tone and the pace of movements. Thus electrical activation of the ZI is very effective for control of tremors, bradykinesia, and rigidity in patients with Parkinson disease (Plaha et al. 2006). Using nonspecific gene transfer of the arousal peptide orexin, we found that orexin release from all kinds of incerto-lateral hypothalamic neurons prevents sudden bouts of muscle paralysis known as cataplexy (Liu et al. 2011). Cataplexy and sleep attacks are cardinal symptoms of narcolepsy. In contrast, we observed that transferring the orexin gene specifically into MCH neurons exacerbated cataplexy and sleep attacks suggesting MCH neurons exert negative control over the skeletal muscle tone and levels of arousal.

Our contrasting results using gene transfer vividly underscore the need for attaining cell specificity during experiments of complex networks. Electrical currents are limited in this regard. Electrophysiological studies have been done almost always using ex vivo preparations, i.e., neurons in slices or in culture. Glass pipette microelectrodes in whole-cell patch clamp are exquisite tools for controlling the excitability of individual neurons. Neurons in these preparations can also be identified post hoc either by intracellular injection of dyes or a priori through genetic tagging. Alas behaviors like sleep/wake engaging multiple neurons and being displayed only in vivo settings are an unsurmountable challenge for using the whole-cell patch-clamp approach. To achieve intracellular stimulation in vivo, the animal must be anesthetized and even though it is still a very challenging study (Brecht et al. 2004). Even the enthusiasm to study the role of discrete brain regions using very fine microelectrodes (5–30 microns) has significantly waned. By the end of the twenty-first century, it became unavoidable the realization that electrical fields produce activation of the neural tissue by a myriad of factors making extremely hard to interpret the results using a cellular framework. Type of neural tissue (gray vs. white matter), type of electrode configuration (monopolar vs. bipolar), type of current (anodal, cathodal, biphasic), size of electrode tip, distance to the excitable unit, orientation of electrical field with respect to axons or cell bodies, whether the neuron has or not myelinated axons, and the chronoaxie signature all influence whether neurons or axons [or both] will be activated by electrical currents (Ranck 1975; Iggo 1978; Brocker and Grill 2013). Artificial electrical currents can even activate axons instead of neuronal perikarya (Nowak and Bullier 1998a, b). Even small electrical currents are able to stimulate multiple neurons and often activate thousands of them (Tehovnik 1996). In the case of chronic electrical stimulation, its efficacy diminishes with time, based on how brain tissue reacts to the electrode material (Biran et al. 2005). These many drawbacks motivated researchers to develop new approaches so as to unravel the role of specific neurons on behavior or its electrophysiological correlates. These novel approaches should overcome the caveats of electrical stimulation but still keep its advantages, i.e., small size and precise time resolution. The solution came combining molecular biology tools with photonics, i.e., optogenetics. Now it follows reviewing the optogenetic experiments done to control the excitability of MCH neurons.

5 Optogenetic Control of MCH Neurons and Sleep Regulation

5.1 Stimulation of MCH Neurons

Findings from numerous previous studies led to the proposition that activity of MCH neurons plays a role regulating the sleep states. Yet any of those studies directly tested this hypothesis. Therefore we reasoned that if MCH neurons are sleep active (Hassani et al. 2009), its stimulation should push the awake brain to transit into NREM sleep and when in NREM sleep to transit into REM sleep. Thus we were the first research group to apply optogenetics to selectively stimulate MCH neurons while measuring its effects on sleep/wake stages (Konadhode et al. 2013). Because of the genetic advantages that afford studying mice, we chose mice as our first animal model, i.e., WT C57/BL. In the past we used a very specific and efficient MCH promoter to target the expression of heterologous proteins solely in MCH neurons (Liu et al. 2011). Our MCH promoter was donated and validated by one of our collaborators (van den Pol et al. 2004). Using the same gene expression cassette, we inserted the ChR2$_{H134R}$-YFP gene (donated by K. Deisseroth) replacing the orexin gene. The re-engineered plasmid was packed into recombinant adeno-associated virus (stereotype 5) at a titer of 5 × 10^{12GC} /μl (University of North

Carolina, Chapel Hill, NC). As a control we used our previous rAAV-MCH promoter-GFP vector (Liu et al. 2011). To transfect a large number of MCH neuron vectors, we injected the vectors at two different loci within the LH. Likewise the injected volume was large (0.75 μl/hemisphere). Histological analysis confirmed that our MCH promoter produced a eutopic expression of YFP solely in MCH neurons. For instance, no orexin neurons expressed YFP even though orexin neurons are closely intermingled with MCH neurons. The YFP expression was strong and restricted to the plasma membrane particularly in perikarya but it was less dense in proximal dendrites. Roughly 53% of MCH neurons expressed EYFP.

Whole-cell patch-clamp studies indicated that our vector efficiently drove the expression of the ChR2 gene. In current clamp mode, the YFP-positive-labeled neurons showed APs in responses to 10 ms blue light pulses delivered up to 30 Hz. As already reported by others using $ChR2_{H134R}$ (Gunaydin et al. 2010), light pulses sometimes evoked spikes in doublet or even triplets. Under constant illumination transfected neurons responded with depolarization current of 30 mV. Photocurrents were evoked in a light intensity manner. At subthreshold light intensity, transfected neurons only showed depolarization currents. Transfected neurons were also able to follow the light pulses when stimulated chronically at 10 Hz for 1 min repeated every other minute.

The in vitro findings gave us the confidence to test optogenetic stimulation in vivo. To test the hypnotic effect of MCH, we chose to stimulate MCH neurons over a 24 h period starting when lights turned off, i.e., mouse active period. Instead of stimulating MCH neurons within a particular vigilance state, we chose an unbiased approach stimulating neurons for 1 min every 4 min so stimulation would fall in all vigilance states over the course of the experiment. A LED blue light produced 10 ms duration pulses with at an output power of 1 mW (at the tip of fiber optic) that shone the transfected neurons via bilateral 200 μm O.D. optic fiber probes. We stimulated MCH neurons at three different frequencies, 5, 10, and 30 Hz, keeping 72 h apart between experiments. During the first 12 h (night), stimulation at 5 Hz was ineffective to change sleep amounts; however, stimulation at 10 and 30 Hz significantly increased both NREM sleep and REM sleep. The hypnotic effects started immediately and lasted for the next 6 h. Stimulation at 10 Hz was the most effective increasing NREMS by 40% and REMS by 70%. This finding was expected since in vitro recordings indicated 10 Hz stimulation evoked APs with the highest amplitude and even elicited double or triple spiking. Higher sleep amounts were caused by a significant reduction in the number of long-duration wake bouts. As such during stimulation, mice could not stay awake longer than 8 min. Yet sleep bout average duration remained unchanged. During stimulation mice fell more often asleep as well as transited more often into REM sleep. Transitions out of sleep stages also increased suggesting that activation of MCH neurons facilitates all sorts of state transitions. Increase in sleep drive was also observed measuring delta power during NREM sleep.

When stimulation continued over the next 12 h (day), sleep amounts remained unchanged likely due to the ceiling effects. Remarkably NREM sleep delta power analysis revealed a paradoxical increase during daytime stimulation. Sleep amounts remained unchanged over the 24 h of stimulation in mice infected with the control vector. To assess whether ceiling effects observed during the day time could have been caused by exhaustion of MCH neurons, MCH neurons were stimulated for 6 h starting at noon. Protocol of stimulation was the same as before, that is, 10 ms of pulse duration, at 10 Hz for 1 min every 4 min. Unlike stimulation started at night, stimulation started at noon failed to increase sleep amounts or change the sleep architecture. Our conclusion was then that the hypnotic effects of optogenetic stimulation of MCH neurons depend on the time of day manifesting very strongly during the mouse's active period.

To corroborate the hypnotic effect of stimulation of MCH neurons, we repeated the study in rats (Blanco-Centurion et al. 2016). We followed closely the same protocol of transfection and stimulation except we delivered more viruses (3 μl/ hemisphere) as well as we used a wider optic fiber (rat = 400 μm O.D. vs. mouse = 200 μm O.D.). The larger number of MCH neurons and brain size in rats made these adjustments compulsory. First we set out to confirm the effects of optogenetic stimulation in vitro. As reported earlier for mouse MCH neurons, rat MCH neurons did not show spontaneous firing activity and had almost identical membrane resting potential (≈ −60 mV). Rat transfected MCH neurons responded with APs when illuminated with blue light pulses of 10 ms of duration. However unlike mouse MCH neurons, rat MCH neurons could not fire in synchrony with light pulses delivered faster than 10 Hz. Voltage-clamp recordings showed that fast-frequency stimulation evoked a single spike first but then only evoked subthreshold depolarization currents. Histological analysis in rats indicated that our vector specifically transfected MCH neurons as we observed in mice. We observed very intense YFP labeling within the LH, ZI, and PeF. YFP was also exclusively present along neuronal plasma membranes, and it was particularly dense along dendrites unlike what occurred in mice. Similar to what was measured in mice, 52% of MCH neurons were transfected. We then tested stimulation of MCH neurons in vivo at three different frequencies: 5, 10, and 30 Hz. Again stimulation lasted for 24 h starting at lights off. Similar to previous findings in mice during the first 12 h of stimulation at 10 Hz, NREM sleep and REM sleep were significantly increased. The magnitude of NREM sleep increase was similar to that in mice (≈ 50%); however, in rats REM sleep was further augmented (≈ 200%). Another difference from the observations in mice is that stimulation at 5 Hz significantly increased REM sleep, whereas at 30 Hz, it did not. Optogenetic stimulation of MCH neurons in rats also had an immediate hypnotic effect that lasted for the next 9 h. During the peak of effects, REM sleep was augmented by 300% and NREM sleep by 100%. During the next 12 h of stimulation (day), all frequencies of stimulation significantly increased REM sleep amounts. Particularly stimulation at 5 Hz enhanced REM sleep amounts by 50% breaking the ceiling effects seen before in mice. Analysis of the sleep architecture confirmed that the hypnotic effect was mainly due to a dramatic reduction of long-duration wake bouts. Stimulation also increased the number of short duration NREM and REM sleep bouts instead of lengthening them. It corroborates our prior hypothesis that activation of MCH neurons is meant to facilitate brain transitions from waking into NREM sleep and from NREM sleep into REM sleep. This function was clearly revealed when we analyzed NREM EEG delta power over the 24 h of the stimulation. Normally NREM sleep delta power waxes and wanes across the 24 h, i.e., waxing during the active period and waning during the rest period. However during stimulation delta power daily oscillation became flat. The simplest explanation of it is that in order to discharge delta power during the resting phase, NREM sleep bouts should remain consolidated, i.e., plenty of long-duration bouts. Because during the day stimulation forced the brain to switch regularly into REM sleep, it caused a reduction of the long-duration bouts of NREM sleep blunting the delta power discharge. At night, constant intrusion of NREM sleep into the waking also produced a noticeable trend to elevate delta power so it too failed to wax. EEG spectral analysis in REM sleep revealed that theta power was also significantly higher for almost the entire 24 h of the stimulation. It indicates that activation of MCH neurons significantly builds up REM sleep pressure as well. In conclusion, optogenetic stimulation of MCH neurons had a potent hypnotic effect that was capable of counteracting circadian wake drive and, in the case of REM sleep, breaking the ceiling effect during daytime.

Other two research groups have also used optogenetic tools to investigate the role of activity of MCH neurons on sleep stages. Just a few months after we published our study in mice, Jego et al. published another study using a Tg line of mice expressing Cre in MCH neurons (Jego et al. 2013). A Cre-dependent DIO virus was microinjected into the hypothalamus of the MCH-Cre mice for targeting $ChETA_A$-YFP expression in MCH neurons. Around 87% of MCH neurons expressed YFP which was almost double from what we achieved in mice and rats. Whole-cell patch-clamp recordings from transfected neurons indicated blue light pulses also elicited spiking. Similar to transfected neurons in rats, transfected neurons in MCH-Cre mice showed 100% fidelity on their response to light pulses delivered between 1 and 20 Hz. However when stimulated at faster frequencies, it failed to evoke spiking. The fact that MCH neurons in the Cre mice could not be driven to fire faster than 20 Hz, even with $ChETA_A$, indicates that rodent MCH's neurons feature strong spike adaptation. Transfected neurons did not show spontaneous firing either.

Unlike our stimulation paradigm of unbiased timing, Jego et al. stimulated MCH neurons across sleep states during the second half of the resting phase (afternoon). They stimulated at the two opposite fringes of the response curve: 1 and 20 Hz. Unlike us, they used a blue laser light as source of illumination. Since lasers emit photons in a coherent way, the photon energy in laser light is higher than the LED energy (30–40 mW vs. 1 mW) and is also non-divergent. Stimulation at either frequency across NREM sleep episodes did not lengthen the episodes but significantly increased the number of transition into REM sleep. When stimulation was given at 20 Hz across the REM sleep episodes, it significantly prolonged the bout average duration. Yet when stimulated at 1 Hz, it failed to modify any of the sleep states. Stimulation at any rate during REM sleep did not change the EEG power spectra. Subsequently, they transfected MCH neurons with a ChR2 SFO confirming that 50 ms blue light pulses delivered every 10 s evoked steady depolarization currents until it was stopped by a 50 ms yellow pulse. Spiking was also produced under those conditions upon activation of postsynaptic excitatory inputs. They used the SFO to stimulate MCH neurons in vivo. Similar to previous findings with $ChETA_A$, stimulation across NREM sleep did not change this stage, whereas stimulation across REM sleep episodes prolonged its duration. Jego et al.'s experiments both confirmed and contradicted ours. Both studies found activation of MCH neurons significantly increased REM sleep by increasing transitions from NREM sleep. However, Jego et al. found that it also prolonged REM sleep bout duration. We found no evidence of MCH neurons activation prolonged bout duration.

It is possible to explain the discrepancies. Jego et al. observed effects on REM sleep during daytime stimulation, whereas we did not simply because they transfected the double of neurons than us. Also, blue light energy used to activate ChR2 was more powerful and concentrated in Jego et al.'s study than in ours (Laser vs. LED). When we used a thicker optic fiber and produced very dense expression of ChR2 on dendrites in the rat study, REM sleep amount was potently increased across a spectrum of stimulation frequencies. The fact that we did not observe a lengthening of the REM sleep bout might have to do with the duration of the stimulation. Whereas we stimulated for 1 min, Jego et al. stimulated along the entire REM sleep episode either intermittently (20 Hz) or continuously (SFO). Our analysis of the sleep architecture revealed that REM sleep bouts distribution was phase-locked to the duration of the stimulation; i.e. 1 min. Since 1 Hz stimulation did not evoke longer REM sleep episodes, it suggests this effect only occurs when the MCH neurons are forced to stay highly active across the REM sleep episode. Jego et al. found that MCH neuronal activation did not increase NREM sleep. When we stimulated during daytime, we did not observed this effect either. We only observed it during the nighttime stimulation. Since Jego et al. did not stimulate MCH neurons at that time, the controversy could not be resolved then.

The following year another study from Tsunematsu et al. (2014) attempted to tackle it. These researchers engineered a double Tg line of mice to conditionally express the SFO $ChR2_{E123T/T159C}$ in MCH neurons. For that purpose they used the tetracycline-controlled transcriptional activation system. Crossing a mouse endowed with the tetracycline transactivator protein (tTA) downstream of the MCH promoter gene with another mouse endowed with the tetracycline operator (TetO) upstream of the $ChR2_{E123T/T159C}$-YFP gene resulted in offspring having MCH-tTA/TetO. Histology of these MCH-tTA/TetO revealed that almost 90% of the MCH neurons expressed the reporter gene YFP. Its expression was exclusive to MCH neurons and was observed in somata and dendrites. It is worthy to note that antibodies against GFP were needed to see YFP; hence it is inferred that the endogenous gene expression had to be weak. Voltage-clamp mode recordings showed the YFP-expressing neurons responded with depolarizing currents. The response was elicited by either sustained blue light, or stronger yet, by 10 ms duration light pulses given at 10 Hz. Blue light pulses at 10 Hz for 1 min every 5 min also produced consistent spiking with 100% of fidelity at each stimulation bout. Tsunematsu et al. stimulated the MCH-tTA/TetO mice both during daytime and nighttime. During the first experiment, the daytime stimulation lasted for 3 h at 10 Hz (pulse = 10 ms, light power = 26 mW). Their findings closely replicated our results in the sense that REM sleep amount doubled. This effect was caused by a higher number of REM sleep transitions associated with fragmentation of NREM sleep and waking as well as increased NREM sleep delta power. The average duration of REM sleep remained unchanged despite mice had constant stimulation for 3 h. When 10 Hz stimulation was given across the night for 1 min every 5 min, transitions into REM sleep were also significantly augmented. The authors stated that REM sleep transitions always occurred from NREM sleep, never from waking, i.e., no SOREMPs. Thus it can be surely inferred that the number of NREM sleep bouts had to increase as well. These researchers did not measure the amounts of vigilance states because they were only interested in transitions into REM sleep. The hypnogram presented on the paper to illustrate this experiment clearly shows more transitions into NREM sleep during the stimulation. Then in a second set of experiments, stimulation at 10 Hz for 1 min was given at the beginning of either wake or NREM sleep episodes. Experiments were done during night and day. Authors claimed that only stimulation at the beginning of NREM sleep caused changes in vigilance states. However they only quantified the state transitions that occurred during the minute of stimulation and the next minute of post-stimulation. These conditions did not reproduce our experiments, and their conclusion should be taken with caution due to sampling bias. Even their state transition plots show a clear trend to increase the number of animals who had NREM sleep as stimulation time went on. We claimed that activation of MCH neurons during waking and during the active period will not put the animal to sleep unless it is given repetitively. At this time the circadian drive to stay awake is too strong to be counteracted by only 1 min of stimulation.

Findings from a recent study using DREADDs to activate MCH neurons highlight our point that effects on the two main sleep states will depend on strength and pattern of stimulation. MCH-Cre mice were transfected with DIO vector containing the excitatory DREADD hM3-DGq-mCherry (Vetrivelan et al. 2016). More than 90% of the mCherry-positive neurons co-labeled with MCH. When recorded in vitro mCherry-labeled neurons responded with continuous spiking following a bath application of the DREADD ligand CNO. c-Fos activation was also observed in mCherry-labeled neurons after CNO administration. c-Fos usually signals the continuous activation of neurons for at least 90 min. In vivo between 24% and 56% of MCH neurons were activated by CNO. DREADD-mediated activation of MCH neurons caused a selective increase of REM sleep both during day and night. A higher number of REM sleep bouts caused this effect. REM sleep average bout duration remained unchanged again confirming that MCH neurons normally do not play a role in REM sleep maintenance. It can be said that continuous activation of a small number of MCH neurons seems to be insufficient to significantly change the full spectrum of sleep stages. In our study in mice, the activation of 50% of those neurons did not yield meaningful changes in any of the sleep states during the day likely because we could not break the ceiling effect recruiting only half of the population. Likewise, the Vetrivelan et al.'s study where only one third of the total MCH neuronal population was steadily activated, it did not observe effects in NREM sleep because it did not have the cadence and strength to counteract the strong circadian drive to stay awake. Whereas CNO activates MCH neurons for several hours, our protocol of intermittent stimulation only activated these neurons for 2% of the total time. As mentioned earlier MCH neurons show prominent spike adaptation when stimulated continuously (Gao et al. 2003). Tsunematsu et al. also noticed how, when MCH neurons are stimulated at 10 Hz, the evoked depolarizing currents were larger than when stimulated continuously for 1 s (Tsunematsu et al. 2014). Because of this intrinsic property of the MCH neurons, we chose to stimulate them intermittently. Neurons encode their messages to other neurons by timing its APs. In vivo MCH neurons have both tonic and phasic firing, and both modes are quite dissimilar (Hassani et al. 2009). Tonic firing is only 1 Hz, whereas instantaneous firing is 22 Hz more or less 8 Hz. In lieu of stimulating MCH neurons at its tonic rhythm, we assumed that the message of MCH neurons is mainly encoded through their phasic firing. Hence we tested stimulation frequencies from 5 to 30 Hz but delivered intermittently. Thus in our rat's study combining the appropriate phasic stimulation with enough recruitment of stimulated neurons (more light + more ChR2), we then were able to evoke strong effects on both types of sleep at any time of the day counteracting the circadian constraints. Other researchers have also suggested that depending on the mode of optogenetic stimulation, stimulated neurons will release either amino acid (like GABA, glutamate), neurotransmitters, or neuropeptides (like MCH or orexin). They proposed that tonic mild stimulation releases fast amino acid neurotransmitters, whereas neuropeptides may require sustained strong stimulation (Arrigoni and Saper 2014). This concept perfectly would explain why our protocol of stimulation likely resulted in strong MCH release and effects on both types of sleep.

Now it follows to review experiments where optogenetic tools were used to inhibit the activity of MCH neurons.

5.2 *Inhibition of MCH Neurons*

Both Jego et al. (2013) and Tsunematsu et al. (2014) used optogenetic tools to silence MCH neurons and study its effect on sleep. Jego et al. transfected MCH-cre mice with a DIO viral vector containing either the eNpHR3.0-YFP or Arch HR gene. In vitro slice recordings of those transfected neurons corroborated that yellow light illumination for 30 s caused a complete silencing of spikes as well as an outward current when those neurons were voltage clamped. Yellow light 30 s duration pulses were then used to test the in vivo effect on REM sleep and theta power. Inhibition either by eNpHR3.0 or Arch significantly decreased the 6–9 Hz EEG theta power bandwidth at the expense of increasing the 3–5 Hz bandwidth. These results corroborate ours when we found stimulation of MCH neurons significantly increased EEG theta power during REM sleep.

Remarkably yellow light inhibition of MCH neurons neither decreased the average bout duration nor the amount of REM sleep. This result contradicts their hypothesis claiming activity of the MCH neurons is necessary for REM sleep initiation or maintenance. Since these researchers did not publish any histology for those particular experiments, it is not possible to determine whether the lack of effect had to do with technical issues. We claim that it is not sufficient to rely on the in vitro data to assure the data in vivo had the same degree of validity. Placement of the optic fiber with respect to the targeted cells, the extent of transfection (or selective expression), and strength of transgene expression judged by the reporter gene fluorescent signal all are critical factors that must be reported to interpret correctly the data from optogenetic experiments. It is important that researchers produce maps depicting optic probe placement in relation to transfected cells. The reader must keep in mind that optogenetic tools demand that great number of photons reach the opsins as well as a robust expression of the opsins among a relatively large population of transfected neurons. One important question is how to assess which of the total transfected neurons are actually being stimulated. One way of answering this should be using a marker of activity during the stimulation, like c-Fos, but could be another one more sensitive like GCAMP6. In any case it can be said that the strong results we measured in rats can be explained because rats have more MCH neurons in the first place. However in addition to this species intrinsic difference, the remarkable expression of ChR2 we observed and correct placement of a thicker optic probe that transmitted more photons all likely caused enough neuronal recruitment resulting in strong effects on both types of sleep.

Tsunematsu et al. also inhibited MCH neurons and studied its effects on REM sleep. They again used the TtA/TtO system to express ArchT in MCH neurons. Unlike Jego et al., Tsunematsu et al. published their histology analysis corroborating ArchT only expressed in MCH neurons. Again the reporter signal must have been weak since IF was needed to visualize the transfected neurons. In any case almost all MCH neurons (97.2%) showed colocalization with YFP. In vitro recordings indicated that green light illumination produced 100% of inhibition of APs induced by current injection. Despite the outstanding expression of ArchT and complete inhibition of APs recorded in vitro, 3 h of inhibition failed to change any of the vigilance states. Total amount, average bout duration, or number of episodes of all three states remained unchanged. The conclusion derived from both studies indicates that acute silencing MCH neuronal activity does not affect any of the sleep states.

Tsunematsu et al. also made selective ablations of MCH neurons with the diphtheria toxin A gene under the control of the TtA/TtO gene expression system. One week after doxycycline was removed from the diet, MCH neurons started to die. By the fourth week, only 2.4% of MCH neurons remained. Contrary to the findings obtained with acute silencing of MCH neurons, its ablation had significant impacts on both wake and NREM sleep daily amounts. Thus wake amounts started to significantly increase by the second week and continued along that trend all the way to the fourth week. The increase in wake amounts was due to the expense of a significant decrease of NREM sleep. Reduction in NREM sleep was measured both during the active and resting phases although it was more remarkable during daytime. The reduction of NREM sleep during daytime was mainly caused by decrease in the number of episodes, whereas during nighttime NREM sleep bout duration significantly decreased. No changes in the EEG spectral power were observed after the loss of MCH neurons. REM sleep amounts were not altered by the complete loss of MCH neurons.

Thus the results obtained after a total loss of MCH neurons corroborate our hypothesis concerning the role of these neurons in regulating daily amounts of NREM sleep. Also they negate the hypothesis that these neurons are critical to generate REM sleep. MCH neurons seem to execute the NREM sleep regulation differently depending on the time of day. This hypothesis is supported by our results where stimulation had different effects depending on the phase of the diurnal cycle. Previously we measured the highest level of MCH in the CSF of rats around noon, a timing associated with high amounts of NREM sleep and REM sleep (Pelluru et al. 2013). Likewise in rats the ICV infusion of MCH at night onset was followed by a significant increase of NREM sleep (+70%) and REM sleep (+200%) (Verret et al. 2003). At night onset MCH levels in the CSF is waning and stay low during the active period (Pelluru et al. 2013). In humans the release of MCH was highest after sleep onset which is also normally associated with higher amounts of delta sleep (Blouin et al. 2013). Thus forced stimulation of MCH neurons, or ICV infusion of MCH after night onset, yielded higher levels of MCH receptor activation recapitulating the sleep facilitation role of MCH occurred otherwise at the onset of the resting phase.

Pharmacological and gene knockout studies provide additional strong support for the role of MCH neurons as modulators of both NREM and REM sleep. In rats selective pharmacological blocking of the MCH1 receptors is followed by a remarkable decrease of both delta sleep and REM sleep (Ahnaou et al. 2008). This effect is dose-dependent, not followed by a sleep rebound and caused by a reduction of bouts duration. MCHR1 KO mice also show significantly less daily amounts of NREM sleep and higher core temperature at rest (Ahnaou et al. 2011). Similar effects were reported by Jego et al. who measured sleep in double MCH-Cre/MCH1R KO mice (Jego et al. 2013). These double transgenic mice had significant reductions in both NREM and REM sleep during daytime.

Recent studies making selective ablations of MCH neurons have put forward an interesting twist on the role of MCH neurons regulating REM sleep during the resting phase. One study replicated the Tsunematsu et al. lesion approach by using diphtheria toxin (DT) to kill selectively MCH neurons (Vetrivelan et al. 2016). MCH-Cre mice were crossed with floxed-P Tg mice harboring the DT receptor. MCH neurons were killed in the offspring with systemic administrations of DT. During the active period, lesioned mice had higher core body temperature, hyperactivity, as well as a significant loss of body mass but no changes in the amounts of sleep states. Remarkably during the resting period, the number of REM sleep bouts paradoxically increased. Interestingly this effect was linked to a shortening of wake episodes. On the second study, researchers made partial lesions of the MCH neurons inserting the Ataxin-3 lethal gene upstream of the MCH promoter (Varin et al. 2016). Adult mice showed $\approx$30% less MCH neurons. The surviving MCH neurons had clear signs of degeneration such as severe reduction in projection terminals as well as 80% reduction in hypothalamic MCH mRNAs. Similar to the previous lesion study during daytime, MCH Ataxin-3 mice had higher number and longer duration of the REM sleep bouts.

In addition MCH Ataxin-3 mice showed sleep fragmentation. After 12 h of total sleep deprivation, MCH Ataxin-3 mice also had a significant deficit to consolidate NREM sleep rebound including a decrease in delta power. Declining and recovering slopes for REM sleep amounts were also significantly decreased in lesioned mice. Another study in MCHR1 KO mice also found higher amounts of REM sleep during the day (Adamantidis et al. 2008).

It is obvious from these three studies that MCH neuronal activity cannot be critical for generation of REM sleep because it is hard to explain the increase of REM sleep without MCH neurons or MCH receptors. Since the pioneering studies of brain transections, it is known that REM sleep is generated by the brainstem. However it is also evident that absence of MCH neurons does impact how the brain shifts among vigilance states, a function that is particularly important to achieve during sleep homeostasis. Deficit in sleep homeostasis is tightly linked to disturbances of metabolism/body temperature. As pointed out extensively already, loss of MCH neurons or MCH receptors is followed by nighttime hyperactivity, loss of lean body mass, and high core body temperature in addition to loss of appetite. It is conceivable that all these changes have led to abnormal sleepiness during the resting period. The persistent higher core temperature can lead to increase of entries into REM sleep because higher core/brain temperature is known to occur during the transition from NREM to REM sleep (Obal et al. 1985; Alfoldi et al. 1990; Gao et al. 1995). Higher basal/brain metabolism during the rest period also leads to sleep fragmentation (Nofzinger et al. 2004).

6 Conclusion: Ying and Yang Interplay Between MCH and Orexin Neurons

Next, we propose a model where orexin and MCH neuronal activity counterbalance each other. The main assumption of our model demands that both populations of neurons show firing activity in close synchrony. Under the prevalent view, this is unlikely because MCH neurons are considered strictly as sleep active, whereas orexin is considered strictly as wake active. A recent study, however, has debunked the dogma that MCH neurons are completely silent during waking (Hassani et al. 2009). Using fiber photometry and GCAMP6 as calcium sensor in MCH-Cre mice, this study found that MCH neurons showed strong population bursts during novelty exploration (Gonzalez et al. 2016). MCH neuronal activation bursts were inversely correlated with activity bursts from orexin neurons. During a novel object presentation, orexin neurons became very active, while MCH neurons did not, and then MCH neurons became active, while orexin neurons quieted down. During the entry into the exploration area, MCH neurons also turned suddenly active, while orexin neurons remained inactive. The calcium proxy of neuronal activity between orexin and MCH corroborates what in vivo juxtacellular recordings reported earlier, that is, orexin and MCH neurons fire in opposite phases (Hassani et al. 2009). However, in this case, counterbalancing is occurring during waking. It is conceivable that activity of MCH neurons during waking is geared toward balancing the activity of orexin neurons. In vitro studies have shown MCH blocks the increase of firing in orexin neurons driven by positive feedback (Rao et al. 2008). Without MCH receptors, the glutamatergic inputs into orexin neurons evoke larger excitatory postsynaptic potentials and stronger firing. During novelty exploration, orexin release within the LH is the highest (Kiyashchenko et al. 2002) and so is orexin firing (Mileykovskiy et al. 2005). Thus it is conceivable that MCH neurons become active to offset the positive feedback of orexin neurons preventing a state of hyperarousal, hyperactivity, and anxiety. Overexpression of orexin leads to this abnormal state (Willie et al. 2011). Highest level of activity of orexin neurons is also associated with behaviors characterized by strong and positive emotions like playing or exploration (Mileykovskiy et al. 2005). These behaviors feature high levels of arousal, memory retrieval, and attention. By contrast, orexin neurons are moderately active during repetitive motor behaviors like treadmill running, ingesting, or grooming which are less emotionally charged. Perhaps this is so because MCH tone is counterbalancing the orexin tone. When we stimulated repetitively MCH neurons during waking, we putatively set up an intermittent restrain over the burst of activity in orexin neurons. In other words, we facilitated NREM sleep during the active period by creating a relaxed state of mind conducive for sleep onset.

MCH neuronal activity may be assisting other inhibitory pathways like GABA in setting up a restraining tone for stable vigilance state switching. During NREM, orexin neurons are already under restrain by GABAergic sleep-active neurons (Alam et al. 2005) and fire very occasionally. Orexin neurons have multiple GABAergic synapses contacting them (Henny and Jones 2006). When GABA receptors on orexin neurons are removed, there is a decrease in sensitivity of orexin neurons to both excitatory and inhibitory inputs. As a result these mice show sleep fragmentation switching frequently among vigilance states (Matsuki et al. 2009). Frequent switching among vigilance states is what we also produced by stimulating MCH neurons. We propose that, during NREM sleep, extra inhibition from MCH neurons to orexin neurons would create the right conditions for REM sleep to initiate. However, when MCH tone is abnormally high, orexin neurons become desensitized creating instability among vigilance states. This hypothesis would explain squarely why optogenetic activation of MCH neurons during NREM sleep is so effective facilitating transition into REM sleep but also why we observed frequent transitions among all vigilance states.

Narcolepsy with cataplexy can also be understood as an imbalance between orexin and MCH neurons. Narcolepsy features disturbance of all vigilances states. Narcoleptics show deficits staying awake, staying asleep, and maintaining proper muscle tone during waking and sleep. In narcolepsy with cataplexy, orexin neurons have died, but MCH neurons are still present in normal numbers creating an imbalance (Thannickal et al. 2000). Normally orexin inhibits voltage-dependent calcium channels on MCH neurons (Gao et al. 2003). Furthermore optogenetic stimulation of orexin neurons inhibits MCH firing burst by exciting GABAergic synapsis onto MCH neurons (Apergis-Schoute et al. 2015). Hence in narcoleptics the lack of orexin inhibition onto MCH neurons may trigger abnormal activity bursts across all vigilance states. This burst of activity in MCH neurons will, in turn, break down arousal (i.e., sleep attacks), and sometimes, when the emotional tone becomes very high, it will bring down muscle tone resulting in cataplexy attacks or direct transitions into REM sleep, i.e., SOREMPs. It also creates sleep fragmentation manifested as constant switches among waking, NREM, and REM sleep, particularly transitions into REM sleep.

Narcolepsy also illustrates how deficits to stay fully alert and maintain skeletal muscle tone are tightly connected to disturbances regulating emotional tone (Bayard and Dauvilliers 2013). We claim that the connection of cataplexy with unrestrained emotional tone can be also explained by an imbalance between orexin and MCH neuronal activity. In narcoleptic rodents or dogs, cataplexy attacks are usually preceded by emotionally laden motor repetitive behaviors such as eating palatable food (Gerashchenko et al. 2001; Clark et al. 2009) play, and copulation (Nishino and Mignot 1997) or other less emotive like grooming or running (Chemelli et al. 1999; Beuckmann et al. 2004; Espana et al. 2007). Infusion of orexin (Mieda et al. 2004) or orexin gene therapy within the dorsolateral pons, a REM sleep generator (Blanco-Centurion et al. 2013); or within the central nucleus of the amygdala, a strong emotional center (Liu et al. 2016); or in the ZI/LH (Liu et al. 2011), where MCH neurons are located, all blocks cataplexy. However we found that when orexin is expressed in MCH neurons only, cataplexy and sleep fragmentation actually worsen. Thus when the timing of orexin and MCH release becomes simultaneous, emotionally driven cataplexy and hyperarousal during sleep occur more often than when MCH is released without any counterbalance from orexin tone. Repetitive pattern motor behaviors like eating are precisely the behaviors that we know kindle MCH release in the amygdala of humans (Blouin et al. 2013). In the amygdala every behavior where orexin tone turns high is negatively mirrored by low MCH tone and vice versa. We propose this counterbalance scenario within the main center of emotional regulation is not epiphenomena, but it reflects an essential counterbalance modulation.

Burst of MCH activity is critical to integrate caloric value, ingesting behavior and level of locomotor activity. It is hard to understand this function if MCH neurons would only fire during sleep. Thus MCH neurons were optogenetically stimulated during a drinking choice preference test where animals had to choose between drinking an artificial sweetener or sucrose (Domingos et al. 2013). In the control group, animals preferred sucrose because of its true caloric content and rewarding effect assessed by dopamine release. Stimulation of MCH neurons at 20 Hz, or continuously for 1 s, given during the choice test completely reversed the preference for sucrose. Stimulation at 5 Hz was ineffective. Effective stimulation also caused a peak of dopamine release, whereas the ineffective stimulation did not. Selective ablation of MCH neurons abolished the sucrose preference as well as the peak in dopamine release elicited by sucrose intake. An antagonist of the MCH receptor also blocks the preference for ingesting highly palatable food (Morens et al. 2005). Remarkably in humans the second high level of MCH release occurs after eating (Blouin et al. 2013). In this context years ago direct evidence emerged that activation of MCH receptors is key to regulate caloric intake. MCH receptor KO mice are lean, eat less, and show high basal metabolism (Shimada et al. 1998). When MCH receptor KO mice are fed with high caloric diet, animals also become hyperactive (Kokkotou et al. 2005). Hyperactivity in MCH receptor KO mice fed with fat diet is especially prominent during the active period and also occurs in MCH KO mice (Zhou et al. 2005) and mice with selective genetic ablation of MCH neurons (Alon and Friedman 2006). By contrast, overexpression of MCH receptors produced a phenotype of obesity and insulin resistance (Ludwig et al. 2001). Same results occurred in mice that had continuous ICV infusions of MCH (Gomori et al. 2003). MCH neurons are indeed strongly activated by glucose (Burdakov et al. 2005) or by cannabinoids which are well known for its orexinergic effect (Huang et al. 2007).

Glucose has the opposite effect on orexin neurons causing strong inhibition of its firing (Burdakov et al. 2005). In contrast to the lean mice whose MCH neurons were ablated, mice without orexin neurons are obese despite eating less. These mice are obese because they have significantly low locomotor activity during the active period so calories are not burned (Hara et al. 2001). On top of that, sleepiness lower even further both the metabolism and the energy expenditure. The effects on large body weight with low activity and metabolism are not surprising because the selective activation of orexin neurons causes an increase of eating, drinking, locomotor activity, and metabolism (Inutsuka et al. 2014).

As we already mentioned, electrical stimulation experiments indicated that the LH is a brain hub for integration of feeding and reward behaviors. The link between MCH neuronal activation and relaxed emotions is rooted in several studies that firmly established the anxiolytic effects of MCH (Monzon et al. 2001; Borowsky et al. 2002; Carlini et al. 2006). Unlike the calming effects of MCH neuronal activation, optogenetic activation of orexin increases anxiety in mice (Heydendael et al. 2014). The orexin pathway regulating anxiety and fear involves connection from the amygdala and the locus coeruleus (Sears et al. 2013). By contrast, GABAergic neurons from the amygdala make monosynaptic connections with MCH neurons inhibiting them (Gonzalez et al. 2016). Altogether the picture that emerges depicts the MCH neurons becoming active in response to biological critically goal-directed behaviors like eating. Ingestion is primarily driven by high orexinergic tone that if unchecked could become excessive. Therefore it is conceivable MCH neurons would activate, counterbalancing the orexin drive and generating a calming/pleasant sensation that leads to the end of the appetitive behavior. The satisfaction of the appetitive drive may eventually lead to sleep onset. This balance/counterbalance model of orexin/MCH may also apply to understand other essential appetitive behavior like drinking or mating.

Stress and anxiety inhibit activity of MCH neurons (Gonzalez et al. 2016) through an orexin and amygdala jointed pathway. This effect of stress would explain why during the juxtacellular recordings of these neurons firing activity during waking was barely detected, i.e., only during transitions to NREM sleep (Hassani et al. 2009). Juxtacellular recordings demand the rat to be head fixed and held lifted onto a hammock. Even though daily habituation makes the rat to eventually fall asleep during the recording, it is obvious that the animal must be somewhat stressed while awake. Stress will prevent the MCH neurons to fire giving a bias toward sleep activity.

The counterbalance between orexin and MCH neurons also can take place during REM sleep. Contrary to the prevalent view of orexin neurons being silent during REM sleep, a microdialysis study found that orexin release, within the LH and basal forebrain, is highest during REM sleep (Kiyashchenko et al. 2002). This finding is supported by the fact that, in unrestrained animals, orexin neurons produce burst of activity linked to the phasic movements occurring during REM sleep (Mileykovskiy et al. 2005). Jego et al. observed that 20 Hz optogenetic stimulation of the MCH projections to the posterior hypothalamus, LH, and medial septum significantly increased the duration of REM sleep episodes (Jego et al. 2013). All these regions receive strong projections from orexin neurons as well (Peyron et al. 1998). We claim that MCH projections to those areas are such to counterbalance the burst of phasic activity from orexin neurons during the phasic events of REM sleep. Recent unpublished data from our laboratory indicate that the pathway connecting MCH neurons to noradrenergic neurons in the locus coeruleus also mediate the increase in REM sleep produced by the optogenetic stimulation. A previous study found projections from MCH to the REM sleep generator zone in the pons. Local infusion of MCH into this pontine area increased REM sleep (Torterolo et al. 2009). Needless to add that orexin neurons also project to the locus coeruleus and the pontine reticular formation. Orexin input into both areas is critically involved in the control of muscle tone during REM sleep and cataplexy (Wu et al. 1999; Kiyashchenko et al. 2001). Thus our hypothesis neatly explains why burst of activity of MCH neurons occurs phase-locked to the phasic motor events of REM sleep (Hassani et al. 2009). Whether the optogenetic stimulation of MCH neurons would lengthen REM sleep, as Jego et al. observed, must depend on how effective counteracting the burst of activity of orexin neurons is. Yet if the latter is mild, optogenetic activation of MCH neurons could be ineffective prolonging REM sleep episode as we and other researchers have observed. As we sustain the timing between orexin and MCH burst of neuronal activity is what will determine the outcome of behavior. An out-of-phase interplay between these two hypothalamic types of neurons is what ultimately governs the expression of the vigilance states.

7 Future Directions

Our hypothesis can be empirically tested in two ways. One is to monitor the in vivo activity of both MCH and orexin neurons using state-of-the-art imaging techniques. Nowadays it is possible to observe intracellular calcium transients in specific neuronal phenotypes occurring in real time and to single-cell spatial resolution level. Hundreds of neurons can be imaged in this way. Less than a second time-resolution scale observations are possible. For instance, a recent imaging study focused on observing how GABAergic neurons within the LH get activated during either the consummatory or appetitive phases of the feeding behavior (Jennings et al. 2015). Calcium imaging in vivo also has been applied to study neuronal activity of both GABAergic and glutamatergic dorsolateral pontine neurons occurring during the sleep states (Cox et al. 2016). If our counterbalance hypothesis is correct and the out-of-phase interplay regulates the states of vigilance, it should be observed as distinctive patterns of activity burst in orexin and MCH neurons. Furthermore, using optogenetics coupled to in vivo micro-endoscopic imaging, we can ask hypothesis-driven experimental questions. Either stimulating or inhibiting orexin and MCH neurons and at the same time imaging those neurons, we should be able to predict changes in the neuronal activity patterns and sleep behavior.

Another aspect that is critical to explore is how different projections of both orexin and MCH are integrating sleep states. Using clarity to make the brain transparent, we discovered MCH neurons are organized in three different clusters according to the direction of its projections (Shiromani and Peever 2017). One cluster projects mostly to the forebrain and the other toward the brainstem, and a third one seems to link both clusters within the LH. We believe this arrangement may reflect the way MCH neurons counterbalance orexin neurons along different aspects of sleep–wake behavior. Perhaps forebrain projections counterbalance orexin signal in terms of memory, attention, and emotion and may regulate the transition from wake into NREM sleep. Projections to the brainstem may integrate the counterbalance of orexin tone to restrain sympathetic and muscle tone that may underlie the transitions into REM sleep. In any case what is needed is to stimulate or inhibit in a specific projection manner and observe its impact on neuronal activity and behavior. Using retrograde virus carrying either opsins or GCAMP6 Cre-dependent gene, we could address this type of important questions.

References

Adamantidis A, Salvert D, Goutagny R, Lakaye B, Gervasoni D, Grisar T, Luppi PH, Fort P (2008) Sleep architecture of the melanin-concentrating hormone receptor 1-knockout mice. Eur J Neurosci 27(7):1793–1800

Ahnaou A, Drinkenburg WH, Bouwknecht JA, Alcazar J, Steckler T, Dautzenberg FM (2008) Blocking melanin-concentrating hormone MCH1 receptor affects rat sleep-wake architecture. Eur J Pharmacol 579(1–3):177–188

Ahnaou A, Dautzenberg FM, Huysmans H, Steckler T, Drinkenburg WH (2011) Contribution of melanin-concentrating hormone (MCH1) receptor to thermoregulation and sleep stabilization: evidence from MCH1 (−/−) mice. Behav Brain Res 218(1):42–50

Alam MN, Kumar S, Bashir T, Suntsova N, Methippara MM, Szymusiak R, McGinty D (2005) GABA-mediated control of hypocretin- but not melanin-concentrating hormone-immunoreactive neurones during sleep in rats. J Physiol 563(Pt 2):569–582

Alfoldi P, Rubicsek G, Cserni G, Obal F Jr (1990) Brain and core temperatures and peripheral vasomotion during sleep and wakefulness at various ambient temperatures in the rat. Pflugers Arch 417(3):336–341

Al-Juboori SI, Dondzillo A, Stubblefield EA, Felsen G, Lei TC, Klug A (2013) Light scattering properties vary across different regions of the adult mouse brain. PLoS One 8(7):e67626

Alon T, Friedman JM (2006) Late-onset leanness in mice with targeted ablation of melanin concentrating hormone neurons. J Neurosci 26(2):389–397

Anand BK, Brobeck JR (1951) Hypothalamic control of food intake in rats and cats. Yale J Biol Med 24(2):123–140

Apergis-Schoute J, Iordanidou P, Faure C, Jego S, Schone C, Aitta-Aho T, Adamantidis A, Burdakov D (2015) Optogenetic evidence for inhibitory signaling from orexin to MCH neurons via local microcircuits. J Neurosci 35(14):5435–5441

Aravanis AM, Wang LP, Zhang F, Meltzer LA, Mogri MZ, Schneider MB, Deisseroth K (2007) An optical neural interface: in vivo control of rodent motor cortex with integrated fiberoptic and optogenetic technology. J Neural Eng 4(3):S143–S156

Arenkiel BR, Peca J, Davison IG, Feliciano C, Deisseroth K, Augustine GJ, Ehlers MD, Feng G (2007) In vivo light-induced activation of neural circuitry in transgenic mice expressing channelrhodopsin-2. Neuron 54(2):205–218

Arrigoni E, Saper CB (2014) What optogenetic stimulation is telling us (and failing to tell us) about fast neurotransmitters and neuromodulators in brain circuits for wake-sleep regulation. Curr Opin Neurobiol 29C:165–171

Aziz A, Fronczek R, Maat-Schieman M, Unmehopa U, Roelandse F, Overeem S, van Duinen S, Lammers GJ, Swaab D, Roos R (2008) Hypocretin and melanin-concentrating hormone in patients with Huntington disease. Brain Pathol 18(4):474–483

Baker BI, Bird DJ (2002) Neuronal organization of the melanin-concentrating hormone system in primitive actinopterygians: evolutionary changes leading to teleosts. J Comp Neurol 442(2): 99–114

Bamberg E, Tittor J, Oesterhelt D (1993) Light-driven proton or chloride pumping by halorhodopsin. Proc Natl Acad Sci USA 90(2):639–643

Banghart M, Borges K, Isacoff E, Trauner D, Kramer RH (2004) Light-activated ion channels for remote control of neuronal firing. Nat Neurosci 7(12):1381–1386

Bayard S, Dauvilliers YA (2013) Reward-based behaviors and emotional processing in human with narcolepsy-cataplexy. Front Behav Neurosci 7:50

Berndt A, Schoenenberger P, Mattis J, Tye KM, Deisseroth K, Hegemann P, Oertner TG (2011) High-efficiency channelrhodopsins for fast neuronal stimulation at low light levels. Proc Natl Acad Sci USA 108(18):7595–7600

Berthoud HR, Munzberg H (2011) The lateral hypothalamus as integrator of metabolic and environmental needs: from electrical self-stimulation to opto-genetics. Physiol Behav 104(1): 29–39

Beuckmann CT, Sinton CM, Williams SC, Richardson JA, Hammer RE, Sakurai T, Yanagisawa M (2004) Expression of a poly-glutamine-ataxin-3 transgene in orexin neurons induces narcolepsy-cataplexy in the rat. J Neurosci 24(18):4469–4477

Biran R, Martin DC, Tresco PA (2005) Neuronal cell loss accompanies the brain tissue response to chronically implanted silicon microelectrode arrays. Exp Neurol 195(1):115–126

Bittencourt JC (2011) Anatomical organization of the melanin-concentrating hormone peptide family in the mammalian brain. Gen Comp Endocrinol 172(2):185–197

Blanco-Centurion C, Liu M, Konadhode R, Pelluru D, Shiromani PJ (2013) Effects of orexin gene transfer in the dorsolateral pons in orexin knockout mice. Sleep 36(1):31–40

Blanco-Centurion C, Liu M, Konadhode RP, Zhang X, Pelluru D, van den Pol AN, Shiromani PJ (2016) Optogenetic activation of melanin-concentrating hormone neurons increases non-rapid eye movement and rapid eye movement sleep during the night in rats. Eur J Neurosci 44 (10):2846–2857

Blouin AM, Fried I, Wilson CL, Staba RJ, Behnke EJ, Lam HA, Maidment NT, Karlsson KAE, Lapierre JL, Siegel JM (2013) Human hypocretin and melanin-concentrating hormone levels are linked to emotion and social interaction. Nat Commun 4:1547

Borowsky B, Durkin MM, Ogozalek K, Marzabadi MR, DeLeon J, Lagu B, Heurich R, Lichtblau H, Shaposhnik Z, Daniewska I, Blackburn TP, Branchek TA, Gerald C, Vaysse PJ, Forray C (2002) Antidepressant, anxiolytic and anorectic effects of a melanin-concentrating hormone-1 receptor antagonist. Nat Med 8(8):825–830

Boyden ES, Zhang F, Bamberg E, Nagel G, Deisseroth K (2005) Millisecond-timescale, genetically targeted optical control of neural activity. Nat Neurosci 8(9):1263–1268

Brecht M, Schneider M, Sakmann B, Margrie TW (2004) Whisker movements evoked by stimulation of single pyramidal cells in rat motor cortex. Nature 427(6976):704–710
Brocker DT, Grill WM (2013) Principles of electrical stimulation of neural tissue. Handb Clin Neurol 116:3–18
Burdakov D, Gerasimenko O, Verkhratsky A (2005) Physiological changes in glucose differentially modulate the excitability of hypothalamic melanin-concentrating hormone and orexin neurons in situ. J Neurosci 25(9):2429–2433
Callaway EM, Katz LC (1993) Photostimulation using caged glutamate reveals functional circuitry in living brain slices. Proc Natl Acad Sci USA 90(16):7661–7665
Cardot J, Fellmann D, Bugnon C (1994) Melanin-concentrating hormone-producing neurons in reptiles. Gen Comp Endocrinol 94(1):23–32
Carlini VP, Schioth HB, de Barioglio SR (2006) Melanin-concentrating hormone (MCH) reverts the behavioral effects induced by inescapable stress. Peptides 27(9):2300–2306
Chambers J, Ames RS, Bergsma D, Muir A, Fitzgerald LR, Hervieu G, Dytko GM, Foley JJ, Martin J, Liu WS, Park J, Ellis C, Ganguly S, Konchar S, Cluderay J, Leslie R, Wilson S, Sarau HM (1999) Melanin-concentrating hormone is the cognate ligand for the orphan G-protein-coupled receptor SLC-1. Nature 400(6741):261–265
Chemelli RM, Willie JT, Sinton CM, Elmquist JK, Scammell T, Lee C, Richardson JA, Williams SC, Xiong Y, Kisanuki Y, Fitch TE, Nakazato M, Hammer RE, Saper CB, Yanagisawa M (1999) Narcolepsy in orexin knockout mice: molecular genetics of sleep regulation. Cell 98(4): 437–451
Chow BY, Han X, Dobry AS, Qian X, Chuong AS, Li M, Henninger MA, Belfort GM, Lin Y, Monahan PE, Boyden ES (2010) High-performance genetically targetable optical neural silencing by light-driven proton pumps. Nature 463(7277):98–102
Chuhma N, Tanaka KF, Hen R, Rayport S (2011) Functional connectome of the striatal medium spiny neuron. J Neurosci 31(4):1183–1192
Clark EL, Baumann CR, Cano G, Scammell TE, Mochizuki T (2009) Feeding-elicited cataplexy in orexin knockout mice. Neuroscience 161(4):970–977
Clark KL, Armstrong KM, Moore T (2011) Probing neural circuitry and function with electrical microstimulation. Proc Biol Sci 278(1709):1121–1130
Conductier G, Brau F, Viola A, Langlet F, Ramkumar N, Dehouck B, Lemaire T, Chapot R, Lucas L, Rovere C, Maitre P, Hosseiny S, Petit-Paitel A, Adamantidis A, Lakaye B, Risold PY, Prevot V, Meste O, Nahon JL, Guyon A (2013) Melanin-concentrating hormone regulates beat frequency of ependymal cilia and ventricular volume. Nat Neurosci 16(7):845–847
Courseaux A, Nahon JL (2001) Birth of two chimeric genes in the Hominidae lineage. Science 291(5507):1293–1297
Cox J, Pinto L, Dan Y (2016) Calcium imaging of sleep-wake related neuronal activity in the dorsal pons. Nat Commun 7:10763
Denk W (1994) Two-photon scanning photochemical microscopy: mapping ligand-gated ion channel distributions. Proc Natl Acad Sci USA 91(14):6629–6633
Dhawale AK, Hagiwara A, Bhalla US, Murthy VN, Albeanu DF (2010) Non-redundant odor coding by sister mitral cells revealed by light addressable glomeruli in the mouse. Nat Neurosci 13(11):1404–1412
Domingos AI, Sordillo A, Dietrich MO, Liu ZW, Tellez LA, Vaynshteyn J, Ferreira JG, Ekstrand MI, Horvath TL, de Araujo IE, Friedman JM (2013) Hypothalamic melanin concentrating hormone neurons communicate the nutrient value of sugar. Elife 2:e01462
Espana RA, McCormack SL, Mochizuki T, Scammell TE (2007) Running promotes wakefulness and increases cataplexy in orexin knockout mice. Sleep 30(11):1417–1425
Farber IC, Grinvald A (1983) Identification of presynaptic neurons by laser photostimulation. Science 222(4627):1025–1027
Filadelfi AM, Castrucci AM (1994) Melatonin desensitizing effects on the in vitro responses to MCH, alpha-MSH, isoproterenol and melatonin in pigment cells of a fish (S. marmoratus), a toad (B. ictericus), a frog (R. pipiens), and a lizard (A. carolinensis), exposed to varying photoperiodic regimens. Comp Biochem Physiol A Physiol 109(4):1027–1037

Gao BO, Franken P, Tobler I, Borbely AA (1995) Effect of elevated ambient temperature on sleep, EEG spectra, and brain temperature in the rat. Am J Phys 268(6 Pt 2):R1365–R1373

Gao XB, Ghosh PK, van den Pol AN (2003) Neurons synthesizing melanin-concentrating hormone identified by selective reporter gene expression after transfection in vitro: transmitter responses. J Neurophysiol 90(6):3978–3985

Gerashchenko D, Kohls MD, Greco M, Waleh NS, Salin-Pascual R, Kilduff TS, Lappi DA, Shiromani PJ (2001) Hypocretin-2-saporin lesions of the lateral hypothalamus produce narcoleptic-like sleep behavior in the rat. J Neurosci 21(18):7273–7283

Gomori A, Ishihara A, Ito M, Mashiko S, Matsushita H, Yumoto M, Ito M, Tanaka T, Tokita S, Moriya M, Iwaasa H, Kanatani A (2003) Chronic intracerebroventricular infusion of MCH causes obesity in mice. Melanin-concentrating hormone. Am J Physiol Endocrinol Metab 284 (3):E583–E588

Gonzalez JA, Iordanidou P, Strom M, Adamantidis A, Burdakov D (2016) Awake dynamics and brain-wide direct inputs of hypothalamic MCH and orexin networks. Nat Commun 7:11395

Govorunova EG, Spudich EN, Lane CE, Sineshchekov OA, Spudich JL (2011) New channelrhodopsin with a red-shifted spectrum and rapid kinetics from Mesostigma viride. MBio 2(3): e00115-00111

Gradinaru V, Thompson KR, Deisseroth K (2008) eNpHR: a Natronomonas halorhodopsin enhanced for optogenetic applications. Brain Cell Biol 36(1–4):129–139

Gradinaru V, Zhang F, Ramakrishnan C, Mattis J, Prakash R, Diester I, Goshen I, Thompson KR, Deisseroth K (2010) Molecular and cellular approaches for diversifying and extending optogenetics. Cell 141(1):154–165

Gunaydin LA, Yizhar O, Berndt A, Sohal VS, Deisseroth K, Hegemann P (2010) Ultrafast optogenetic control. Nat Neurosci 13(3):387–392

Hagglund M, Borgius L, Dougherty KJ, Kiehn O (2010) Activation of groups of excitatory neurons in the mammalian spinal cord or hindbrain evokes locomotion. Nat Neurosci 13(2):246–252

Han X, Boyden ES (2007) Multiple-color optical activation, silencing, and desynchronization of neural activity, with single-spike temporal resolution. PLoS One 2(3):e299

Han X, Chow BY, Zhou H, Klapoetke NC, Chuong A, Rajimehr R, Yang A, Baratta MV, Winkle J, Desimone R, Boyden ES (2011) A high-light sensitivity optical neural silencer: development and application to optogenetic control of non-human primate cortex. Front Syst Neurosci 5:18

Hara J, Beuckmann CT, Nambu T, Willie JT, Chemelli RM, Sinton CM, Sugiyama F, Yagami K, Goto K, Yanagisawa M, Sakurai T (2001) Genetic ablation of orexin neurons in mice results in narcolepsy, hypophagia, and obesity. Neuron 30(2):345–354

Hassani OK, Lee MG, Jones BE (2009) Melanin-concentrating hormone neurons discharge in a reciprocal manner to orexin neurons across the sleep-wake cycle. Proc Natl Acad Sci USA 106 (7):2418–2422

Henny P, Jones BE (2006) Innervation of orexin/hypocretin neurons by GABAergic, glutamatergic or cholinergic basal forebrain terminals evidenced by immunostaining for presynaptic vesicular transporter and postsynaptic scaffolding proteins. J Comp Neurol 499(4):645–661

Hess WR, Akert K (1955) Experimental data on role of hypothalamus in mechanism of emotional behavior. AMA Arch Neurol Psychiatry 73(2):127–129

Heydendael W, Sengupta A, Beck S, Bhatnagar S (2014) Optogenetic examination identifies a context-specific role for orexins/hypocretins in anxiety-related behavior. Physiol Behav 130: 182–190

Hoebel BG, Teitelbaum P (1962) Hypothalamic control of feeding and self-stimulation. Science 135(3501):375–377

Hou SY, Govorunova EG, Ntefidou M, Lane CE, Spudich EN, Sineshchekov OA, Spudich JL (2012) Diversity of Chlamydomonas channelrhodopsins. Photochem Photobiol 88(1):119–128

Huang H, Acuna-Goycolea C, Li Y, Cheng HM, Obrietan K, van den Pol AN (2007) Cannabinoids excite hypothalamic melanin-concentrating hormone but inhibit hypocretin/orexin neurons: implications for cannabinoid actions on food intake and cognitive arousal. J Neurosci 27(18): 4870–4881

Iggo A (1978) The physiological interpretation of electrical stimulation of the nervous system. Electroencephalogr Clin Neurophysiol Suppl(34):335–341

Inutsuka A, Inui A, Tabuchi S, Tsunematsu T, Lazarus M, Yamanaka A (2014) Concurrent and robust regulation of feeding behaviors and metabolism by orexin neurons. Neuropharmacology 85:451–460

Ishizuka T, Kakuda M, Araki R, Yawo H (2006) Kinetic evaluation of photosensitivity in genetically engineered neurons expressing green algae light-gated channels. Neurosci Res 54(2):85–94

Jego S, Glasgow SD, Herrera CG, Ekstrand M, Reed SJ, Boyce R, Friedman J, Burdakov D, Adamantidis AR (2013) Optogenetic identification of a rapid eye movement sleep modulatory circuit in the hypothalamus. Nat Neurosci 16(11):1637–1643

Jennings JH, Ung RL, Resendez SL, Stamatakis AM, Taylor JG, Huang J, Veleta K, Kantak PA, Aita M, Shilling-Scrivo K, Ramakrishnan C, Deisseroth K, Otte S, Stuber GD (2015) Visualizing hypothalamic network dynamics for appetitive and consummatory behaviors. Cell 160(3): 516–527

Johns DC, Marx R, Mains RE, O'Rourke B, Marban E (1999) Inducible genetic suppression of neuronal excitability. J Neurosci 19(5):1691–1697

Kelly L, Bielajew C (1996) Short-term stimulation-induced decreases in brown fat temperature. Brain Res 715(1–2):172–179

Kiyashchenko LI, Mileykovskiy BY, Lai YY, Siegel JM (2001) Increased and decreased muscle tone with orexin (hypocretin) microinjections in the locus coeruleus and pontine inhibitory area. J Neurophysiol 85(5):2008–2016

Kiyashchenko LI, Mileykovskiy BY, Maidment N, Lam HA, Wu MF, John J, Peever J, Siegel JM (2002) Release of hypocretin (orexin) during waking and sleep states. J Neurosci 22(13): 5282–5286

Kleinlogel S, Terpitz U, Legrum B, Gokbuget D, Boyden ES, Bamann C, Wood PG, Bamberg E (2011) A gene-fusion strategy for stoichiometric and co-localized expression of light-gated membrane proteins. Nat Methods 8(12):1083–1088

Kokkotou E, Jeon JY, Wang X, Marino FE, Carlson M, Trombly DJ, Maratos-Flier E (2005) Mice with MCH ablation resist diet-induced obesity through strain-specific mechanisms. Am J Physiol Regul Integr Comp Physiol 289(1):R117–R124

Konadhode RR, Pelluru D, Blanco-Centurion C, Zayachkivsky A, Liu M, Uhde T, Glen WB Jr, van den Pol AN, Mulholland PJ, Shiromani PJ (2013) Optogenetic stimulation of MCH neurons increases sleep. J Neurosci 33(25):10257–10263

Lanyi JK (1990) Halorhodopsin, a light-driven electrogenic chloride-transport system. Physiol Rev 70(2):319–330

Lechner HA, Lein ES, Callaway EM (2002) A genetic method for selective and quickly reversible silencing of Mammalian neurons. J Neurosci 22(13):5287–5290

Lerchner W, Xiao C, Nashmi R, Slimko EM, van Trigt L, Lester HA, Anderson DJ (2007) Reversible silencing of neuronal excitability in behaving mice by a genetically targeted, ivermectin-gated Cl-channel. Neuron 54(1):35–49

Li X, Gutierrez DV, Hanson MG, Han J, Mark MD, Chiel H, Hegemann P, Landmesser LT, Herlitze S (2005) Fast noninvasive activation and inhibition of neural and network activity by vertebrate rhodopsin and green algae channelrhodopsin. Proc Natl Acad Sci USA 102(49): 17816–17821

Lin JY, Lin MZ, Steinbach P, Tsien RY (2009) Characterization of engineered channelrhodopsin variants with improved properties and kinetics. Biophys J 96(5):1803–1814

Liu M, Blanco-Centurion C, Konadhode R, Begum S, Pelluru D, Gerashchenko D, Sakurai T, Yanagisawa M, van den Pol AN, Shiromani PJ (2011) Orexin gene transfer into zona incerta neurons suppresses muscle paralysis in narcoleptic mice. J Neurosci 31(16):6028–6040

Liu M, Blanco-Centurion C, Konadhode RR, Luan L, Shiromani PJ (2016) Orexin gene transfer into the amygdala suppresses both spontaneous and emotion-induced cataplexy in orexin knockout mice. Eur J Neurosci 43(5):681–688

Ludwig DS, Tritos NA, Mastaitis JW, Kulkarni R, Kokkotou E, Elmquist J, Lowell B, Flier JS, Maratos-Flier E (2001) Melanin-concentrating hormone overexpression in transgenic mice leads to obesity and insulin resistance. J Clin Invest 107(3):379–386

Madisen L, Mao T, Koch H, Zhuo JM, Berenyi A, Fujisawa S, Hsu YW, Garcia AJ 3rd, Gu X, Zanella S, Kidney J, Gu H, Mao Y, Hooks BM, Boyden ES, Buzsaki G, Ramirez JM, Jones AR, Svoboda K, Han X, Turner EE, Zeng H (2012) A toolbox of Cre-dependent optogenetic transgenic mice for light-induced activation and silencing. Nat Neurosci 15(5):793–802

Matsuki T, Nomiyama M, Takahira H, Hirashima N, Kunita S, Takahashi S, Yagami K, Kilduff TS, Bettler B, Yanagisawa M, Sakurai T (2009) Selective loss of GABA(B) receptors in orexin-producing neurons results in disrupted sleep/wakefulness architecture. Proc Natl Acad Sci USA 106(11):4459–4464

Mattis J, Tye KM, Ferenczi EA, Ramakrishnan C, O'Shea DJ, Prakash R, Gunaydin LA, Hyun M, Fenno LE, Gradinaru V, Yizhar O, Deisseroth K (2011) Principles for applying optogenetic tools derived from direct comparative analysis of microbial opsins. Nat Methods 9(2):159–172

McGregor R, Shan L, Wu MF, Siegel JM (2017) Diurnal fluctuation in the number of hypocretin/orexin and histamine producing: implication for understanding and treating neuronal loss. PLoS One 12(6):e0178573

Mieda M, Willie JT, Hara J, Sinton CM, Sakurai T, Yanagisawa M (2004) Orexin peptides prevent cataplexy and improve wakefulness in an orexin neuron-ablated model of narcolepsy in mice. Proc Natl Acad Sci USA 101(13):4649–4654

Mikrouli E, Wortwein G, Soylu R, Mathe AA, Petersen A (2011) Increased numbers of orexin/hypocretin neurons in a genetic rat depression model. Neuropeptides 45(6):401–406

Mileykovskiy BY, Kiyashchenko LI, Siegel JM (2005) Behavioral correlates of activity in identified hypocretin/orexin neurons. Neuron 46(5):787–798

Monzon ME, Varas MM, De Barioglio SR (2001) Anxiogenesis induced by nitric oxide synthase inhibition and anxiolytic effect of melanin-concentrating hormone (MCH) in rat brain. Peptides 22(7):1043–1047

Morens C, Norregaard P, Receveur JM, van Dijk G, Scheurink AJ (2005) Effects of MCH and a MCH1-receptor antagonist on (palatable) food and water intake. Brain Res 1062(1–2):32–38

Nagel G, Ollig D, Fuhrmann M, Kateriya S, Musti AM, Bamberg E, Hegemann P (2002) Channelrhodopsin-1: a light-gated proton channel in green algae. Science 296(5577):2395–2398

Nagel G, Szellas T, Huhn W, Kateriya S, Adeishvili N, Berthold P, Ollig D, Hegemann P, Bamberg E (2003) Channelrhodopsin-2, a directly light-gated cation-selective membrane channel. Proc Natl Acad Sci USA 100(24):13940–13945

Nagel G, Brauner M, Liewald JF, Adeishvili N, Bamberg E, Gottschalk A (2005) Light activation of channelrhodopsin-2 in excitable cells of Caenorhabditis elegans triggers rapid behavioral responses. Curr Biol 15(24):2279–2284

Nahon JL (1994) The melanin-concentrating hormone: from the peptide to the gene. Crit Rev Neurobiol 8(4):221–262

Nishino S, Mignot E (1997) Pharmacological aspects of human and canine narcolepsy. Prog Neurobiol 52(1):27–78

Nofzinger EA, Buysse DJ, Germain A, Price JC, Miewald JM, Kupfer DJ (2004) Functional neuroimaging evidence for hyperarousal in insomnia. Am J Psychiatry 161(11):2126–2128

Nowak LG, Bullier J (1998a) Axons, but not cell bodies, are activated by electrical stimulation in cortical gray matter. I. Evidence from chronaxie measurements. Exp Brain Res 118(4):477–488

Nowak LG, Bullier J (1998b) Axons, but not cell bodies, are activated by electrical stimulation in cortical gray matter. II. Evidence from selective inactivation of cell bodies and axon initial segments. Exp Brain Res 118(4):489–500

Obal F Jr, Rubicsek G, Alfoldi P, Sary G, Obal F (1985) Changes in the brain and core temperatures in relation to the various arousal states in rats in the light and dark periods of the day. Pflugers Arch 404(1):73–79

Pelluru D, Konadhode R, Shiromani PJ (2013) MCH neurons are the primary sleep-promoting group. Sleep 36(12):1779–1781

Peyron C, Tighe DK, van den Pol AN, de Lecea L, Heller HC, Sutcliffe JG, Kilduff TS (1998) Neurons containing hypocretin (orexin) project to multiple neuronal systems. J Neurosci 18(23): 9996–10015

Plaha P, Ben Shlomo Y, Patel NK, Gill SS (2006) Stimulation of the caudal zona incerta is superior to stimulation of the subthalamic nucleus in improving contralateral parkinsonism. Brain 129(Pt 7):1732–1747

Raimondo JV, Kay L, Ellender TJ, Akerman CJ (2012) Optogenetic silencing strategies differ in their effects on inhibitory synaptic transmission. Nat Neurosci 15(8):1102–1104

Ranck JB Jr (1975) Which elements are excited in electrical stimulation of mammalian central nervous system: a review. Brain Res 98(3):417–440

Rao Y, Lu M, Ge F, Marsh DJ, Qian S, Wang AH, Picciotto MR, Gao XB (2008) Regulation of synaptic efficacy in hypocretin/orexin-containing neurons by melanin concentrating hormone in the lateral hypothalamus. J Neurosci 28(37):9101–9110

Sailer AW, Sano H, Zeng Z, McDonald TP, Pan J, Pong SS, Feighner SD, Tan CP, Fukami T, Iwaasa H, Hreniuk DL, Morin NR, Sadowski SJ, Ito M, Ito M, Bansal A, Ky B, Figueroa DJ, Jiang Q, Austin CP, MacNeil DJ, Ishihara A, Ihara M, Kanatani A, Van der Ploeg LH, Howard AD, Liu Q (2001) Identification and characterization of a second melanin-concentrating hormone receptor, MCH-2R. Proc Natl Acad Sci USA 98(13):7564–7569

Saito Y, Nothacker HP, Wang Z, Lin SH, Leslie F, Civelli O (1999) Molecular characterization of the melanin-concentrating-hormone receptor. Nature 400(6741):265–269

Schuz A, Palm G (1989) Density of neurons and synapses in the cerebral cortex of the mouse. J Comp Neurol 286(4):442–455

Sears RM, Fink AE, Wigestrand MB, Farb CR, de Lecea L, Ledoux JE (2013) Orexin/hypocretin system modulates amygdala-dependent threat learning through the locus coeruleus. Proc Natl Acad Sci USA 110(50):20260–20265

Shimada M, Tritos NA, Lowell BB, Flier JS, Maratos-Flier E (1998) Mice lacking melanin-concentrating hormone are hypophagic and lean. Nature 396(6712):670–674

Shiromani PJ, Peever JH (2017) New neuroscience tools that are identifying the sleep-wake circuit. Sleep 40(4)

Sinnamon HM (1984) Forelimb and hindlimb stepping by the anesthetized rat elicited by electrical stimulation of the diencephalon and mesencephalon. Physiol Behav 33(2):191–199

Sinnamon HM (1993) Preoptic and hypothalamic neurons and the initiation of locomotion in the anesthetized rat. Prog Neurobiol 41(3):323–344

Slimko EM, McKinney S, Anderson DJ, Davidson N, Lester HA (2002) Selective electrical silencing of mammalian neurons in vitro by the use of invertebrate ligand-gated chloride channels. J Neurosci 22(17):7373–7379

Smith OA, DeVito JL, Astley CA (1990) Neurons controlling cardiovascular responses to emotion are located in lateral hypothalamus-perifornical region. Am J Phys 259(5 Pt 2):R943–R954

Sohal VS, Zhang F, Yizhar O, Deisseroth K (2009) Parvalbumin neurons and gamma rhythms enhance cortical circuit performance. Nature 459(7247):698–702

Tehovnik EJ (1996) Electrical stimulation of neural tissue to evoke behavioral responses. J Neurosci Methods 65(1):1–17

Thannickal TC, Moore RY, Nienhuis R, Ramanathan L, Gulyani S, Aldrich M, Cornford M, Siegel JM (2000) Reduced number of hypocretin neurons in human narcolepsy. Neuron 27(3): 469–474

Thannickal TC, Lai YY, Siegel JM (2007) Hypocretin (orexin) cell loss in Parkinson's disease. Brain 130(Pt 6):1586–1595

Thomas RK, Young CD (1993) A note on the early history of electrical stimulation of the human brain. J Gen Psychol 120(1):73–81

Tittor J, Oesterhelt D, Maurer R, Desel H, Uhl R (1987) The photochemical cycle of halorhodopsin: absolute spectra of intermediates obtained by flash photolysis and fast difference spectra measurements. Biophys J 52(6):999–1006

Toossi H, Del Cid-Pellitero E, Jones BE (2016) GABA receptors on orexin and melanin-concentrating hormone neurons are differentially homeostatically regulated following sleep deprivation. eNeuro 3(3)
Torterolo P, Sampogna S, Chase MH (2009) MCHergic projections to the nucleus pontis oralis participate in the control of active (REM) sleep. Brain Res 1268:76–87
Tsunematsu T, Ueno T, Tabuchi S, Inutsuka A, Tanaka KF, Hasuwa H, Kilduff TS, Terao A, Yamanaka A (2014) Optogenetic manipulation of activity and temporally controlled cell-specific ablation reveal a role for MCH neurons in sleep/wake regulation. J Neurosci 34(20): 6896–6909
Vallarino M, Bruzzone F, Vaudry H (2009) Neuroanatomical distribution of MCH in the brain and pituitary of submammalian vertebrates. Peptides 30(11):1973–1978
van den Pol AN, Acuna-Goycolea C, Clark KR, Ghosh PK (2004) Physiological properties of hypothalamic MCH neurons identified with selective expression of reporter gene after recombinant virus infection. Neuron 42(4):635–652
van der Plas J, Wiersinga-Post JE, Maes FW, Bohus B (1995) Cardiovascular effects and changes in midbrain periaqueductal gray neuronal activity induced by electrical stimulation of the hypothalamus in the rat. Brain Res Bull 37(6):645–656
Varin C, Arthaud S, Salvert D, Gay N, Libourel PA, Luppi PH, Leger L, Fort P (2016) Sleep architecture and homeostasis in mice with partial ablation of melanin-concentrating hormone neurons. Behav Brain Res 298(Pt B):100–110
Verret L, Goutagny R, Fort P, Cagnon L, Salvert D, Leger L, Boissard R, Salin P, Peyron C, Luppi PH (2003) A role of melanin-concentrating hormone producing neurons in the central regulation of paradoxical sleep. BMC Neurosci 4:19
Vetrivelan R, Kong D, Ferrari LL, Arrigoni E, Madara JC, Bandaru SS, Lowell BB, Lu J, Saper CB (2016) Melanin-concentrating hormone neurons specifically promote rapid eye movement sleep in mice. Neuroscience 336:102–113
Wang H, Peca J, Matsuzaki M, Matsuzaki K, Noguchi J, Qiu L, Wang D, Zhang F, Boyden E, Deisseroth K, Kasai H, Hall WC, Feng G, Augustine GJ (2007) High-speed mapping of synaptic connectivity using photostimulation in Channelrhodopsin-2 transgenic mice. Proc Natl Acad Sci USA 104(19):8143–8148
Wen L, Wang H, Tanimoto S, Egawa R, Matsuzaka Y, Mushiake H, Ishizuka T, Yawo H (2010) Opto-current-clamp actuation of cortical neurons using a strategically designed channelrhodopsin. PLoS One 5(9):e12893
Wietek J, Prigge M (2016) Enhancing Channelrhodopsins: an overview. Methods Mol Biol 1408:141–165
Willie JT, Takahira H, Shibahara M, Hara J, Nomiyama M, Yanagisawa M, Sakurai T (2011) Ectopic overexpression of orexin alters sleep/wakefulness states and muscle tone regulation during REM sleep in mice. J Mol Neurosci 43(2):155–161
Wu MF, Gulyani SA, Yau E, Mignot E, Phan B, Siegel JM (1999) Locus coeruleus neurons: cessation of activity during cataplexy. Neuroscience 91(4):1389–1399
Yaroslavsky AN, Schulze PC, Yaroslavsky IV, Schober R, Ulrich F, Schwarzmaier HJ (2002) Optical properties of selected native and coagulated human brain tissues in vitro in the visible and near infrared spectral range. Phys Med Biol 47(12):2059–2073
Yizhar O, Fenno LE, Prigge M, Schneider F, Davidson TJ, O'Shea DJ, Sohal VS, Goshen I, Finkelstein J, Paz JT, Stehfest K, Fudim R, Ramakrishnan C, Huguenard JR, Hegemann P, Deisseroth K (2011) Neocortical excitation/inhibition balance in information processing and social dysfunction. Nature 477(7363):171–178
Zemelman BV, Lee GA, Ng M, Miesenbock G (2002) Selective photostimulation of genetically chARGed neurons. Neuron 33(1):15–22
Zemelman BV, Nesnas N, Lee GA, Miesenbock G (2003) Photochemical gating of heterologous ion channels: remote control over genetically designated populations of neurons. Proc Natl Acad Sci USA 100(3):1352–1357

Zhang F, Wang LP, Brauner M, Liewald JF, Kay K, Watzke N, Wood PG, Bamberg E, Nagel G, Gottschalk A, Deisseroth K (2007) Multimodal fast optical interrogation of neural circuitry. Nature 446(7136):633–639

Zhang F, Prigge M, Beyriere F, Tsunoda SP, Mattis J, Yizhar O, Hegemann P, Deisseroth K (2008) Red-shifted optogenetic excitation: a tool for fast neural control derived from Volvox carteri. Nat Neurosci 11(6):631–633

Zhao S, Ting JT, Atallah HE, Qiu L, Tan J, Gloss B, Augustine GJ, Deisseroth K, Luo M, Graybiel AM, Feng G (2011) Cell type-specific channelrhodopsin-2 transgenic mice for optogenetic dissection of neural circuitry function. Nat Methods 8(9):745–752

Zhou D, Shen Z, Strack AM, Marsh DJ, Shearman LP (2005) Enhanced running wheel activity of both Mch1r- and Pmch-deficient mice. Regul Pept 124(1–3):53–63

The Effects of Melanin-Concentrating Hormone on Neurotransmitter Systems Involved in the Generation and Maintenance of Wakefulness

Jaime M. Monti, Seithikurippu R. Pandi-Perumal, and Pablo Torterolo

Abstract Within the central nervous system (CNS), melanin-concentrating hormone (MCH) participates in a number of functions including sleep-wake behavior. In this respect, MCHergic neurons project widely throughout the central nervous system (CNS) to neural structures involved in the regulation of wakefulness (W). An enhancement of REM sleep time has been described following the microinjection of MCH into the dorsal raphe nucleus (serotonergic neurons), locus coeruleus nucleus (noradrenergic neurons), and basal forebrain [(horizontal limb of the diagonal band of Broca) glutamatergic and cholinergic (W-on) neurons] of rodents. In addition, optogenetic stimulation of MCH terminals in the tuberomammillary nucleus (histaminergic neurons) is followed by an increase in the duration of REM sleep episodes. Moreover, the finding that the neuropeptide negatively modulates the mesolimbic dopaminergic function tends to indicate that the inhibition of nucleus accumbens and ventral tegmental nucleus dopaminergic neurons by MCH could facilitate the occurrence of REM sleep. Thus, the REM sleep-inducing and sleep-facilitating effect of MCH is at least partly related to the deactivation of monoaminergic, glutamatergic, and cholinergic (W-on) neurons.

J. M. Monti (✉)
School of Medicine, University of the Republic, Montevideo, Uruguay
e-mail: jmonti@mednet.org.uy

S. R. Pandi-Perumal
Somnogen Canada Inc., Toronto, ON, Canada

P. Torterolo
Department of Physiology, School of Medicine, University of the Republic, Montevideo, Uruguay

S. R. Pandi-Perumal et al. (eds.), *Melanin-Concentrating Hormone and Sleep*,
https://doi.org/10.1007/978-3-319-75765-0_5

1 Neural Structures and Neurotransmitter Systems Involved in the Generation and Maintenance of Wakefulness

The brain regions involved in the promotion of the waking state are found in the brainstem, hypothalamus, and basal forebrain (BFB). The nuclei located in the brainstem include:

Neurons containing serotonin [5-HT—dorsal raphe nucleus (DRN), median raphe nucleus (MRN)]
Norepinephrine [NE—locus coeruleus]
Dopamine [DA—ventral tegmental area (VTA), substantia nigra pars compacta (SNc), ventral periaqueductal gray matter (vPAG)]
Acetylcholine [ACh (wakefulness-on)—laterodorsal and pedunculopontine tegmental nuclei (LDT/PPT)]

The hypothalamic component of the arousal system includes cells containing orexin [OX; posterior lateral hypothalamus (LH) around the fornix] and histamine [HA; tuberomammillary nucleus (TMN)]. The cholinergic and glutamatergic neurons of the BFB involved in the regulation of wakefulness (W) are located predominantly in the medial septum, diagonal band of Broca, and substantia innominata (Pace-Schott and Hobson 2002; Jones 2003). The 5-HT-, NE-, HA-, and ACh-containing neurons that participate in the regulation of W give rise to mainly ascending projections to (1) the thalamus (dorsal route) which in turn projects to the cerebral cortex and (2) the BFB (ventral route) where cells in turn project to the cerebral cortex and the hippocampus. The DA-containing cells of the VTA and SNc project to the striatum and the prefrontal cortex, while those corresponding to the vPAG project predominantly to the BFB and midline thalamus. Furthermore, OX-containing neurons carry projections to the entire forebrain and brainstem arousal systems as well as to the thalamus and cortex (Brown et al. 2012). Additionally, most of these neural structures send descending projections to the brainstem and spinal cord regions that modulate muscle tone.

2 Neural Structures and Neurotransmitter Systems Involved in the Generation and Maintenance of NREM Sleep

Neurons of the preoptic area and adjacent BFB constitute the sleep-inducing system (Szymusiak et al. 2007). Sleep active neurons of the preoptic area are mainly located in the ventrolateral preoptic area (VLPO). A majority of these neurons contain γ-aminobutyric acid (GABA) and galanin and project to brainstem and hypothalamic areas involved in the promotion of W including the DRN, LC, LDT/PPT, vPAG, and

LH neurons. More recently, melanin-concentrating hormone (MCH) neurons located in the lateral hypothalamus and zona incerta have been proposed to participate also in the regulation of non-rapid-eye movement (NREM) sleep and REM sleep (Torterolo et al. 2011).

3 Neural Structures and Neurotransmitter Systems Involved in the Generation and Maintenance of REM Sleep

It has been proposed that the sublaterodorsal tegmental nucleus (SLD) is the brain structure necessary and sufficient to induce and maintain REM sleep in the rat, and the effect would depend upon the activation of glutamatergic neurons located in this neuroanatomical structure (Luppi et al. 2013). Its equivalent in the cat is called the subcoeruleus nucleus or nucleus pontis oralis. In favor of the proposal is the finding that SLD glutamatergic neurons increase their firing rate during REM sleep. Furthermore, microiontophoretic administration of the glutamate agonist kainic acid into the SLD induces a REM sleep-like state (Clément et al. 2011). Luppi et al. (2013) have posed, in addition, that the activation of SLD glutamatergic neurons would depend upon the removal of an inhibitory tone present during W and NREM sleep arising from GABAergic REM-off neurons located in the ventrolateral periaqueductal gray (vlPAG) and the deep mesencephalic nucleus (DPMe) and monoaminergic neurons located in the brainstem and hypothalamus. Accordingly, local microinjection into the SLD of the $GABA_A$ receptor antagonists bicuculline or gabazine produced an increase of REM sleep in laboratory animals (Boissard et al. 2002). It has been contended also that inhibition of vlPAG and DPMe REM-off neurons would depend upon the activation of GABAergic REM-on neurons located in the vlPAG and dorsal gigantocellular nucleus (DPGi) and MCH neurons of the lateral hypothalamus (Sapin et al. 2009; Monti et al. 2016). On the other hand, the reciprocal-interaction hypothesis of REM sleep generation originally proposed by McCarley and Hobson (Hobson et al. 1975) identifies interconnected populations of REM-on and REM-off neurons compatible with reciprocal interactions as a physiological basis of sleep cycle oscillation. In the updated version of the reciprocal-interaction model (McCarley 2007), cholinergic neurons of the LDT/PPT are identified as promoting REM sleep and interact with serotonergic and noradrenergic neurons of the DRN and LC that inhibit REM sleep. In addition, McCarley (2007) includes GABAergic mechanisms that deactivate neurotransmitter systems responsible for the inhibition of LDT/PPT cholinergic (REM-on) neurons. This would lead to the activation of pontine reticular formation glutamatergic neurons and the occurrence of REM sleep.

Thus, it can be proposed that there is much agreement with respect to the neural structures and mechanisms of action involved in the regulation of W and NREM sleep. However, discrepancies still exist in relation to the neuroanatomical systems and synaptic processes involved in the facilitation and inhibition of REM sleep.

4 Structure, Mechanism of Action, and Projections of MCH-Containing Neurons

In mammals MCH is a cyclic neuropeptide with 19 amino acids. It is generated by the cleavage of a precursor of 165 amino acids, the prepro-MCH. MCH is confined to a group of neurons in the lateral hypothalamus and incerto-hypothalamic area and acts through its G-protein-coupled receptors named MCHR1 and MCHR2. Rodents present only the MCHR1 (Lembo et al. 1999). The binding of MCH to MCHR1 activates diverse intracellular signaling pathways by coupling to G_i, G_q, and G_o proteins, while MCHR2 is known to couple to the G_q protein (Hawes et al. 2000; Sailer et al. 2001). Studies of MCH neurons in vitro have shown a predominantly inhibitory effect of the neuropeptide both at pre- and postsynaptic levels (Gao and van den Pol 2001). Sekiya et al. (1988) determined the distribution of MCH-like immunoreactivity by radio immunoassay in the central nervous system (CNS) of the rat, guinea pig, and man. The highest concentrations of MCH were found in the hypothalamus of all these species. Later on, Bittencourt et al. (1992) characterized the organization of the system using antisera raised against rat MCH. It was shown that medium-sized and multipolar to fusiform MCH-containing cells were localized predominantly in the lateral hypothalamic area and zona incerta of the rat. In addition, monosynaptic fibers stained for MCH were broadly distributed throughout the CNS. In this respect, MCH-immunoreactive (MCH-ir) axons innervate several neuroanatomical structures located in the telencephalon, diencephalon, mesencephalon, and rhombencephalon that are involved in the regulation of the behavioral state. In addition, Bittencourt and Elias (1998) established that the origin of MCH-ir projections in the medial septal nucleus, vertical and horizontal limbs of the diagonal band of Broca, and spinal cord resides mainly in the lateral hypothalamus. Moreover, dense MCH innervation was reported in the cerebral motor cortex and LDT/PPT of the rat, the latter originating mainly from the dorsal half of the lateral hypothalamus (Elias et al. 2008; Hong et al. 2011).

5 Local Brain Delivery of MCH into CNS Structures Involved in the Regulation of the Sleep-Wake Cycle

5.1 *Dorsal Raphe Nucleus*

Bittencourt et al. (1992) described for the first time a descending component of MCH fibers that innervated the DRN in the rat. More recently, Yoon and Lee (2013) examined the projections from MCH neurons in the medial and lateral subdivision of the LH to the DRN of rodents. The authors found that MCH axon terminals from both subdivisions of the LH made contact with DRN somata at both rostral and caudal levels. A further study by Urbanavicius et al. (2016) characterized in detail the distribution and density of the MCHergic fibers along the rostro-caudal axis of

Table 1 The effects of melanin-concentrating hormone microinjection into neural structures involved in the occurrence of wakefulness on sleep variables in rodents

Neural structure	W	SWS	REMS	Reference
Dorsal raphe nucleus (serotonergic neurons)	–	+	+	Lagos et al. (2009)
Locus coeruleus nucleus (noradrenergic neurons)	n.s.	n.s.	+	Monti et al. (2015)
Lateral basal forebrain (glutamatergic and cholinergic neurons)	–	n.s.	+	Lagos et al. (2012)
Tuberomammillary nucleus (histaminergic neurons)	n.a.	n.a.	+	Jego et al. (2013)[a]

Abbreviations: *W* wakefulness, *SWS* slow wave sleep, *REMS* rapid-eye movement sleep, + significant increase, – significant decrease, *n.s.* nonsignificant, *n.a.* not available
[a]Jego et al. (2013) stimulated the MCHergic afferents to the tuberomammillary nucleus

the rat DRN and their anatomical relationship with 5-HT and GABA-containing neurons. In addition, the authors evaluated the MCH effects on the 5-HT extracellular levels. Accordingly, MCHergic axons reached all the rostro-caudal levels of the DRN, although their density was lower at the most caudal level. Of note, MCH fibers were found to be in apposition with both 5-HT- and GABA-containing cells. The local microinjection of a relatively low dose (30 μM) of the neuropeptide was followed by a significant and long-lasting decrease of stimulated 5-HT levels. The finding of particular dense MCHergic projections to the DRN by Bittencourt et al. (1992) led to the analysis of the effect of microinjections of MCH into this neuroanatomical structure on sleep variables in the rat. To this purpose MCH (50 and 100 ng) and vehicle were microinjected into the DRN of rodents prepared for chronic sleep recordings. Compared with the control vehicle, MCH 100 ng significantly increased REM sleep time during 6-hour polysomnographic recordings (Table 1) (Lagos et al. 2009). The increment of REM sleep time amounted to 70.7% of the control value and was related to a greater number of REM sleep episodes. MCH 100 ng produced, in addition, a small but significant increase in SWS (9.2% increment of the control value). Besides, administration of the neuropeptide significantly reduced the time spent in W and light sleep (33.8% and 26% decrement of the control values, respectively). The analysis of REM sleep values in 2-hour blocks showed that the increase of REM sleep after MCH 100 ng microinjection was maintained up to 6 hours. During a second step, the effect of immunoneutralization of MCH in the DRN on sleep and W was determined in the rat (Lagos et al. 2011). Compared to the control solution, microinjection of anti-MCH antibodies (1/100 solution in 0.2 μl) induced a significant reduction of REM sleep time and the number of REM sleep episodes, while REM sleep latency was increased. Additionally, there was an increase in total W time. Light sleep and SWS remained almost unchanged. These findings strongly support the proposal that MCH released in the DRN facilitates the occurrence of predominantly REM sleep. Because MCH is an inhibitory neuropeptide, it can be hypothesized that MCH would facilitate the generation of REM sleep by inhibiting serotonergic neurons. In support of the suggestion,

Devera et al. (2015) described that the microinjection of MCH into lateral ventricle resulted in a significant decrease in the firing rate of 59% of the neurons recorded in the DRN. Moreover, the juxtacellular administration of MCH reduced the discharge in 80% of these cells. Based on the electrophysiological and pharmacological properties of the neurons affected by MCH, including action potential average duration, basal firing rate, and suppression of the discharge following the systemic administration of the selective 5-HT_{1A} receptor agonist 8-OH-DPAT, it was concluded that these neurons were likely serotonergic.

5.2 *Locus Coeruleus Nucleus*

Harthoorn (2007) and Bittencourt (2011) initially reported the existence of MCH terminals within the LC. More recently, Yoon and Lee (2013) examined the distribution to the LC of fibers from MCH neurons located within medial and lateral subdivisions of the lateral hypothalamus. The authors established that MCH projections to the nuclear core of the LC exhibit differential distribution with a predominance of the lateral subdivision over the medial one. Accordingly, the proportions of retrograde-labeled MCH neurons over the total retrograde-labeled cells amounted to 4.4% ± 0.5% at the medial subdivision and 7.4% ± 0.6% at the lateral one. It has been proposed, in addition, that MCH neurons might utilize GABA as a co-transmitter, since their soma contain the enzyme glutamic acid decarboxylase (Gao and van den Pol 2001; Elias et al. 2001; Sapin et al. 2010). In support of the proposal, Del Cid-Pellitero and Jones (2012) found that a small number of MCH varicosities present at the LC in the rat were immunopositive for the vesicular transporter for GABA. Furthermore, the MCH varicosities containing the vesicular transporter for GABA contacted the tyrosine hydroxylase immunostained neurons of the LC.

The effects of MCH microinjection into the right LC on sleep variables during 6-h recording sessions have been characterized in the rat (Table 1). Compared with the control vehicle, MCH 200 ng significantly increased REM sleep from a control value of 27.7 ± 4.1 min (7.7% of the total recording time) to 39.6 ± 3.5 min (11.0% of the total recording time). The increment of REM sleep time was related to a greater number of REM sleep episodes. SWS and REM sleep latency were not significantly modified. Analysis of sleep variables in 2 h blocks showed that MCH 200 ng significantly augmented REM sleep during the first, second, and third 2 h of recording. Additionally, MCH 100 ng induced a significant increase of REM sleep during the first 2 h period after treatment. W, light sleep, and SWS showed slight but inconsistent changes that did not attain significance (Monti et al. 2015).

Thus, it can be proposed that under normal conditions MCH and to a smaller extent GABA released by MCH-containing neurons would inhibit the noradrenergic cells located in the LC and increase REM sleep values.

Interestingly, Bayer et al. (2005) examined the effects of NE and carbachol, a cholinergic agonist, on identified MCH neurons in rat hypothalamic slices. It was found that both receptor agonists hyperpolarized MCH cells by direct postsynaptic actions. The authors concluded that MCH neurons would be inhibited by NE and ACh (W-on cells) during W while disinhibited and active during SWS and REM sleep. Thus, the study by Bayer et al. (2005) further supports the proposal that MCH-induced inhibition of LC noradrenergic neurons favors the occurrence of REM sleep in laboratory animals.

5.3 *Lateral Basal Forebrain*

The effects of bilateral microinjection of MCH into the horizontal limb of the diagonal band of Broca (HDB) on sleep variables during the light phase of the light-dark cycle have been examined in the rat. The microinjection of MCH was aimed at the HBD because this is where choline-acetyltransferase (ChAT)-immunoreactive cells, glutamic acid decarboxylase (GAD)-immunoreactive cells, and, as judged by the presence of vesicular glutamate transporter 2 (VGLUT2), glutamatergic neurons are present in great numbers (Semba 2000).

Bilateral microinjection of MCH 100 ng into the HDB significantly reduced W during the 6-hour recording sessions from a control value of 81.7 $\pm$ 10.8 min (22.6% of the total recording time) to 64.7 $\pm$ 9.3 min (17.9% of the total recording time) (Table 1). REM sleep time showed a slight increase that did not attain significance. Notwithstanding this, MCH 100 ng significantly decreased REM sleep latency and augmented the number of REM episodes during the first 2-h period after treatment (Table 1). Analysis of sleep variables in 2 h blocks showed that MCH 50 and 100 ng significantly reduced W during the first 2 h of recording. In contrast, only the 100 ng dose of the neuropeptide significantly increased REM sleep during the first 2 h after treatment. Light sleep and SWS showed slight but inconsistent changes that did not attain significance (Lagos et al. 2012). MCHergic projections from the lateral hypothalamus and zona incerta have been described to the nuclei of the HDB (Bittencourt et al. 1992; Hervieu et al. 2000). In addition, the MCHR1 receptor is densely expressed in the basal forebrain (Saito et al. 2001). Cholinergic, glutamatergic, and GABAergic neurons of the basal forebrain have been shown to be involved in the promotion of W (Jones 2005; Lee et al. 2005). The former two neurotransmitter systems influence directly cortical and hippocampal activities, whereas GABAergic neurons target inhibitory interneurons and through disinhibition activate pyramidal cells (Freund and Gulyas 1991; Deurveilheur and Semba 2011). In addition, BFB GABAergic and non-GABAergic, mainly cholinergic, neurons have been proposed to contribute to the modulation of REM sleep via descending projections acting on REM sleep-regulatory neurons located in the brainstem (Rodrigo-Angulo et al. 2008; Deurveilheur and Semba 2011; Semba 2011). The former would induce a direct inhibition of REM sleep generating and maintaining neurons, while the latter would exert an indirect inhibition through the activation of local GABAergic cells. It can be proposed that MCH microinjected into the HDB would inhibit cholinergic, glutamatergic, and GABAergic neurons that express MCHR1R. Inhibition of the acetylcholine- and glutamate-containing cell groups involved in the occurrence of W would be followed by its reduction. Moreover, the deactivation of REM sleep-off GABAergic and non-GABAergic cells that project to the brainstem would lead to the disinhibition of neuroanatomical structures involved in the induction and maintenance of REM sleep. As a result, values corresponding to this behavioral state would be increased. Thus, it can be proposed that under normal conditions, the MCH-induced deactivation of HDB neurons would contribute to the regulation of W and REM sleep.

5.4 Dopaminergic Mesocorticolimbic System

The DAergic neuronal system relevant to sleep and W is located in the upper mesencephalon. Dopamine neurons project to several brain areas via a number of tracts. One group of DA neurons arises in the SNc and projects via the nigrostriatal tract to the dorsal striatum (caudate-putamen). A second group of DA neurons arises in the VTA and projects via the mesolimbic and the mesocortical tract to limbic areas [septal area, olfactory tubercles, nucleus accumbens (ACb), amygdaloid complex, hippocampus, and piriform cortex] and the cerebral cortex [medial prefrontal, cingulate, and entorhinal areas], respectively (Monti and Jantos 2009). Additionally, DA neurons have been characterized in the vPAG of the rat (Lu et al. 2006). Dopamine D1 and D2 receptors are present within these structures. Behavioral arousal is impaired in DA D1 receptor knockout mice. On the other hand, systemic administration of a selective DA D1 agonist induces an increase of W and a reduction of SWS and REM sleep in laboratory animals. Mice with genetically induced lesions that target the D2 receptor show reduced levels of spontaneous locomotor activity. Similar results have been observed following systemic, i.c.v., or intra-accumbens injection of a selective DA D2 autoreceptor agonist. Systemic administration of DA D2 agonists induces biphasic effects such that low doses reduce W and increase SWS and REM sleep, whereas large doses induce the opposite effects (Monti and Monti 2007).

Although there is no direct evidence regarding the interaction between MCHergic and DAergic neurons in the control of the behavioral state, indirect evidence strongly suggests an interplay between both systems in the control of W and sleep. In this respect, MCH-containing fibers and MCHR1 expression are abundant in the ACb and VTA (Bittencourt et al. 1992; Hervieu et al. 2000; Saito et al. 2001). Pissios et al. (2008) evaluated the striatal-dependent functions in MCH knockout (MCH−/−) mice and distinguished a number of functional changes in the ACb shell, including increased electrically evoked DA release and augmented reuptake of the neurotransmitter in the presynaptic terminals. The latter was consistent with the increased expression of the DA transporter in the ACb. Moreover, MCH−/− mice showed a significantly greater increase in locomotor activity following repeated d-amphetamine injection compared to wild-type animals, which led the authors to propose that chronic absence of MCH enhanced the sensitizing effects of d-amphetamine in mice. The selective DA D1 receptor agonist SKF81297 induced also a significantly greater enhancement of locomotor activity as compared to the MCH+/+ animals. Furthermore, Smith et al. (2005) determined whether basal and evoked tissue levels of DA within the ACb and locomotor activity were altered in MCHR1−/− as compared to controls. No significant differences in basal or evoked tissue levels of DA were observed in MCHR1−/− mice compared with wild-type animals. Concerning the mice behavior, it was observed that MCHR1−/− animals became hyperactive when exposed to a novel environment. In addition, they were supersensitive to the locomotor stimulating effect of d-amphetamine and the selective DA D1 receptor agonist SKF38393. Thus, presently available evidence tends to indicate that deletion of MCH or MCHR1 results in an upregulation of mesolimbic DA receptors, indicating that the neuropeptide may negatively modulate mesolimbic DA function. This outcome could tentatively result in a reduction of W and increase of NREM sleep and REM sleep following the microinjection of MCH into the ACb.

Related to our topic, Alberto et al. (2011) and Conductier et al. (2011) investigated the effect of DA on MCH neurons using whole-cell patch-clamp recordings in hypothalamic mouse brain slices. It was observed that DA hyperpolarizes MCH cells through activation of α2 NE receptors and the opening of G-protein-activated inward rectifier K^+ (GIRK) channels. The finding by van den Pol et al. (2004) that α2 NE receptors are present on MCH neurons further supports the proposal that DA indirectly regulates MCH neurons. Thus, it can be proposed that DA induces a reduction in MCH neuron excitability and decreases MCH release. Interestingly, Yang et al. (2014) described that DA induces a hyperpolarization of the membrane potentials in the rat SLD glutamatergic neurons and a decrease of their firing, which is mediated also by α2-noradrenergic receptors.

5.5 Histaminergic Tuberomammillary Nucleus

Skofitsch et al. (1985) characterized the distribution of MCH-like immunoreactivity in the rat brain and concluded that only a moderate number of axons reach the TMN. In contrast, Bittencourt and Elias (1998) found that the TMN, mainly the medial mammillary nucleus and the supramammillary nucleus, receive a large number of MCH axons in the rat. On the other hand, a very low number of MCH fibers occurred in the lateral mammillary nucleus.

Jego et al. (2013) tested the hypothesis that MCH neurons induce an inhibitory effect on W-promoting HA-containing neurons in the TMN of mice. In this respect, it was found that in vitro activation of MCH neuron terminals induces inhibitory postsynaptic currents in histaminergic neurons of the TMN. Moreover, in vivo stimulation of TMN MCH terminals by means of optogenetic tools augmented the duration of REM sleep episodes in mice prepared for sleep recordings (Table 1). It can be concluded that MCH-induced inhibition of W-promoting histaminergic neurons of the TMN facilitates REM sleep occurrence in mice.

6 Conclusions

Presently available evidence tends to indicate that MCH inhibition of neurotransmitter systems involved in the occurrence of W, including the serotonergic, noradrenergic, glutamatergic, cholinergic (W-on cells), and histaminergic ones, is followed by an increase of REM sleep.

References

Alberto CO, Trask RB, Hirasawa M (2011) Dopamine acts as a partial agonist for α2A adrenoceptor in melanin-concentrating hormone neurons. J Neurosci 31:10671–10676

Bayer L, Eggermann E, Serafin M, Grivel J, Machard D, Muhlethaler M et al (2005) Opposite effects of noradrenaline and acetylcholine upon hypocretin/orexin versus melanin concentrating hormone neurons in rat hypothalamic slices. Neuroscience 130:807–811

Bittencourt JC (2011) Anatomical organization of the melanin-concentrating hormone peptide family in the mammalian brain. Gen Comp Endocrinol 172:185–197

Bittencourt JC, Elias CF (1998) Melanin-concentrating hormone and neuropeptide EI projections from the lateral hypothalamic area and zona incerta to the medial septal nucleus and spinal cord: a study using multiple neural tracers. Brain Res 805:1–19

Bittencourt JC, Press F, Arias C, Peto C, Vaughan J, Nahon JL et al (1992) The melanin-concentrating hormone system of the rat brain: an immuno- and hybridization histochemical characterization. J Comp Neurol 319:218–245

Boissard R, Gervasoni D, Schmidt MH, Barbagli B, Fort P, Luppi PH (2002) The rat ponto-medullary network responsible for paradoxical sleep onset and maintenance: a combined microinjection and functional neuroanatomical study. Eur J Neurosci 16:1959–1973

Brown RE, Basheer R, McKenna JT, Strecker RE, McCarley RW (2012) Control of sleep and wakefulness. Physiol Rev 92:1087–1187

Clément O, Sapin E, Bérod A, Fort P, Luppi PH (2011) Evidence that neurons of the sublaterodorsal tegmental nucleus triggering paradoxical (REM) sleep are glutamatergic. Sleep 34:419–423

Conductier G, Nahon JL, Guyon A (2011) Dopamine depresses melanin-concentrating hormone neuronal activity through multiple effects on α2-noradrenergic, D1 and D2-like dopaminergic receptors. Neuroscience 178:89–100

Del Cid-Pellitero E, Jones BE (2012) Immunohistochemical evidence for synaptic release of GABA from MCH-containing varicosities in the locus coeruleus. Neuroscience 223:269–276

Deurveilheur S, Semba L (2011) Basal forebrain regulation of cortical activity and sleep-wake states: roles of cholinergic and non-cholinergic neurons. Sleep Biol Rhythms 9(Suppl. 1):65–70

Devera A, Pascovich C, Lagos P, Falconi A, Sampogna S, Chase MH et al (2015) Melanin-concentrating hormone (MCH) modulates the activity of dorsal raphe nucleus. Brain Res 1598:114–128

Elias CF, Lee CE, Kelly JF, Ahima RS, Kuhar M, Saper CB et al (2001) Characterization of CART neurons in the rat and human hypothalamus. J Comp Neurol 432:1–19

Elias CF, Sita LV, Zambon BK, Oliveira ER, Vasconcelos LAP, Bittencourt JC (2008) Melanin-concentrating hormone projections to areas involved in somatomotor responses. J Chem Neuroanat 35:188–201

Freund TF, Gulyas AI (1991) GABAergic interneurons containing calbindin D28K or somatostatin are major targets of GABAergic basal forebrain afferents in the rat neocortex. J Comp Neurol 314:187–199

Gao XB, van den Pol AN (2001) Melanin-concentrating hormone depresses synaptic activity of glutamate and GABA neurons from rat lateral hypothalamus. J Physiol 55:237–252

Harthoorn LF (2007) Projection-dependent differentiation of melanin-concentrating hormone-containing neurons. Cell Mol Neurobiol 27:49–55

Hawes BE, Kil E, Green B, O'Neill K, Fried S, Graziano MP (2000) The melanin-concentrating hormone receptor couples to multiple G proteins to activate diverse intracellular signaling pathways. Endocrinology 141:4524–4532

Hervieu GJ, Cluderay JE, Harrison D, Meakin J, Maycox P, Nasir S et al (2000) The distribution of the mRNA and protein products of the melanin-concentrating hormone (MCH) receptor gene, sic-1, in the central nervous system of the rat. Eur J Neurosci 12:1194–1216

Hobson JA, McCarley RW, Wyzinski PW (1975) Sleep cycle oscillation: reciprocal discharge by two brainstem neuronal groups. Science 189:55–58

Hong EY, Yoon YS, Lee HS (2011) Differential distribution of melanin-concentrating hormone (MCH) and hypocretin (Hcrt)-immunoreactive neurons projecting to the mesopontine cholinergic complex in the rat. Brain Res 1424:20–31

Jego S, Glasgow SD, Gutierrez-Herrera C, Ekstrand M, Reed SJ, Boyce R et al (2013) Optogenetic identification of a rapid eye movement sleep modulatory circuit in the hypothalamus. Nat Neurosci 16:1637–1644

Jones BE (2003) Arousal systems. Front Bioci 8:438–451

Jones BE (2005) From waking to sleeping: neuronal and chemical substrates. TIPS 26:578–586

Lagos P, Torterolo P, Jantos H, Chase MH, Monti JM (2009) Effects on sleep of melanin-concentrating hormone (MCH) microinjections into the dorsal raphe nucleus. Brain Res 1265:103–110

Lagos P, Torterolo P, Jantos H, Monti JM (2011) Immunoneutralization of melanin-concentrating hormone (MCH) in the dorsal raphe nucleus: effects on sleep and wakefulness. Brain Res 1369:112–111

Lagos P, Monti JM, Jantos H, Torterolo P (2012) Microinjections of the melanin-concentrating hormone into the lateral basal forebrain increases REM sleep and reduces wakefulness in the rat. Life Sci 90:895–899

Lee M, Hassani O, Alonso A, Jones BE (2005) Cholinergic basal forebrain neurons burst with theta during waking and paradoxical sleep. J Neurosci 25:4365–4369

Lembo PM, Grazzini E, Cao J, Hubatsch DA, Pelletier M, Hoffert C et al (1999) The receptor for the orexigenic peptide melanin-concentrating hormone is a G-protein-coupled receptor. Nat Cell Biol 1:267–271

Lu J, Jhou TC, Saper CB (2006) Identification of wake-active dopaminergic neurons in the ventral periaqueductal gray matter. J Neurosci 26:191–202

Luppi PH, Clément O, Fort P (2013) Paradoxical (REM) sleep genesis by the brainstem is under hypothalamic control. Curr Opin Neurobiol 23:1–7

McCarley RW (2007) Neurobiology of REM and NREM sleep. Sleep Med 8:302–330

Monti JM, Jantos H (2009) The roles of dopamine and serotonin, and of their receptors, in regulating sleep and waking. Progr Brain Res 172:625–646

Monti JM, Monti D (2007) The involvement of dopamine in the modulation of sleep and waking. Sleep Med Rev 11:113–133

Monti JM, Lagos P, Jantos H, Torterolo P (2015) Increased REM sleep after intra-locus coeruleus nucleus microinjection of melanin-concentrating hormone (MCH) in the rat. Prog Neuro-Psychopharmacol Biol Psychiatry 56:185–188

Monti JM, Torterolo P, Jantos H, Lagos P (2016) Microinjection of the melanin-concentrating hormone into the sublaterodorsal tegmental nucleus inhibits REM sleep in the rat. Neurosci Lett 630:66–69

Pace-Schott EF, Hobson JA (2002) Basic mechanisms of sleep: new evidence on the neuroanatomy and neuromodulation of the NREM-REM cycle. In: Charney D, Nemeroff C (eds) Neuropsychopharmacology—the fifth generation of progress. Lippincott, Williams and Wilkins, Philadelphia, pp 1859–1877

Pissios P, Frank L, Kennedy AR, Porter DR, Marino FE, Liu FF et al (2008) Dysregulation of the mesolimbic dopamine system and reward in MCH−/− mice. Biol Psychiatry 64:184–191

Rodrigo-Angulo M, Heredero S, Rodríguez-Veiga E, Reinoso-Suárez F (2008) GABAergic and non-GABAergic thalamic, hypothalamic and basal forebrain projections to the oral pontine reticular nucleus: their implication in REM sleep regulation. Brain Res 1210:116–125

Sailer AW, Sano H, Zeng Z, McDonald TP, Pan J, Pong SS et al (2001) Identification and characterization of a second melanin-concentrating hormone receptor MCH-2R. Proc Natl Acad Sci U S A 98:7564–7569

Saito Y, Cheng M, Leslie FM, Civelli O (2001) Expression of the melanin-concentrating hormone (MCH) receptor mRNA in the rat brain. J Comp Neurol 435:26–40

Sapin E, Lapray D, Bérod A, Goutagny R, Léger L, Ravassard P et al (2009) Localization of the brainstem GABAergic neurons controlling paradoxical (REM) sleep. PLoS One 4:1–12

Sapin E, Bérod A, Léger L, Herman PA, Luppi PH, Peyron C (2010) A very large number of GABAergic neurons are activated in the tuberal hypothalamus during paradoxical (REM) sleep hypersomnia. PLoS One 5(7):e11766. https://doi.org/10.1371/journal.pone.0011766

Sekiya K, Ghatei MA, Lacoumenta S, Burnet PW, Zamir N, Burrin JM et al (1988) The distribution of melanin-concentrating hormone-like immunoreactivity in the central nervous system of rat, guinea-pig and man. Neuroscience 25:925–930

Semba K (2000) Multiple output pathways of the basal forebrain: organization, chemical heterogeneity, and roles in vigilance. Behav Brain Res 115:117–141

Semba K (2011) Preoptic and basal forebrain modulation of REM sleep. In: Mallick BN, Pandi-Perumal SR, McCarley R, Morrison AR (eds) REM sleep: regulation and function. Cambridge University Press, Cambridge, pp 99–109

Skofitsch G, Jacobowitz DM, Zamir N (1985) Immunohistochemical localization of melanin-concentrating hormone-like peptide in the rat brain. Brain Res Bull 15:653–649

Smith DG, Tzavara ET, Shaw J, Luecke S, Wade M, Davis R et al (2005) Mesolimbic dopamine super-sensitivity in melanin-concentrating hormone-1 receptor-deficient mice. J Neurosci 25:914–922

Szymusiak R, Gvilia I, McGinty D (2007) Hypothalamic control of sleep. Sleep Med 8:291–301

Torterolo P, Lagos P, Monti JM (2011) Melanin-concentrating hormone: a new sleep factor? Front Neurol 2:2–12

Urbanavicius J, Lagos P, Torterolo P, Abin-Carriquiry JA (2016) Melanin-concentrating hormone projections to the dorsal raphe nucleus: an immunofluorescence and in vivo microdialysis study. J Chem Neuroanat 72:16–24

van den Pol AN, Acuna-Goycolea C, Clark KR, Ghosh PK (2004) Physiological properties of hypothalamic MCH neurons identified with selective expression of reporter gene after recombinant virus infection. Neuron 42:635–652

Yang N, Zhang KY, Wang FF, Hu ZA, Zhang J (2014) Dopamine inhibits neurons from the rat dorsal subcoeruleus nucleus through the activation of α2-adrenergic receptors. Neurosci Lett 559:61–66

Yoon YS, Lee HS (2013) Projections from melanin-concentrating hormone (MCH) neurons to the dorsal raphe or the nuclear core of the locus coeruleus in the rat. Brain Res 1490:72–82

Cannabinoids, Sleep, and the MCH System

Eric Murillo-Rodríguez, Daniela Morales-Lara, José Carlos Pastrana-Trejo, Lorena Macías-Triana, Karen Romero-Cordero, Miriel de-la-Cruz, and Clelia De-la-Peña

Abstract Marijuana is a colloquial name given to *Cannabis sativa*, which has been used for diverse purposes, including as a therapeutical element for multiple health issues. The neurobiological effects of *C. sativa* involve a complex biological machinery including receptors, named CB_1 and CB_2 cannabinoid receptors. These receptors recognize endogenous cannabinoid-like compounds, such as anandamide and 2-arachinonolglycerol which seems to display sleep-inducing properties. Along decades, the study of the putative role of exogenous and endogenous cannabinoids in sleep modulation has brought critical data. Since endocannabinoids have been described in sleep-related brain areas, intriguing issues regarding whether hypothalamic substrates, such as MHC, may be interacting with the endocannabinoids have been raised.

1 Introduction

Cannabinoids are a group of terpenophenolic molecules present in *Cannabis* (*Cannabis sativa*). It is important to note that a broader definition of cannabinoids also includes a group of compounds that are structurally related to the principal active molecule of *Cannabis*, delta-9-tetrahydrocannabinol (Δ^9-THC; Gertsch et al. 2010; Schafroth and Carreira 2017; ElSohly et al. 2017; Fig. 1).

E. Murillo-Rodríguez (✉) · D. Morales-Lara · J. C. Pastrana-Trejo · L. Macías-Triana · K. Romero-Cordero · M. de-la-Cruz
Laboratorio de Neurociencias Moleculares e Integrativas, Escuela de Medicina, División Ciencias de la Salud, Universidad Anáhuac Mayab, Mérida, Yucatán, Mexico

Grupo de Investigación en Envejecimiento, División Ciencias de la Salud, Universidad Anáhuac Mayab, Mérida, Yucatán, Mexico

Intercontinental Neuroscience Research Group, Mérida, Yucatán, Mexico
e-mail: eric.murillo@anahuac.mx

C. De-la-Peña
Centro de Investigación Científica de Yucatán. A.C, Unidad de Biotecnología, Mérida, Yucatán, Mexico

S. R. Pandi-Perumal et al. (eds.), *Melanin-Concentrating Hormone and Sleep*,
https://doi.org/10.1007/978-3-319-75765-0_6

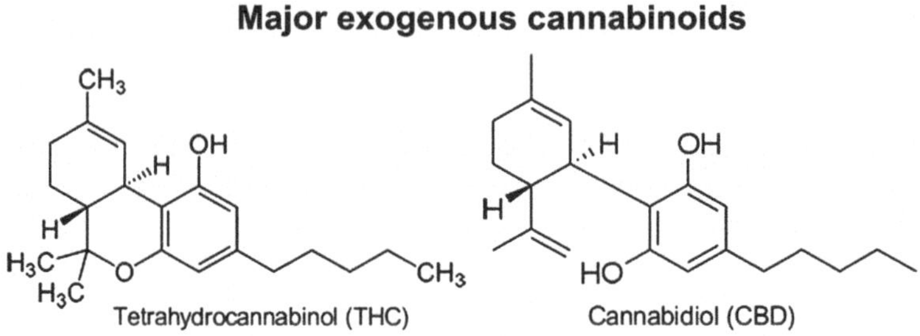

Fig. 1 Molecular structure of two major exogenous cannabinoids: delta-9-tetrahydrocannabinol and cannabidiol

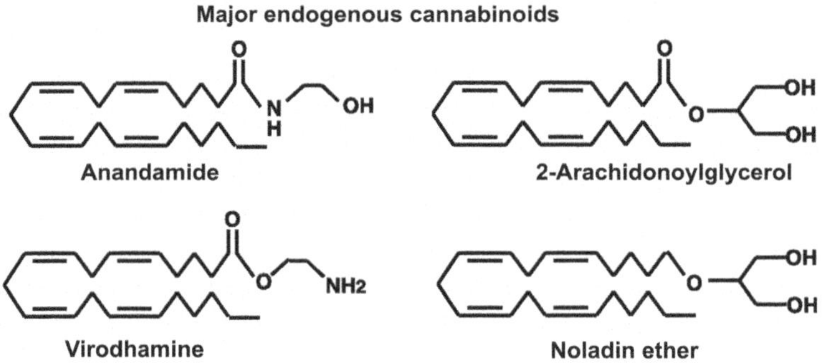

Fig. 2 Molecular structure of major endogenous cannabinoids: anandamide, 2-arachidonoylglycerol, virodhamine, and noladin ether

The physiological effects of cannabinoids are produced by the involvement of specific membrane-bound receptors. In general consensus, the receptors that recognize Δ^9-THC have been denominated as CB_1 and CB_2 cannabinoid receptors which are present in several species such as mammals, birds, fish, and invertebrates (Abood 2005; McPartland et al. 2006; Gilbert and Soderstrom 2013; Krug and Clark 2015; Laprairie et al. 2016).

Intriguing questions regarding what might be the natural molecule that binds to the cannabinoid receptors are raised. Later, lipids that bind and activate the CB_1/CB_2 cannabinoid receptors (Cascio and Marini 2015; Pertwee 2015) were discovered. Moreover, multiple pharmacological experiments demonstrated that administration of these lipids induced marihuana-like effects, such as hyperphagia, hypomobility, hypothermia, etc. (Gerra et al. 2010; Mechoulam and Parker 2013). Then, the endogenous compounds were named endocannabinoids (Elphick and Egertová 2001; Iversen 2003; Fig. 2).

Among the most studied endocannabinoids, 2-arachidonoylglycerol (2-AG) and anandamide (ANA) have shown that they modulate diverse neurobiological functions, such as learning and memory, feeding, pain perception, and sleep generation

(Basavarajappa et al. 2009; Bermudez-Silva et al. 2010; Murillo-Rodríguez et al. 2011a, b; Guindon and Hohmann 2011; Woodhams et al. 2015; Morena et al. 2016).

Since cannabinoid receptors and endogenous cannabinoids are present in several species, including humans, an active neurobiological role of the endocannabinoid system in diverse functions has been proposed (Gatta-Cherifi and Cota 2015; Katona 2015; Rubino et al. 2015; Fernández-Ruiz et al. 2015; Sido et al. 2015). The current knowledge has shown that the endocannabinoid system that comprises receptors, enzymes, and ligands is linked with human diseases such as Parkinson's disease, Alzheimer's disease, and obesity, among other health issues (Bedse et al. 2015; More and Choi 2015; Jourdan et al. 2016; Gruden et al. 2016; Fakhoury 2017).

2 Phytocannabinoids and Sleep

2.1 Δ9-THC and Sleep

Marijuana is a common name given to the plant *Cannabis sativa* (*C. sativa*) which has been used by diverse cultures for distinct purposes, including therapeutic aims (Carlini et al. 2017; Walsh et al. 2017). Multiple pieces of evidence have demonstrated that *C. sativa* or Δ^9-THC modulates the sleep-wake cycle. Historical reports have shown that during the 1960s, 1970s, and 1980s, several experiments in animal models and humans were carried out in order to evaluate the effects of the cannabinoids on sleep. The main outcome from these reports is that *C. sativa* or Δ^9-THC increases sleep (Buonamici et al. 1982) including in humans (Pivik et al. 1972; Feinberg et al. 1975, 1976; Freemon 1972). For example, dosage of 70–210 mg of Δ^9-THC induced sleep in humans (Feinberg et al. 1975, 1976). In summary, consumption of *C. sativa* promotes sleep (Haney et al. 2007; Schierenbeck et al. 2008).

2.2 CBD and Sleep

It is worthy to mention that administration of cannabidiol (CBD), a non-psychotropic constituent of marijuana, induces opposite effects that are observed for Δ^9-THC. For example, Monti (1977) observed a reduction in sleep after systemic administration of CBD. Similar results were obtained later by Nicholson et al. (2004) since they reported that administrations of CBD enhanced waking in humans. In recent years, our group has demonstrated that intracerebroventricular administration of CBD (10 μg/5 μL) to rats increased alertness and decreased rapid eye movement sleep (REMS) during the resting period of the animals. In addition, it was found that CBD enhanced the extracellular levels of dopamine collected from nucleus accumbens, whereas an increase in *c*-Fos expression was detected in waking-related brain areas, such as hypothalamus and dorsal raphe nucleus (Murillo-Rodríguez et al. 2006a, b). Similar findings were observed when CBD was injected into the lateral hypothalamus of rats during the lights-on period (Murillo-Rodríguez et al. 2008, 2011a, b). Recently, Hsiao et al. (2012) reported that CBD blocked anxiety-induced REMS suppression. Lastly, the wake-promoting effects of CBD might engage neurochemicals related to control of waking such as adenosine (Mijangos-Moreno et al. 2014). Taken together, these data suggest that CBD promotes alertness. Despite this fascinating result, the mechanism of action of this phytocannabinoid in sleep control remains unclear.

3 Cannabinoids, Sleep, and the MCH System

Several reports have indicated that ANA is present in multiple brain areas, including the hypothalamus (Valenti et al. 2004; Murillo-Rodríguez et al. 2006a, b; Egertová et al. 2008). Interestingly, this endocannabinoid increases its concentration in the hypothalamus during the lights-on period (Murillo-Rodríguez et al. 2006a, b) suggesting its putative role in modulating hypothalamic neurobiological functions, such as sleep (Li et al. 2016; Gao and Sun 2016; Chowen et al. 2016; Blais et al. 2017; Al Massadi et al. 2017).

The role of the hypothalamus in the control of sleep has been widely studied (Fraigne et al. 2014; de Lecea 2015; Schwartz and Kilduff 2015; Di Cristoforo et al. 2015; Mehta et al. 2015). Among the neurochemicals that hypothalamus synthesizes, the melanin-concentrating hormone (MCH) is a compound of peptidergic nature. This molecule exerts a critical role in sleep modulation. Briefly, the MCHergic neurons increase their electrophysiological firing during REMS. Moreover, pharmacological experiments in laboratory animals have demonstrated that administration of MCH into brain areas linked to the sleep modulation promotes REMS (Konadhode et al. 2015; Devera et al. 2015; Apergis-Schoute et al. 2015; Dias Abdo Agamme et al. 2015; Torterolo et al. 2015; González et al. 2016; Urbanavicius et al. 2016; Chometton et al. 2016; Blanco-Centurion et al. 2016; Vetrivelan et al. 2016).

Since MCH neurons are located in the lateral hypothalamus and ANA has been described to be present in this brain area, it is likely to suspect a possible relationship between these two neuromodulators. Huang et al. (2007) demonstrated that cannabinoid agonists activated MCH cells by increasing their spike frequency. These results show that cannabinoids modulate the activity of MCH neurons. Further report showed that administrations of 2-AG increased *c*-Fos expression in MCH neurons (Pérez-Morales et al. 2013). Whether MCH system interacts with the endocannabinoid elements (ligands, receptors, enzymes, transporters, etc.) lacks of fully study. Moreover, the role of exogenous, endogenous, and synthetic cannabinoids modulating the activity of MCH neurons remains to be elucidated (Table 1).

Table 1 Effects on sleep and MCH system of exogenous, endogenous, and synthetic cannabinoids

	Effects on Sleep	Effects on MCH
	Exogenous cannabinoids	
delta-9-tetrahydrocannabino	Increases sleep Decreases waking	Unknown
Cannabidiol	Increases waking Decreases sleep	Unknown
	Effects on Sleep	Effects on MCH
	Endogenous cannabinoids	
Anandamide	Increases sleep Decreases waking	Unknown
2-Arachidonylglycerol	Increases sleep Decreases waking	Activates neurons in MCH
Virodhamine	Unknown	Unknown
Noladin Eter	Unknown	Unknown
	Synthetic cannabinoid agonists	
WIN 55, 212,2	Increases sleep Decreases waking	Activates neurons in MCH
	Synthetic cannabinoid antagonists	
AM251	Increases waking Decreases sleep	Deactivates neurons in MCH
ABD459	Increases sleep	Unknown

4 Conclusions

The sleep-wake cycle is controlled by multiple neurochemical substrates, including exogenous and endogenous cannabinoids (Buonamici et al. 1982; Carlini and Cunha 1981; Feinberg et al. 1975, 1976; Freemon 1972; Monti 1977; Murillo-Rodríguez et al. 2006a, b, 2008, 2011a, b; Nicholson et al. 2004; Pivik et al. 1972). In this chapter, we revised that phytocannabinoids modulate the sleep-wake cycle.

Since the endocannabinoids have been localized in sleep-related brain areas, such as hypothalamus (Valenti et al. 2004; Murillo-Rodríguez et al. 2006a, b; Egertová et al. 2008), intriguing issues regarding whether hypothalamic substrates, such as MHC (Konadhode et al. 2015; Devera et al. 2015; Apergis-Schoute et al. 2015; Dias Abdo Agamme et al. 2015; Torterolo et al. 2015; González et al. 2016; Urbanavicius et al. 2016; Chometton et al. 2016; Blanco-Centurion et al. 2016; Vetrivelan et al. 2016), might be interacting with the endocannabinoids have been raised. The demonstration that exogenous ligands for CB_1 cannabinoid receptor activate MCH neurons (Huang et al. 2007) opens the possibility that both neuromodulator systems might display a cross taking in the sleep control. Despite these fascinating discoveries, the full understanding of the relationship between exogenous or endogenous cannabinoids and MCH system awaits further investigation.

References

Abood ME (2005) Molecular biology of cannabinoid receptors. Handb Exp Pharmacol (168):81–115

Al Massadi O, López M, Tschöp M, Diéguez C, Nogueiras R. Current understanding of the hypothalamic ghrelin pathways inducing appetite and adiposity. Trends Neurosci. 2017. pii: S0166–2236(16)30190–4. doi: https://doi.org/10.1016/j.tins.2016.12.003

Apergis-Schoute J, Iordanidou P, Faure C, Jego S, Schöne C, Aitta-Aho T, Adamantidis A, Burdakov D (2015) Optogenetic evidence for inhibitory signaling from orexin to MCH neurons via local microcircuits. J Neurosci 35(14):5435–5441. https://doi.org/10.1523/JNEUROSCI.5269-14.2015

Basavarajappa BS, Nixon RA, Arancio O (2009) Endocannabinoid system: emerging role from neurodevelopment to neurodegeneration. Mini Rev Med Chem 9(4):448–462

Bedse G, Romano A, Lavecchia AM, Cassano T, Gaetani S (2015) The role of endocannabinoid signaling in the molecular mechanisms of neurodegeneration in Alzheimer's disease. J Alzheimers Dis 43(4):1115–1136. https://doi.org/10.3233/JAD-141635

Bermudez-Silva FJ, Viveros MP, McPartland JM, Rodriguez de Fonseca F (2010) The endocannabinoid system, eating behavior and energy homeostasis: the end or a new beginning? Pharmacol Biochem Behav 95(4):375–382. https://doi.org/10.1016/j.pbb.2010.03.012

Blais A, Drouin G, Chaumontet C, Voisin T, Couvelard A, Even PC, Couvineau A (2017) Impact of Orexin-A treatment on food intake, energy metabolism and body weight in mice. PLoS One 12 (1):e0169908. https://doi.org/10.1371/journal.pone.0169908

Blanco-Centurion C, Liu M, Konadhode RP, Zhang X, Pelluru D, van den Pol AN, Shiromani PJ (2016) Optogenetic activation of melanin-concentrating hormone neurons increases non-rapid eye movement and rapid eye movement sleep during the night in rats. Eur J Neurosci 44 (10):2846–2857. https://doi.org/10.1111/ejn.13410

Buonamici M, Young GA, Khazan N (1982) Effects of acute delta 9-THC administration on EEG and EEG power spectra in the rat. Neuropharmacology 21(8):825–829
Carlini EA, Cunha JM (1981) Hypnotic and antiepileptic effects of cannabidiol. J Clin Pharmacol. 21(8–9 Suppl):417S–427S
Carlini BH, Garrett SB, Carter GT (2017) Medicinal cannabis. Am J Hosp Palliat Care 34(1):85–91. https://doi.org/10.1177/1049909115604669
Cascio MG, Marini P (2015) Biosynthesis and fate of endocannabinoids. Handb Exp Pharmacol 231:39–58. https://doi.org/10.1007/978-3-319-20825-1_2
Chometton S, Croizier S, Fellmann D, Risold PY (2016) The MCH neuron population as a model for the development and evolution of the lateral and dorsal hypothalamus. J Chem Neuroanat 75 (Pt A):28–31. https://doi.org/10.1016/j.jchemneu.2015.09.004
Chowen JA, Argente-Arizón P, Freire-Regatillo A, Frago LM, Horvath TL, Argente J (2016) The role of astrocytes in the hypothalamic response and adaptation to metabolic signals. Prog Neurobiol 144:68–87. https://doi.org/10.1016/j.pneurobio.2016.03.001
de Lecea L (2015) Optogenetic control of hypocretin (orexin) neurons and arousal circuits. Curr Top Behav Neurosci 25:367–378. https://doi.org/10.1007/7854_2014_364
Devera A, Pascovich C, Lagos P, Falconi A, Sampogna S, Chase MH, Torterolo P (2015) Melanin-concentrating hormone (MCH) modulates the activity of dorsal raphe neurons. Brain Res 1598:114–128. https://doi.org/10.1016/j.brainres.2014.12.032
Di Cristoforo A, Cerri M, Del Vecchio F, Hitrec T, Luppi M, Perez E, Zamboni G, Amici R (2015) Wake-sleep, thermoregulatory, and autonomic effects of cholinergic activation of the lateral hypothalamus in the rat: a pilot study. Arch Ital Biol 153(2–3):67–76. https://doi.org/10.12871/000398292015232
Dias Abdo Agamme AL, Aguilar Calegare BF, Fernandes L, Costa A, Lagos P, Torterolo P, D'Almeida V (2015) MCH levels in the CSF, brain preproMCH and MCHR1 gene expression during paradoxical sleep deprivation, sleep rebound and chronic sleep restriction. Peptides 74:9–15. https://doi.org/10.1016/j.peptides.2015.10.001
Egertová M, Simon GM, Cravatt BF, Elphick MR (2008) Localization of N-acyl phosphatidylethanolamine phospholipase D (NAPE-PLD) expression in mouse brain: a new perspective on N-acylethanolamines as neural signaling molecules. J Comp Neurol 506(4):604–615
Elphick MR, Egertová M (2001) The neurobiology and evolution of cannabinoid signalling. Philos Trans R Soc Lond Ser B Biol Sci 356(1407):381–408
ElSohly MA, Radwan MM, Gul W, Chandra S, Galal A (2017) Phytochemistry of Cannabis sativa L. Prog Chem Org Nat Prod 103:1–36. https://doi.org/10.1007/978-3-319-45541-9_1
Fakhoury M (2017) Role of the endocannabinoid system in the pathophysiology of schizophrenia. Mol Neurobiol 54(1):768–778. https://doi.org/10.1007/s12035-016-9697-5
Feinberg I, Jones R, Walker JM, Cavness C, March J (1975) Effects of high dosage delta-9-tetrahydrocannabinol on sleep patterns in man. Clin Pharmacol Ther 17(4):458–466
Feinberg I, Jones R, Walker J, Cavness C, Floyd T (1976) Effects of marijuana extract and tetrahydrocannabinol on electroencephalographic sleep patterns. Clin Pharmacol Ther 19 (6):782–794
Fernández-Ruiz J, Romero J, Ramos JA (2015) Endocannabinoids and neurodegenerative disorders: Parkinson's disease, Huntington's chorea, Alzheimer's disease, and others. Handb Exp Pharmacol 231:233–259. https://doi.org/10.1007/978-3-319-20825-1_8
Fraigne JJ, Grace KP, Horner RL, Peever J (2014) Mechanisms of REM sleep in health and disease. Curr Opin Pulm Med 20(6):527–532. https://doi.org/10.1097/MCP.0000000000000103
Freemon FR (1972) Effects of marijuana on sleeping states. JAMA 220(10):1364–1365
Gao Y, Sun T (2016) Molecular regulation of hypothalamic development and physiological functions. Mol Neurobiol 53(7):4275–4285. https://doi.org/10.1007/s12035-015-9367-z
Gatta-Cherifi B, Cota D (2015) Endocannabinoids and metabolic disorders. Handb Exp Pharmacol 231:367–391. https://doi.org/10.1007/978-3-319-20825-1_13

Gerra G, Zaimovic A, Gerra ML, Ciccocioppo R, Cippitelli A, Serpelloni G, Somaini L (2010) Pharmacology and toxicology of Cannabis derivatives and endocannabinoid agonists. Recent Pat CNS Drug Discov 5(1):46–52

Gertsch J, Pertwee RG, Di Marzo V (2010) Phytocannabinoids beyond the cannabis plant—do they exist? Br J Pharmacol 160(3):523–529. https://doi.org/10.1111/j.1476-5381.2010.00745.x

Gilbert MT, Soderstrom K (2013) Novel song-stimulated dendritic spine formation and Arc/Arg3.1 expression in zebra finch auditory telencephalon are disrupted by cannabinoid agonism. Brain Res 1541:9–21. https://doi.org/10.1016/j.brainres.2013.10.012

González JA, Iordanidou P, Strom M, Adamantidis A, Burdakov D (2016) Awake dynamics and brain-wide direct inputs of hypothalamic MCH and orexin networks. Nat Commun 7:11395. https://doi.org/10.1038/ncomms11395

Gruden G, Barutta F, Kunos G, Pacher P (2016) Role of the endocannabinoid system in diabetes and diabetic complications. Br J Pharmacol 173(7):1116–1127. https://doi.org/10.1111/bph.13226

Guindon J, Hohmann AG (2011) The endocannabinoid system and cancer: therapeutic implication. Br J Pharmacol 163(7):1447–1463. https://doi.org/10.1111/j.1476-5381.2011.01327.x

Haney M, Gunderson EW, Rabkin J, Hart CL, Vosburg SK, Comer SD, Foltin RW (2007) Dronabinol and marijuana in HIV-positive marijuana smokers. Caloric intake, mood, and sleep. J Acquir Immune Defic Syndr 45(5):545–554

Hsiao YT, Yi PL, Li CL, Chang FC (2012) Effect of cannabidiol on sleep disruption induced by the repeated combination tests consisting of open field and elevated plus-maze in rats. Neuropharmacology 62(1):373–384. https://doi.org/10.1016/j.neuropharm.2011.08.013

Huang H, Acuna-Goycolea C, Li Y, Cheng HM, Obrietan K, van den Pol AN (2007) Cannabinoids excite hypothalamic melanin-concentrating hormone but inhibit hypocretin/orexinneurons: implications for cannabinoid actions on food intake and cognitive arousal. J Neurosci 27 (18):4870–4881

Iversen L (2003) Cannabis and the brain. Brain 126(Pt 6):1252–1270

Jourdan T, Godlewski G, Kunos G (2016) Endocannabinoid regulation of β-cell functions: implications for glycemic control and diabetes. Diabetes Obes Metab 18(6):549–557. https://doi.org/10.1111/dom.12646

Katona I (2015) Cannabis and endocannabinoid signaling in epilepsy. Handb Exp Pharmacol 231:285–316. https://doi.org/10.1007/978-3-319-20825-1_10

Konadhode RR, Pelluru D, Shiromani PJ (2015) Neurons containing orexin or melanin concentrating hormone reciprocally regulate wake and sleep. Front Syst Neurosci 8:244. https://doi.org/10.3389/fnsys.2014.00244. Review

Krug RG 2nd, Clark KJ (2015) Elucidating cannabinoid biology in zebrafish (Danio rerio). Gene 570(2):168–179. https://doi.org/10.1016/j.gene.2015.07.036

Laprairie RB, Bagher AM, Denovan-Wright EM (2016) Cannabinoid receptor ligand bias: implications in the central nervous system. Curr Opin Pharmacol 32:32–43. https://doi.org/10.1016/j.coph.2016.10.005

Li SB, Giardino WJ, de Lecea L (2016) Hypocretins and arousal. Curr Top Behav Neurosci. https://doi.org/10.1007/7854_2016_58

McPartland JM, Agraval J, Gleeson D, Heasman K, Glass M (2006) Cannabinoid receptors in invertebrates. J Evol Biol 19(2):366–373

Mechoulam R, Parker LA (2013) The endocannabinoid system and the brain. Annu Rev Psychol 64:21–47. https://doi.org/10.1146/annurev-psych-113011-143739

Mehta R, Khanday MA, Mallick BN (2015) REM sleep loss associated changes in orexin-A levels in discrete brain areas in rats. Neurosci Lett 590:62–67. https://doi.org/10.1016/j.neulet.2015.01.067

Mijangos-Moreno S, Poot-Aké A, Arankowsky-Sandoval G, Murillo-Rodríguez E (2014) Intrahypothalamic injection of cannabidiol increases the extracellular levels of adenosine in nucleus accumbens in rats. Neurosci Res 84:60–63

Monti JM (1977) Hypnoticlike effects of cannabidiol in the rat. Psychopharmacology 55 (3):263–265

More SV, Choi DK (2015) Promising cannabinoid-based therapies for Parkinson's disease: motor symptoms to neuroprotection. Mol Neurodegener 10:17. https://doi.org/10.1186/s13024-015-0012-0
Morena M, Patel S, Bains JS, Hill MN (2016) Neurobiological interactions between stress and the endocannabinoid system. Neuropsychopharmacology 41(1):80–102. https://doi.org/10.1038/npp.2015.166
Murillo-Rodríguez E, Millán-Aldaco D, Palomero-Rivero M, Mechoulam R, Drucker-Colín R (2006a) Cannabidiol, a constituent of Cannabis sativa, modulates sleep in rats. FEBS Lett 580:4337–4345
Murillo-Rodríguez E, Desarnaud F, Prospéro-Garcia O (2006b) Diurnal variation of arachidonoylethanolamine, palmitoylethanolamide and oleoylethanolamide in the brain of the rat. Life Sci 79:30–37
Murillo-Rodríguez E, Millán-Aldaco D, Palomero-Rivero M, Mechoulam R, Drucker-Colín R (2008) The non-psychoactive cannabis constituent cannabidiol is a wake-inducing agent. Behav Neurosci 122:1378–1382
Murillo-Rodríguez E, Poot-Ake A, Arias-Carrion O, Pacheco-Pantoja E, Fuente-Ortegon Ade L, Arankowsky-Sandoval G (2011a) The emerging role of the endocannabinoid system in the sleep-wake cycle modulation. Cent Nerv Syst Agents Med Chem 11(3):189–196
Murillo-Rodríguez E, Palomero-Rivero M, Millán-Aldaco D, Mechoulam R, Drucker-Colín R (2011b) Effects on sleep and dopamine levels of microdialysis perfusion of cannabidiol into the lateral hypothalamus of rats. Life Sci 88:504–511
Nicholson AN, Turner C, Stone BM, Robson PJ (2004) Effect of delta-9-tetrahydrocannabinol and cannabidiol on nocturnal sleep and early-morning behavior in young adults. J Clin Psychopharmacol 24(3):305–313
Pérez-Morales M, De La Herrán-Arita AK, Méndez-Díaz M, Ruiz-Contreras AE, Drucker-Colín R, Prospéro-García O (2013) 2-AG into the lateral hypothalamus increases REM sleep and cFos expression in melanin concentrating hormone neurons in rats. Pharmacol Biochem Behav 108:1–7. https://doi.org/10.1016/j.pbb.2013.04.006
Pertwee RG (2015) Endocannabinoids and their pharmacological actions. Handb Exp Pharmacol 231:1–37. https://doi.org/10.1007/978-3-319-20825-1_1
Pivik RT, Zarcone V, Dement WC, Hollister LE (1972) Delta-9-tetrahydrocannabinol and synhexl: effects on human sleep patterns. Clin Pharmacol Ther 13(3):426–435
Rubino T, Zamberletti E, Parolaro D (2015) Endocannabinoids and mental disorders. Handb Exp Pharmacol 231:261–283. https://doi.org/10.1007/978-3-319-20825-1_9
Schafroth MA, Carreira EM (2017) Synthesis of phytocannabinoids. Prog Chem Org Nat Prod 103:37–59. https://doi.org/10.1007/978-3-319-45541-9_2
Schierenbeck T, Riemann D, Berger M, Hornyak M (2008) Effect of illicit recreational drugs upon sleep: cocaine, ecstasy and marijuana. Sleep Med Rev 12(5):381–389. https://doi.org/10.1016/j.smrv.2007.12.004
Schwartz MD, Kilduff TS (2015) The neurobiology of sleep and wakefulness. Psychiatr Clin North Am 38(4):615–644. https://doi.org/10.1016/j.psc.2015.07.002
Sido JM, Nagarkatti PS, Nagarkatti M (2015) Role of endocannabinoid activation of peripheral CB1 receptors in the regulation of autoimmune disease. Int Rev Immunol 34(5):403–414. https://doi.org/10.3109/08830185.2014.921165
Torterolo P, Scorza C, Lagos P, Urbanavicius J, Benedetto L, Pascovich C, López-Hill X, Chase MH, Monti JM (2015) Melanin-concentrating hormone (MCH): role in rem sleep and depression. Front Neurosci 9:475. https://doi.org/10.3389/fnins.2015.00475
Urbanavicius J, Lagos P, Torterolo P, Abin-Carriquiry JA, Scorza C (2016) Melanin-concentrating hormone projections to the dorsal raphe nucleus: an immunofluorescence and in vivo microdialysis study. J Chem Neuroanat 72:16–24. https://doi.org/10.1016/j.jchemneu.2015.11.010
Valenti M, Viganò D, Casico MG, Rubino T, Steardo L, Parolaro D, Di Marzo V (2004) Differential diurnal variations of anandamide and 2-arachidonoyl-glycerol levels in rat brain. Cell Mol Life Sci 61(7–8):945–950

Vetrivelan R, Kong D, Ferrari LL, Arrigoni E, Madara JC, Bandaru SS, Lowell BB, Lu J, Saper CB (2016) Melanin-concentrating hormone neurons specifically promote rapid eye movement sleep in mice. Neuroscience 336:102–113. https://doi.org/10.1016/j.neuroscience.2016.08.046

Walsh Z, Gonzalez R, Crosby K, S Thiessen M, Carroll C, Bonn-Miller MO (2017) Medical cannabis and mental health: a guided systematic review. Clin Psychol Rev 51:15–29. https://doi.org/10.1016/j.cpr.2016.10.002

Woodhams SG, Sagar DR, Burston JJ, Chapman V (2015) The role of the endocannabinoid system in pain. Handb Exp Pharmacol 227:119–143. https://doi.org/10.1007/978-3-662-46450-2_7

MCH and Thermoregulation

Marco Luppi

Abstract Homeothermy represents a remarkable step in animal evolution, albeit at a very high cost in terms of metabolic demand. The maintenance of core body temperature in mammals represents one of the prominent physiological components contributing to the basal metabolic rate. Homeostatic thermoregulation is coordinated by the central nervous system by means of different strategies, spanning from behavioral modifications, aimed at finding a better environment, to the activation or inhibition of key regulatory mechanisms, which are mainly driven by the autonomic nervous system. The hypothalamic neuropeptide MCH plays a pivotal role in regulating basal metabolism, and the activation of this system results in a slowing down of the metabolic rate and also stimulates food intake. On the contrary, blocking the MCH system, in animal models, promotes a lean phenotype with higher body temperature. Even though MCH is not involved in thermoregulatory processes, modifying MCH activity induces metabolic rate modifications, and thermoregulation is modified accordingly. The activation of the MCH system also leads to the dampening of the normal daily oscillation of body temperature. The well-known involvement of MCH in wake-sleep cycle regulation, by stabilizing sleep, and in particular REM sleep, reinforces the hypothesis that the functions of metabolism, thermoregulation, and sleep regulation are closely linked.

1 Thermoregulation

Thermoregulation represents one of the most cited examples of physiological homeostasis. Homeostatic regulation is a basic conceptual mechanism through which bodily regulations actively control and maintain the constancy of the "internal environment" in the organism, as defined by Walter B. Cannon in the 1920s. There are different ways to drive modifications in order to keep a specific physiological parameter constant. Homeostasis of the body temperature, for instance, is firstly

M. Luppi (✉)
Department of Biomedical and Neuromotor Sciences, University of Bologna, Bologna, Italy
e-mail: marco.luppi@unibo.it

S. R. Pandi-Perumal et al. (eds.), *Melanin-Concentrating Hormone and Sleep*,
https://doi.org/10.1007/978-3-319-75765-0_7

reached by means of behavioral modifications; this strategy is less costly in terms of metabolism than those involving active physiological responses through the intervention of the autonomic nervous system (Cabanac 1996; Satinoff 1996; Willmer et al. 2000). These kinds of homeostatically driven behavioral modifications result in a strategy that leads the animal to the search for a better environment with a more comfortable temperature. In particular, following a cold feeling, the animal looks for a warmer area or a protected niche, where other individuals may be present; on the contrary, when body temperature increases, animals look for more comfortable positions, such as in water, shade, or windy areas. In homeotherms, when this is not sufficient, autonomic thermoregulation is then activated, increasing thermogenic or thermolytic effects, aimed at keeping the core temperature constant. In fact, homeothermy is a characteristic that is shared only by mammals and birds and was a revolutionary step in evolution (Nespolo et al. 2011; Lovegrove 2012). At the cost of a much higher metabolic rate, homeothermy makes it possible for the animal to live in a very broad range of environments, with a wide variety of ambient temperatures, while maintaining the core body temperature within a narrow range (Grigg et al. 2004). The homeothermic organism is able either to dissipate heat, when the ambient temperature is higher than the optimal temperature (defined as thermoneutrality), or to produce heat while minimizing heat dissipation (thermogenesis), in the case of the ambient temperature being lower than thermoneutrality. The constancy of the core body temperature is the main component contributing to the basal metabolism, 66% of which is spent in this process (cf. Clapham 2012). It follows that the link between thermoregulation and metabolism is extremely close, and these mechanisms cannot be modified independently in homeotherms: experimentally, it is not possible to modify core temperature without modifying basal metabolism and, vice versa, is not possible to modify basal metabolism without observing spontaneous temperature modifications.

Recently, thanks to the past two decades of research, light has been shed on the neural pathway controlling thermoregulation (cf. Morrison et al. 2014), even though the dynamics of the pathway itself, as well as the mechanisms underlying thermoregulation, are not yet fully understood. What appears to be clear is that body temperature is continuously monitored both from the periphery and within the brain (Cerri 2017) and is precisely regulated from some neural structures with highly integrative characteristics (Morrison et al. 2014).

2 Thermoregulation and Sleep

One of the first theories on the biological function of sleep is that its aim is to save energy; sleep is a period of non-active behavior, with reduced body temperature and energy expenditure (Brown et al. 2012; Cerri 2017). Clearly, this is now a very simplified point of view, but the strict relationship between sleep and metabolism/thermoregulation still appears peculiar and physiologically intriguing (Amici et al. 2014; Cerri et al. 2017).

At sleep onset, body temperature, as well as brain temperature, drops slightly together with metabolic rate (Krauchi 2007; Cerri 2017). But the most peculiar physiological interplay takes place during REM sleep, when thermoregulation interacts with the occurrence of this sleep state: during REM sleep episodes, central thermoregulation appears to be impaired; consequently, when thermoregulation is stimulated by an adverse ambient condition, REM sleep amount is reduced, and the regularity and stability of episodes appear to be disturbed (Amici et al. 2014; Cerri et al. 2017).

Phylogenetically, the appearance of homeothermy is concomitant with the appearance of REM sleep. This may sustain the hypothesis that there is a marked overlap between the neural pathways controlling thermoregulation and REM sleep occurrence and that the biological function of REM sleep, which is still unknown, may be, for some reason, incompatible with the activation of the thermoregulatory structures (Cerri et al. 2017).

The finding in the late 1990s that the lateral hypothalamic area is important in regulating both metabolism and the stability of the wake-sleep cycle (Eban-Rothschild et al. 2017) opened new possibilities in understanding sleep functions and the relationship between sleep and thermoregulation. The key neurotransmitters in this interplay between physiological regulations are the two neuropeptides that are typically expressed in this area: orexin and MCH (Monti et al. 2013).

3 MCH, Thermoregulation, and Sleep

The main difficulty in studying the involvement of MCH neurotransmission in thermoregulatory processes is the fact that this neurotransmitter acts on almost the same target structures as orexin does. Generally the reciprocal action of these two peptides has an opposite effect, such as in wake-sleep regulation (Luppi et al. 2013; Vetrivelan et al. 2016), but in some other cases, functions are overlapping, for example, in the case of food intake stimulation (Qu et al. 1996). Moreover, the correct organization of their reciprocal functioning is maintained through the reciprocal inhibition between orexinergic and MCHergic hypothalamic neurons (Apergis-Schoute et al. 2015). Experimentally, this very intricate reciprocal interaction makes it difficult to understand whether the effects of MCH observed on thermoregulation are actually the consequence of the neurotransmission itself or, for instance, resulting from the counteraction to the orexinergic action. Nevertheless, the clear involvement of orexin in thermoregulation was experimentally described (Tupone et al. 2011) when this peptide was demonstrated to exert a modulatory effect on thermogenesis: orexinergic input to the pontine Raphe Pallidus amplified an already active thermogenetic process, but when thermogenesis was not activated, the effect of orexin was absent.

Data regarding the direct involvement of the MCH system in thermoregulatory processes are poor and are sometimes inconsistent among experiments. Apparently, the most important role that hypothalamic MCH seems to have in thermoregulation is to determine how energy expenditure takes part in the thermoregulatory process (Pereira-da-Silva et al. 2003). Experimental data show that MCH overexpression is triggered by exposure to a low ambient temperature (Pereira-da-Silva et al. 2003). A thermoregulatory effect observed in this ambient condition was the lack of reduction in rectal temperature when MCH was blocked, even after four days of cold exposure, in contrast to what happens in normal animals whose MCH is functioning normally (Pereira-da-Silva et al. 2003). Another important effect observed is the relationship between thermoregulation and metabolism, especially when thermogenesis is active. When rats are exposed to a low ambient temperature (i.e., +4 °C), they normally show an initial reduction in body weight, which soon recovers to a normal level within 2–3 days. This kind of weight recovery appears to be disturbed when the MCH system is not working properly; when the MCH system is blocked in rats exposed to a cold environment, they are not able to recover their normal body weight following the initial loss (Pereira-da-Silva et al. 2003).

Melanin-concentrating hormone neurons also have a role in controlling energy balance, acting directly on the brown adipose tissue. The intracerebroventricular administration of MCH induces a reduction in body temperature and brown fat uncoupling protein (UCP1; Ito et al. 2004). Consistently with this regulatory function, *ob/ob* mice lacking MCH present an increased activity of UCP1, and their body temperature is regulated and maintained to a higher level compared to *ob/ob* controls (Segal-Lieberman et al. 2003). As previously stated, a direct action of the orexinergic neurons from the lateral hypothalamus to the brown adipose tissue has been described; these neurons project to the Raphe Pallidus, a pontine structure containing the putative premotor autonomic neurons that activate the sympathetic brown adipose tissue nerve (Tupone et al. 2011). As far as MCH neurons are concerned, they do not appear to project directly to the Raphe Pallidus, while a multisynaptic projection to the brown adipose tissue has been described (Oldfield et al. 2002).

Melanin-concentrating hormone may act in neurotransmission through the involvement of two receptors: MCHR1 and MCHR2 (cf. Astrand et al. 2004). The former is present in all the species investigated so far and appears to be widely distributed in the central nervous system, consistently with its fundamental role in energy expenditure (Saito et al. 2001). The second receptor is expressed in humans but not in rodents (Tan et al. 2002).

The complex involvement of the MCH system in energy balance and thermoregulation may be observed in MCHR1 null mice, which eat more than their control littermates but present a lean phenotype. Normally, these coupled characteristics may coexist only when there is a sustained energy expenditure (Astrand et al. 2004). More specifically, the lack of MCHR1 induces a significant rise in core temperature in mice, mostly during only the dark period of the light-dark daily cycle (i.e., the activity period in mice); also, the drop in body temperature that normally follows fasting periods in these mice appears to be lower (Astrand et al. 2004). Consistently, when the daily oscillation of body temperature was analyzed, the difference normally shown was dampened in chronically MCH-infused mice (Martelli et al. 2012; Blessing et al. 2013), and this difference was almost completely the consequence of a temperature reduction during the active period of the animals (Glick et al. 2009).

The role of MCHR1 in the functional relationship between sleep regulation and thermoregulation appears to be significant: mice lacking MCHR1 present higher wakefulness and lower NREM sleep amounts. Regarding thermoregulation specifically in these experiments aimed at studying sleep regulation in MCHR1 knockout mice, body temperature was also seen to be higher compared to that of their wild-type control littermates, but, in contrast to previous data, the higher temperature was detected particularly during the light (rest) period of the day (Ahnaou et al. 2011). In any case, none of these modifications observed in mean body temperature, or in the cyclic changes through the day, affected the efficiency of either regulatory process: in fact, on one hand, the homeostatic rebound of sleep appeared to be normal in knockout mice, and, on the other hand, the increase in body temperature normally observed following a restraint stress protocol was also maintained (Ahnaou et al. 2011).

Antagonizing MCHR1 produces a higher basal metabolic rate, apparently through the increase of heat production. This is supported by the fact that, in animal models, the chronic administration of MCHR1 antagonists reduces body weight, without significantly interfering with food intake and motor activity (Clapham 2012). Therefore, the rise in body temperature observed when the MCHergic system is not working properly, albeit of a very low magnitude, is a process that appears to have important consequences in the basal metabolic regulation, possibly favoring a lean phenotype.

Furthermore, two anesthetics (ketamine and propofol) were studied in relation to MCH brain levels and wake-sleep modifications. For both drugs, an increase in MCH was shown, but there was a contrasting effect on wake-sleep architecture: ketamine augmented post-anesthetic wakefulness, while propofol induced higher levels of NREM sleep amount within the post-anesthetic period (Kushikata et al. 2016). It is well known that anesthetics commonly induce hypothermia (Bornkamp et al. 2016; Clark-Price 2015), and this suggests that the observed effects of these drugs on sleep regulation could be the consequence of the interaction between thermoregulation and sleep regulation (Amici et al. 2014; Cerri et al. 2017). Again, a chemogenetic-specific approach to both the excitation and disruption of MCH signaling showed that MCH induces REM sleep and also the maintenance of the correct circadian distribution of REM sleep episodes (Vetrivelan et al. 2016). Taking into consideration the core body temperature and basal metabolic rate, those results may be also interpreted on the grounds of an indirect effect arising from a disruption in the regulation of metabolism and also the dysregulation of sympathetic effects (Astrand et al. 2004).

To summarize, as shown in Fig. 1, the role of MCH in thermoregulatory processes appears to be of high complexity. First of all, it is important to consider the reciprocal functional interaction between MCH and orexin throughout the whole brain. The action of MCHergic neurotransmission on thermoregulation is mainly expressed through the correct and fine modulation of the basal metabolic rate and, as a consequence, of energy expenditure. When MCH acts on MCHR1, the metabolic rate is kept to a lower level, favoring energy storage and food intake. Due to the fact that most of the basal metabolism is spent for the maintenance of body temperature, MCH may be seen as a modulator that contributes to the thermoregulatory processes

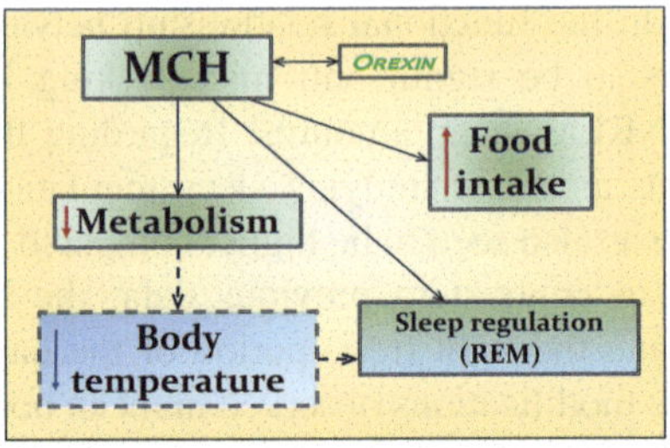

Fig. 1. Schematic representation of the modulatory effects that MCH exerts on thermoregulation. Green color represents direct functional modulation operated by MCH. Light-blue color represents possible indirect effects (also marked with dashed lines and arrows) that MCH may induce through metabolic modulation. See text for explanations

of the autonomic nervous system. Finally, the close link between thermoregulation and sleep regulation, in particular REM sleep regulation, suggests that, in addition to the known direct effect, MCH may also have an indirect role in regulating wake-sleep occurrence, through its action on the regulatory processes of body metabolism.

References

Ahnaou A, Dautzenberg FM, Huysmans H, Steckler T, Drinkenburg WH (2011) Contribution of melanin-concentrating hormone (MCH1) receptor to thermoregulation and sleep stabilization: evidence from MCH1 (−/−) mice. Behav Brain Res 218:42–50

Amici R, Bastianini S, Berteotti C, Cerri M, Del Vecchio F, Lo Martire V, Luppi M, Perez E, Silvani A, Zamboni G, Zoccoli G (2014) Sleep and bodily functions: the physiological interplay between body homeostasis and sleep homeostasis. Arch Ital Biol 152:66–78

Apergis-Schoute J, Iordanidou P, Faure C, Jego S, Schone C, Aitta-Aho T, Adamantidis A, Burdakov D (2015) Optogenetic evidence for inhibitory signaling from orexin to MCH neurons via local microcircuits. J Neurosci 35:5435–5441

Astrand A, Bohlooly-Y M, Larsdotter S, Mahlapuu M, Andersén H, Tornell J, Ohlsson C, Snaith M, Morgan DG (2004) Mice lacking melanin-concentrating hormone receptor 1 demonstrate increased heart rate associated with altered autonomic activity. Am J Physiol Regul Integr Comp Physiol 287: R749–R758

Blessing W, Mohammed M, Ootsuka Y (2013) Brown adipose tissue thermogenesis, the basic rest-activity cycle, meal initiation, and bodily homeostasis in rats. Physiol Behav 121:61–69

Bornkamp JL, Robertson S, Isaza NM, Harrison K, DiGangi BA, Pablo L (2016) Effects of anesthetic induction with a benzodiazepine plus ketamine hydrochloride or propofol on hypothermia in dogs undergoing ovariohysterectomy. Am J Vet Res 77:351–357

Brown RE, Basheer R, McKenna JT, Strecker RE, McCarley RW (2012) Control of sleep and wakefulness. Physiol Rev 92:1087–1187

Cabanac M (1996) The place of behavior in physiology. In: Fregly MJ, Blatteis CM (eds) Handbook of physiology. Environmental physiology (section 4). Oxford University Press, Oxford, pp 1523–1536

Cerri M (2017) The central control of energy expenditure: exploiting torpor for medical applications. Annu Rev Physiol 79:167–186

Cerri M, Luppi M, Tupone D, Zamboni G, Amici R (2017) REM sleep and endothermy: potential sites and mechanism of a reciprocal interference. Front Physiol 8:624
Clapham JC (2012) Central control of thermogenesis. Neuropharmacology 63:111–123
Clark-Price S (2015) Inadvertent perianesthetic hypothermia in small animal patients. Vet Clin North Am Small Anim Pract 45:983–994
Eban-Rothschild A, Giardino WJ, de Lecea L (2017) To sleep or not to sleep: neuronal and ecological insights. Curr Opin Neurobiol 44:132–138
Glick M, Segal-Lieberman G, Cohen R, Kronfeld-Schor N (2009) Chronic MCH infusion causes a decrease in energy expenditure and body temperature, and an increase in serum IGF-1 levels in mice. Endocrine 36:479–485
Grigg GC, Beard LA, Augee ML (2004) The evolution of endothermy and its diversity in mammals and birds. Physiol Biochem Zool 77:982–997
Ito M, Gomori A, Ishihara A, Oda Z, Mashiko S, Matsushita H, Yumoto M, Ito M, Sano H, Tokita S, Moriya M, Iwaasa H, Kanatani A (2004) Characterization of MCH-mediated obesity in mice. Am J Physiol Endocrinol Metab 284:E940–E945
Krauchi K (2007) The thermophysiological cascade leading to sleep initiation in relation to phase of entrainment. Sleep Med Rev 11:439–451
Kushikata T, Sawada M, Niwa H, Kudo T, Kudo M, Tonosaki M, Hirota K (2016) Ketamine and propofol have opposite effects on postanesthetic sleep architecture in rats: relevance to the endogenous sleep-wakefulness substances orexin and melanin-concentrating hormone. J Anesth 30:437–443
Lovegrove BG (2012) The evolution of endothermy in Cenozoic mammals: a plesiomorphic-apomorphic continuum. Biol Rev Camb Philos Soc 87:128–162
Luppi PH, Clément O, Fort P (2013) Paradoxical (REM) sleep genesis by the brainstem is under hypothalamic control. Curr Opin Neurobiol 23:786–792
Martelli D, Luppi M, Cerri M, Tupone D, Perez E, Zamboni G, Amici R (2012) Waking and sleeping following water deprivation in the rat. PLoS One 7:e46116
Monti JM, Torterolo P, Lagos P (2013) Melanin-concentrating hormone control of sleep-wake behavior. Sleep Med Rev 17:293–298
Morrison SF, Madden CJ, Tupone D (2014) Central neural regulation of brown adipose thermogenesis and energy expenditure. Cell Metab 19:741–756
Nespolo RF, Bacigalupe LD, Figueroa CC, Koteja P, Opazo JC (2011) Using new tools to solve an old problem: the evolution of endothermy in vertebrates. Trends Ecol Evol 26:414–423
Oldfield BJ, Giles ME, Watson A, Anderson C, Colvill LM, McKinley MJ (2002) The neurochemical characterisation of hypothalamic pathways projecting polysynaptically to brown adipose tissue in the rat. Neuroscience 110:515–526
Pereira-da-Silva M, Torsoni MA, Nourani HV, Augusto VD, Souza CT, Gasparetti AL, Carvalheira JB, Ventrucci G, Marcondes MC, Cruz-Neto AP, Saad MJ, Boschero AC, Carneiro EM, Velloso LA (2003) Hypothalamic melanin-concentrating hormone is induced by cold exposure and participates in the control of energy expenditure in rats. Endocrinology 144:4831–4840
Qu D, Ludwig DS, Gammeltoft S, Piper M, Pelleymounter MA, Cullen MJ, Mathes WF, Przypek J, Kanarek R, Maratos-Flier E (1996) A role for melanin-concentrating hormone in the central regulation of feeding behaviour. Nature 380:243–247
Saito Y, Cheng M, Leslie FM, Civelli O (2001) Expression of the melanin-concentrating hormone (MCH) receptor mRNA in the rat brain. J Comp Neurol 435:26–40
Satinoff E (1996) Behavioral thermoregulation in the cold. In: Fregly MJ, Blatteis CM (eds) Handbook of physiology. Environmental physiology (section 4). Oxford University Press, Oxford, pp 481–505
Segal-Lieberman G, Bradley RL, Kokkotou E, Carlson M, Trombly DJ, Wang X, Bates S, Myers MG Jr, Flier JS, Maratos-Flier E (2003) Melanin-concentrating hormone is a critical mediator of the leptin-deficient phenotype. Proc Natl Acad Sci U S A 100:10085–10090
Tan CP, Sano H, Iwaasa H, Pan J, Sailer AW, Hreniuk DL, Feighner SD, Palyha OC, Pong SS, Figueroa DJ, Austin CP, Jiang MM, Yu H, Ito J, Ito M, Guan XM, MacNeil DJ, Kanatani A, Van der Ploeg LH, Howard AD (2002) Melanin-concentrating hormone receptor subtypes 1 and 2: species-specific gene expression. Genomics 79:785–792

Tupone D, Madden CJ, Cano G, Morrison SF (2011) An orexinergic projection from perifornical hypothalamus to raphe pallidus increases rat brown adipose tissue thermogenesis. J Neurosci 31:15944–15955
Vetrivelan R, Kong D, Ferrari LL, Arrigoni E, Madara JC, Bandaru SS, Lowell BB, Lu J, Saper CB (2016) Melanin-concentrating hormone neurons specifically promote rapid eye movement sleep in mice. Neuroscience 336:102–113
Willmer P, Stone G, Johnston I (eds) (2000) Environmental physiology of animals. Blackwell Science, Oxford, pp 415–509

MCH, Sleep, and Neuroendocrine Functions

Carlos Eduardo Neves Girardi, Débora Cristina Hipólide, and Vânia D'Almeida

Abstract Melanin-concentrating hormone (MCH) is expressed in the central nervous system, primarily in the lateral hypothalamic area (LHA) (Bittencourt et al., J Comp Neurol, 319(2):218–45, 1992). A number of studies have suggested that MCH acts as an integrative peptide. Although its main function has been long regarded as the regulation of feeding behavior, its widespread fiber distribution throughout the brain suggests a role in a broader range of functions, especially metabolism, and reproductive and parental behavior, besides a putative involvement in the processing of rewarding stimuli. MCH also participates in arousal, being an important hormone related to sleep-wake behavior. In this chapter, we will discuss the role of MCH in behavioral, metabolic, and neuroendocrine functions.

1 MCH and Motivated Behavior Toward Food

A classical experiment showing that self-administration of an electrical current is elicited when electrodes are placed within the lateral hypothalamus (LH) (Olds 1958) has suggested that neurons within this region may regulate behavior toward rewarding stimuli. Recent findings have identified orexin and MCH, intrinsic LH peptides, as important regulators of such reward-directed behavior. Their signaling in the LH has been implicated in several behaviors, including its response to food and drugs of abuse, indicating an important role in the reward pathway of the brain (DiLeone et al. 2003).

Interestingly, a pronounced amount of MCH receptor (MCH-R) mRNA was found in the nucleus accumbens shell (NAccSh) (Hervieu et al. 2000; Saito et al. 2001). The NAccSh is one of the important brain nodes that belong to the motivation

C. E. N. Girardi
Department of Psychobiology, Universidade Federal de São Paulo, São Paulo, Brazil

Programa de Pós-Graduação em Psicossomática, Universidade Ibirapuera, São Paulo, Brazil

D. C. Hipólide · V. D'Almeida (✉)
Department of Psychobiology, Universidade Federal de São Paulo, São Paulo, Brazil

S. R. Pandi-Perumal et al. (eds.), *Melanin-Concentrating Hormone and Sleep*,
https://doi.org/10.1007/978-3-319-75765-0_8

and reward circuitry, which also includes the prefrontal cortex, the striatum, and the ventral tegmental area (VTA). The involvement of MCH in rewarding stimuli is supported by the existence of MCHergic projections from the LH to the nucleus accumbens (Bittencourt et al. 1992; Haemmerle et al. 2015).

MCH is thought to be involved in the control of food intake and to play a potential role in obesity (Della-Zuana et al. 2002). The relevance of MCH to feeding behavior was evidenced by the fact that MCH receptor knockout mice (*Mchr1−/−*), besides being lean, hyperactive, and having altered metabolism, are also hyperphagic when maintained on regular chow (Marsh et al. 2002).

Evidence that feeding behavior depends on MCH was also provided by Georgescu et al. (2005), who showed that MCH injections directly into the NAccSh increase chow consumption, while the MCHR1 antagonist exerts the opposite effect. The contribution of the MCH signal in the NAccSh to the motivational aspects of feeding behavior was shown by Mul et al. (2011) using the *Pmch−/−* rat, which is characterized by a chronic loss of the MCH precursor, Pmch. Rats that chronically lack *Pmch* during development have a decreased food intake due to a reduction in meal size and also show a reduced response to high-fat food reinforcement as adults.

Additionally, self-administration of sucrose, which is caloric and sweet, was reduced by MCHR1 antagonist, while saccharin self-administration, which provides only sweetness but no caloric intake, is not affected by MCHR1 antagonism (Karlsson et al. 2012), indicating that it is the caloric content of food that is signaled by MCH. In the same study, the cue-reinstatement of lever-pressing to obtain sucrose was also decreased by MCH antagonism, confirming the involvement of MCH signaling in the rewarding aspects of sucrose intake.

The specificity for the effects of MCH on the rewarding properties of food intake was further explored using optogenetics in mice. Mice show an innate preference for sucrose over saccharin (Domingos et al. 2011). Stimulation of MCH neurons using optogenetics reverted such innate sucrose preference, stimulating dopaminergic release in the striatum, and this effect is apparently not dependent on taste, since the same effect was observed in mice insensitive to sweetness (Domingos et al. 2013).

From the abovementioned evidence, it becomes clear that the reward pathway plays a fundamental role in the motivational and incentive-related aspects of food consumption. Because feeding is intrinsically related to the control of energy balance, we will further explore the relationship between MCH and metabolic aspects of feeding.

2 MCH, Feeding, and Energy Homeostasis

Additionally to feeding, the potential role of hypothalamic MCH as an important effector of nutritional homeostasis was likewise suggested by the findings from Qu et al. (1996), showing that its expression is increased by fasting and in the *ob/ob* mouse. Maratos-Flier et al. (1998) were the first to develop a deficient MCH mouse model. MCH-deficient mice have reduced body weight due to hypophagia accompanied by decreased energy expenditure despite the reduced mRNA levels of leptin and pro-opiomelanocortin (POMC), which results in obesity when MCH is present. These data support the idea that MCH plays a central role in the control of food intake and energy balance (Maratos-Flier et al. 1998). Subsequently, a number of other studies have suggested a MCH obesogenic effect that is related to its effects on the central control of food intake and in the regulation of energy expenditure (Chen et al. 2002; Marsh et al. 2002; Segal-Lieberman et al. 2003).

The regulatory role of MCH in consummatory behaviors was largely demonstrated by studies in rodents in which knocking out the preproMCH (*Pmch*) gene resulted in reduced body weight following reduced feeding (Chen et al. 2002), and similar results were also found after the deletion of MCH receptor 1 (*Mchr 1*) (Kokkotou et al. 2005; Mashiko et al. 2005), pointing to an important role of MCH as a regulator of feeding and energy balance (Chen et al. 2002).

Meal size is increased after MCH administration and decreased after MCHR1 antagonist injections (Santollo and Eckel 2008; Kowalski et al. 2006). Interestingly, injections of MCH agonists reduce fat oxidation significantly only when administered intracerebrovascularly, supporting the importance of the MCH system in the control of substrate oxidation (Guesdon et al. 2009).

Despite the scarcity of data concerning MCH receptor 2 (MCHR2), since it is not functional in rodents, it has been reported to have a role in the modulation of human feeding behavior (Rodriguez et al. 2001; Wang et al. 2001). However, in a case-control family-based study performed by Ghoussaini et al. (2007) in a French Caucasian population, the authors concluded that MCHR2 is not a major player in the development of obesity in childhood and adulthood. They only found a minor effect of the A76A T/C *MCHR2* single nucleotide polymorphism on food intake abnormalities, probably due to the higher stability of the *MCHR2* mRNA affecting receptor density and the orexigenic effect of MCH.

It is interesting to note that the orexigenic function of MCH is related to inputs in the forebrain, and its response to palatable foods depends on the nutritional value and hedonic characteristics of the food (Naufahu et al. 2013). Moreover, the MCH system is remarkable because it is able to control (or plays a key role in controlling) energy balance despite widely varying levels of food intake. Increased metabolic rate, locomotor activity, and energy expenditure are frequently reported when the MCH system is challenged by dietary, genetic, or pharmacological manipulations (Maratos-Flier et al. 1998; Marsh et al. 2002; Segal-Lieberman et al. 2003; Kokkotou et al. 2005; Jeon et al. 2006; Naufahu et al. 2013).

Different MCH concentrations result in hyperphagia (Chen et al. 2002; Marsh et al. 2002) or hypophagia (Mashiko et al. 2005; Kowalski et al. 2006). In addition, when distinct models have been used, such as acute and chronic rodent models of MCH manipulations, changes in food intake and weight gain have been shown but with important discrepancies in some aspects. Moreover, pharmacological MCHR1 inactivation results in a more pronounced feeding behavior phenotype compared to genetic models in which the energy balance is more affected (Pissios 2009).

It is important to point out that deficiencies in MCH receptor 1 or in MCH precursor (Pmch) result in lean animals, but the mechanisms responsible for these phenotypes are different: *Mchr1−/−* mice are hyperactive whereas *Pmch−/−* mice are hypophagic and have increased metabolic rates (Marsh et al. 2002).

Pissios et al. (2007) demonstrated for the first time the presence of MCH neuropeptide and its receptor, MCHR1, in cultures of mouse and human islets, suggesting a role for this system in the regulation of islet function. In fact, these authors have reported islet hyperplasia in animals that overexpressed MCH, as well as reduced insulin levels and normal glucose tolerance in *Mch−/−* mice, which are lean and hyperactive compared with their wild littermates. Additionally, *Mch−/−* mice presented an increased metabolic rate, which the authors attributed to higher thermogenesis, a consequence of increased activity of the sympathetic nervous system (Jeon et al. 2006). These *Mch−/−* mice also presented an interesting phenotype since their beta cells cannot expand even in the presence of a high-fat diet, a compensatory mechanism that is evident in wild-type animals (Hull et al. 2005). Moreover, the direct effects on beta cells suggest that the MCH system has peripheral actions besides its central effects in regulating glucose metabolism (Pissios 2009; Pissios et al. 2007).

These energy expenditure variations in *Mch −/−* mice were investigated by Astrand et al. (2004). The authors compared energy expenditure to variations in heart rate and blood pressure and the autonomic mechanisms related to them and found that besides the increased locomotor activity which occurs in the dark phase, the increased energy expenditure occurred in the dark and in the light phases, indicating that this increase in energy expenditure is not dependent on locomotor activity and could be a consequence of a higher basal metabolic rate or involuntary energy expenditure. Interestingly, this increase in energy expenditure occurred concomitantly with an increased heart rate in both the light and dark phases and was not related to increased locomotor activity, suggesting that they are regulated by the autonomic nervous system (Astrand et al. 2004).

In a recent study, Naufahu et al. (2017) developed a radioimmunoassay method for the measurement of circulating MCH, in order to evaluate the role of this neuropeptide in human plasma. The method was sufficiently sensitive and specific, but no significant correlations with glucose, insulin, or leptin concentrations were found, indicating that the role of systemic MCH is probably different to that of MCH in the hypothalamus (Naufahu et al. 2017).

It is interesting to note that sleep disruption affects MCH secretion and *Pmch* expression in rodents. The impact of such changes observed in sleep-deprived individuals could be one of the reasons for metabolic alterations (e.g., obesity and metabolic syndrome), **especially considering the role of the hypothalamus as the master integrator of energy imbalance**. These observations highlight MCH as a potential target for the development of therapeutic approaches to metabolic disturbances.

3 MCH and Reproductive Functions

The physiological functions of MCH are not restricted to energy metabolism. MCH also appears to play a role in reproductive function. A great body of evidence has been gathered concerning the reciprocal relationship between MCH and physiological mechanisms of reproductive and parental behavior. Most studies point to a pivotal role of MCH on the modulation of gonadotropin release. Neuroanatomical characterization in rodents and humans further supports the notion that MCH plays an important role in mediating the integration between systems that regulate energy balance and reproductive function. MCH-immunoreactive neurons are found in the hypothalamus, which in turn send generalized projections throughout the central nervous system, including multiple areas relevant to reproductive physiology, such as the anteroventral periventricular nucleus (AVPV), the medial preoptic area (mPOA), the anterior hypothalamic area (AHA), and the median eminence (ME) (Bittencourt et al. 1992).

Integration between metabolic and reproductive function is complex, and many nutritional cues play a role in bringing energy status information to the hypothalamic gonadotrophin-releasing hormone (GnRH) neuronal network, the primary driver for reproductive function (Evans and Anderson 2012). MCH projections abundantly innervate the preoptic region, establishing contact with the cell bodies of GnRH neurons (Williamson-Hughes et al. 2005; Ward et al. 2009; Wu et al. 2009; Skrapits et al. 2015) and the mediobasal hypothalamus, where the hypophysiotropic GnRH terminals are found (Doyle et al. 2015; Ward et al. 2009).

The functional interaction between MCH and GnRH appears to be mediated by MCH-R1 receptors. MCHR1 mRNA is co-expressed in about 50–55% of GnRH neurons (Williamson-Hughes et al. 2005), and the identification of MCHR1 in human and rodent pituitary tissue confirms that MCH exerts a direct function in the pituitary gland (Takahashi et al. 2001; Segal-Lieberman et al. 2003).

Other results support the hypothesis that MCH-GnRH interaction underlies MCH modulation on luteinizing hormone (LH) release. Initially, a few studies have shown that MCH has a stimulatory role in LH secretion when injected into the hypothalamic mPOA or median eminence (Gonzalez et al. 1997; Murray et al. 2000); however, more recently it has been observed that this modulatory effect of MCH on LH release might also be influenced by endocrine status (Naufahu et al. 2013).

Steroidal hormones seem to exert a reciprocal influence on MCH. Experimental estradiol benzoate supplementation in ovariectomized rats impairs the orexigenic effects of MCH, compared to male and to non-supplemented ovariectomized females, an effect which could also be extended to other physiological conditions (Messina et al. 2006; Santollo and Eckel 2008, 2013).

Steroidal influence on MCH is also observed at a structural level. For instance, during diestrus and proestrus, there is an increase in MCH fiber density, estimated by the number of MCH-immunoreactive fibers in the external layer of the ME, which reach a number comparable to males or to females in estrus (Chiocchio et al. 2001; Gallardo et al. 2004).

The involvement of MCH in parental behavior and lactation is also evident. The mPOA is a key region involved in the circuitry of maternal behavior, and MCH is expressed in the mPOA during lactation (Knollema et al. 1992), in a fashion whereby a progressive increase in mPOA MCH expression parallels a progressive decrease in the expression of maternal behavior, suggesting that MCH might exert an inhibitory role on maternal behavior in the late phase of the postpartum period (Benedetto et al. 2014), while a pro-maternal behavior effect is observed during the early postpartum phase (Benedetto et al. 2014; Alachkar et al. 2016).

Additional evidence for the involvement of MCH in maternal care comes from studies using genetic ablation and antagonism of MCHR1. *Mchr1* knockout female mice show impaired nesting, pup retrieval, maternal aggression toward intruder males, milk production, and prolactin mRNA levels, and similar results were obtained with acute pharmacological inhibition of the MCH system (Alachkar et al. 2016). Also, the higher litter postnatal mortality observed in *Pmch −/−* mice has been attributed to diminished maternal capability to integrate olfactory stimuli (Alachkar et al. 2016).

It is clear that MCH exerts its influence on sexual behavior through neuroendocrine activity, and, given the pivotal role of MCH in the sleep-wake cycle and energy balance, we can further speculate about how metabolism and reproduction converge. As the nutritional status of the organism directly influences reproductive success and the ability to respond to energy demands, timing seems to be central to the MCH-sexual behavior relationship, due to the importance of engaging in reproductive activities during periods when energy balance is not negative and, thus, maximizing their chance of surviving and successful reproduction.

References

Alachkar A et al (2016) Inactivation of the melanin concentrating hormone system impairs maternal behavior. Eur Neuropsychopharmacol 26(11):1826–1835. http://linkinghub.elsevier.com/retrieve/pii/S0924977X16301778. Accessed 17 Aug 2017

Astrand A et al (2004) Mice lacking melanin-concentrating hormone receptor 1 demonstrate increased heart rate associated with altered autonomic activity. Am J Physiol Regul Integr Comp Physiol 287(4):R749–R758. http://ajpregu.physiology.org/cgi/doi/10.1152/ajpregu.00134.2004. Accessed 16 Aug 2017

Benedetto L et al (2014) Melanin-concentrating hormone in the medial preoptic area reduces active components of maternal behavior in rats. Peptides 58:20–25. http://linkinghub.elsevier.com/retrieve/pii/S0196978114001582. Accessed 17 Aug 2017

Bittencourt JC et al (1992) The melanin-concentrating hormone system of the rat brain: an immuno- and hybridization histochemical characterization. J Comp Neurol 319(2):218–245. http://doi.wiley.com/10.1002/cne.903190204. Accessed 13 Aug 2017

Chen Y et al (2002) Targeted disruption of the melanin-concentrating hormone receptor-1 results in hyperphagia and resistance to diet-induced obesity. Endocrinology 143(7):2469–2477. https://academic.oup.com/endo/article-lookup/doi/10.1210/endo.143.7.8903. Accessed 13 Aug 2017

Chiocchio SR et al (2001) Melanin-concentrating hormone stimulates the release of luteinizing hormone-releasing hormone and gonadotropins in the female rat acting at both median eminence and pituitary levels. Biol Reprod 64(5):1466–1472. http://www.ncbi.nlm.nih.gov/pubmed/11319153 Accessed 17 Aug 2017

Della-Zuana O et al (2002) Acute and chronic administration of melanin-concentrating hormone enhances food intake and body weight in Wistar and Sprague–Dawley rats. Int J Obes 26 (10):1289–1295. http://www.ncbi.nlm.nih.gov/pubmed/12355323. Accessed 7 Jul 2017

DiLeone RJ, Georgescu D, Nestler EJ (2003) Lateral hypothalamic neuropeptides in reward and drug addiction. Life Sci 73(6):759–768. http://www.ncbi.nlm.nih.gov/pubmed/12801597. Accessed 7 Jul 2017

Domingos AI et al (2011) Leptin regulates the reward value of nutrient. Nat Neurosci 14 (12):1562–1568. http://www.nature.com/doifinder/10.1038/nn.2977. Accessed 13 Aug 2017

Domingos AI et al (2013) Hypothalamic melanin concentrating hormone neurons communicate the nutrient value of sugar. eLife 2:e01462. http://elifesciences.org/lookup/doi/10.7554/eLife.01462. Accessed 13 Aug 2017

Doyle OM et al (2015) The cortical thickness phenotype of individuals with DISC1 translocation resembles schizophrenia. J Clin Investig 125(9):3714–3722. http://www.ncbi.nlm.nih.gov/pubmed/26301809. Accessed 13 Jul 2017

Evans JJ, Anderson GM (2012) Balancing ovulation and anovulation: integration of the reproductive and energy balance axes by neuropeptides. Hum Reprod Update 18(3):313–332. https://academic.oup.com/humupd/article-lookup/doi/10.1093/humupd/dms004. Accessed 16 Aug 2017

Gallardo MGP, Chiocchio SR, Tramezzani JH (2004) Changes of melanin-concentrating hormone related to LHRH release in the median eminence of rats. Brain Res 1030(1):152–158. http://linkinghub.elsevier.com/retrieve/pii/S0006899304016592. Accessed 17 Aug 2017

Georgescu D et al (2005) The hypothalamic neuropeptide melanin-concentrating hormone acts in the nucleus accumbens to modulate feeding behavior and forced-swim performance. J Neurosci 25(11):2933–2940. http://www.ncbi.nlm.nih.gov/pubmed/15772353. Accessed 7 Jul 2017

Ghoussaini M et al (2007) Genetic study of the melanin-concentrating hormone receptor 2 in childhood and adulthood severe obesity. J Clin Endocrinol Metab 92(11):4403–4409. https://academic.oup.com/jcem/article-lookup/doi/10.1210/jc.2006-2316. Accessed 16 Aug 2017

Gonzalez MI, Baker BI, Wilson CA (1997) Stimulatory effect of melanin-concentrating hormone on luteinising hormone release. Neuroendocrinology 66(4):254–262. http://www.ncbi.nlm.nih.gov/pubmed/9349659. Accessed 17 Aug 2017

Guesdon B et al (2009) Effects of intracerebroventricular and intra-accumbens melanin-concentrating hormone agonism on food intake and energy expenditure. Am J Physiol Regul Integr Comp Physiol 296(3):R469–R475. http://ajpregu.physiology.org/cgi/doi/10.1152/ajpregu.90556.2008. Accessed 13 Aug 2017

Haemmerle CAS, Campos AMP, Bittencourt JC (2015) Melanin-concentrating hormone inputs to the nucleus accumbens originate from distinct hypothalamic sources and are apposed to GABAergic and cholinergic cells in the long-Evans rat brain. Neuroscience 289:392–405. http://linkinghub.elsevier.com/retrieve/pii/S0306452215000627. Accessed 13 Aug 2017

Hervieu GJ et al (2000) The distribution of the mRNA and protein products of the melanin-concentrating hormone (MCH) receptor gene, slc-1, in the central nervous system of the rat. Eur J Neurosci 12(4):1194–1216. http://www.ncbi.nlm.nih.gov/pubmed/10762350. Accessed 13 Aug 2017

Hull RL et al (2005) Dietary-fat-induced obesity in mice results in beta cell hyperplasia but not increased insulin release: evidence for specificity of impaired beta cell adaptation. Diabetologia 48(7):1350–1358. http://link.springer.com/10.1007/s00125-005-1772-9. Accessed 16 Aug 2017

Jeon JY et al (2006) MCH−/− mice are resistant to aging-associated increases in body weight and insulin resistance. Diabetes 55(2):428–434. http://www.ncbi.nlm.nih.gov/pubmed/16443777. Accessed 16 Aug 2017

Karlsson C et al (2012) Melanin-concentrating hormone receptor 1 (MCH1-R) antagonism: reduced appetite for calories and suppression of addictive-like behaviors. Pharmacol Biochem Behav 102(3):400–406. http://www.ncbi.nlm.nih.gov/pubmed/22705492. Accessed 7 Jul 2017

Knollema S et al (1992) Novel hypothalamic and preoptic sites of prepro-melanin-concentrating hormone messenger ribonucleic acid and peptide expression in lactating rats. J Neuroendocrinol 4 (6):709–717. http://doi.wiley.com/10.1111/j.1365-2826.1992.tb00222.x. Accessed 17 Aug 2017

Kokkotou E et al (2005) Mice with MCH ablation resist diet-induced obesity through strain-specific mechanisms. Am J Physiol Regul Integr Comp Physiol 289(1):R117–R124. http://ajpregu.physiology.org/cgi/doi/10.1152/ajpregu.00861.2004. Accessed 13 Aug 2017

Kowalski TJ et al (2006) Effects of a selective melanin-concentrating hormone 1 receptor antagonist on food intake and energy homeostasis in diet-induced obese mice. Eur J Pharmacol 535 (1–3):182–191. http://linkinghub.elsevier.com/retrieve/pii/S0014299906000859. Accessed 13 Aug 2017

Maratos-Flier E et al (1998) Mice lacking melanin-concentrating hormone are hypophagic and lean. Nature 396(6712):670–674. http://www.ncbi.nlm.nih.gov/pubmed/9872314. Accessed 13 Aug 2017

Marsh DJ et al (2002) Melanin-concentrating hormone 1 receptor-deficient mice are lean, hyperactive, and hyperphagic and have altered metabolism. Proc Natl Acad Sci 99(5):3240–3245. http://www.ncbi.nlm.nih.gov/pubmed/11867747. Accessed 7 Jul 2017

Mashiko S et al (2005) Antiobesity effect of a melanin-concentrating hormone 1 receptor antagonist in diet-induced obese mice. Endocrinology 146(7):3080–3086. http://www.ncbi.nlm.nih.gov/pubmed/15817671. Accessed 13 Aug 2017

Messina MM et al (2006) Estradiol decreases the orexigenic effect of melanin-concentrating hormone in ovariectomized rats. Physiol Behav 88(4–5):523–528. http://www.ncbi.nlm.nih.gov/pubmed/16793070. Accessed 17 Aug 2017

Mul JD et al (2011) Chronic loss of melanin-concentrating hormone affects motivational aspects of feeding in the rat. PLoS One 6(5):e19600. http://www.ncbi.nlm.nih.gov/pubmed/21573180. Accessed 7 Jul 2017

Murray JF et al (2000) Melanin-concentrating hormone, melanocortin receptors and regulation of luteinizing hormone release. J Neuroendocrinol 12(3):217–223. http://www.ncbi.nlm.nih.gov/pubmed/10718917. Accessed 17 Aug 2017

Naufahu J, Cunliffe AD, Murray JF (2013) The roles of melanin-concentrating hormone in energy balance and reproductive function: are they connected? Reproduction 146(5):R141–R150. http://www.reproduction-online.org/cgi/doi/10.1530/REP-12-0385. Accessed 16 Aug 2017

Naufahu J et al (2017) Melanin-concentrating hormone in peripheral circulation in the human. J Endocrinol 232(3):513–523. http://joe.endocrinology-journals.org/lookup/doi/10.1530/JOE-16-0240. Accessed 16 Aug 2017

OLDS J (1958) Self-stimulation of the brain; its use to study local effects of hunger, sex, and drugs. Science 127(3294):315–324. http://www.ncbi.nlm.nih.gov/pubmed/13506579. Accessed 11 Aug 2017

Pissios P (2009) Animals models of MCH function and what they can tell us about its role in energy balance. Peptides 30(11):2040–2044. http://www.ncbi.nlm.nih.gov/pubmed/19447150. Accessed 16 Aug 2017

Pissios P et al (2007) Melanin concentrating hormone is a novel regulator of islet function and growth. Diabetes 56(2):311–319. http://diabetes.diabetesjournals.org/cgi/doi/10.2337/db06-0708. Accessed 16 Aug 2017

Qu D et al (1996) A role for melanin-concentrating hormone in the central regulation of feeding behaviour. Nature 380(6571):243–247. http://www.ncbi.nlm.nih.gov/pubmed/8637571. Accessed 7 Jul 2017

Rodriguez M et al (2001) Cloning and molecular characterization of the novel human melanin-concentrating hormone receptor MCH2. Mol Pharmacol 60(4):632–639. http://www.ncbi.nlm.nih.gov/pubmed/11562423. Accessed 16 Aug 2017

Saito Y et al (2001) Expression of the melanin-concentrating hormone (MCH) receptor mRNA in the rat brain. J Comp Neurol 435(1):26–40. http://www.ncbi.nlm.nih.gov/pubmed/11370009. Accessed 7 Jul 2017

Santollo J, Eckel LA (2008) The orexigenic effect of melanin-concentrating hormone (MCH) is influenced by sex and stage of the estrous cycle. Physiol Behav 93(4–5):842–850. http://linkinghub.elsevier.com/retrieve/pii/S0031938407004878. Accessed 13 Aug 2017

Santollo J, Eckel LA (2013) Oestradiol decreases melanin-concentrating hormone (MCH) and MCH receptor expression in the hypothalamus of female rats. J Neuroendocrinol 25 (6):570–579. http://doi.wiley.com/10.1111/jne.12032. Accessed 16 Aug 2017

Segal-Lieberman G et al (2003) Melanin-concentrating hormone is a critical mediator of the leptin-deficient phenotype. Proc Natl Acad Sci 100(17):10085–10090. http://www.pnas.org/cgi/doi/10.1073/pnas.1633636100. Accessed 13 Aug 2017

Skrapits K et al (2015) Lateral hypothalamic orexin and melanin-concentrating hormone neurons provide direct input to gonadotropin-releasing hormone neurons in the human. Front Cell Neurosci 9:348. http://www.ncbi.nlm.nih.gov/pubmed/26388735. Accessed 4 Jul 2017

Takahashi K et al (2001) Expression of melanin-concentrating hormone receptor messenger ribonucleic acid in tumor tissues of pheochromocytoma, ganglioneuroblastoma, and neuroblastoma. J Clin Endocrinol Metab 86(1):369–374. https://academic.oup.com/jcem/article-lookup/doi/10.1210/jcem.86.1.7158. Accessed 17 Aug 2017

Wang S et al (2001) Identification and pharmacological characterization of a novel human melanin-concentrating hormone receptor, mch-r2. J Biol Chem 276(37):34664–34670. http://www.jbc.org/lookup/doi/10.1074/jbc.M102601200. Accessed 16 Aug 2017

Ward DR et al (2009) Innervation of gonadotropin-releasing hormone neurons by peptidergic neurons conveying circadian or energy balance information in the mouse. PLoS One 4(4): e5322. http://dx.plos.org/10.1371/journal.pone.0005322. Accessed 17 Aug 2017

Williamson-Hughes PS, Grove KL, Smith MS (2005) Melanin concentrating hormone (MCH): a novel neural pathway for regulation of GnRH neurons. Brain Res 1041(2):117–124. http://linkinghub.elsevier.com/retrieve/pii/S0006899304017950. Accessed 16 Aug 2017

Wu M et al (2009) Melanin-concentrating hormone directly inhibits GnRH neurons and blocks kisspeptin activation, linking energy balance to reproduction. Proc Natl Acad Sci 106 (40):17217–17222. http://www.ncbi.nlm.nih.gov/pubmed/19805188. Accessed 4 Jul 2017

Melanin-Concentrating Hormone: Role in Nursing and Sleep in Mother Rats

Luciana Benedetto, Pablo Torterolo, and Annabel Ferreira

Abstract In mammals, the postpartum female undergoes the most important physiological and behavioral changes in life, which allow orchestrating two essential behaviors for survival: nursing and sleep. Although the melanin-concentrating hormone (MCH) is mainly found within the posterolateral hypothalamus and incerto-hypothalamic area, during lactation this neuropeptide is also expressed in the preoptic area (POA). Remarkably, this brain area controls key components not only of the maternal behavior repertoire but also is involved in the regulation of sleep and wakefulness. In this sense, when MCH is microinjected into the POA, this neuropeptide is capable to reduce the motivational aspects of maternal behavior in postpartum rats while increases sleep in male rats. This effect seems to oppose to one of the dopaminergic systems that promotes wakefulness while in postpartum rats stimulates motivational components of maternal behavior. How the MCHergic system controls maternal behavior and sleep within the POA is still an unresolved question.

In this chapter, we provide neuroanatomical and neurochemical evidences showing that MCHergic and dopaminergic systems interact within the medial POA (mPOA) to regulate maternal behavior and sleep. We suggest that the interplay among these and other neurotransmitter/neuromodulators modulates mother's physiology and behavior assuring not only pups' nutrition and development but also the mother's needs for rest and sleep during this highly demanding period of life. Finally, we discuss some useful directions for future research and some issues yet to be explored.

L. Benedetto (✉) · P. Torterolo
Department of Physiology, School of Medicine, University of the Republic, Montevideo, Uruguay
e-mail: lbenedet@fmed.edu.uy

A. Ferreira
Facultad de Ciencias, Sección de Fisiología y Nutrición, Universidad de la República, Montevideo, Uruguay

S. R. Pandi-Perumal et al. (eds.), *Melanin-Concentrating Hormone and Sleep*,
https://doi.org/10.1007/978-3-319-75765-0_9

1 Introduction

The postpartum period is a highly physically and emotionally demanding period for the mammalian female, characterized by a huge range of physiological and behavioral adaptations that allow the adequate development and survival of the offspring. We recently showed that the mother rat can sleep and nurse the pups at the same time, making possible to fulfill the pups' needs of nutrition together with the maintenance of an adequate homeostasis of her sleep physiology (Benedetto et al. 2017b). Interestingly, the melanin-concentrating hormone (MCH), a neuropeptide synthesized by neurons of the lateral hypothalamus and incerto-hypothalamic areas that project throughout the central nervous system (CNS) (Bittencourt et al. 1992), has been involved in the control of maternal behavior (Alachkar et al. 2016; Benedetto et al. 2014) and sleep (Torterolo et al. 2011). Thus, MCH microinjections into the medial preoptic area (mPOA), an essential area in the control of maternal behavior, reduce the active components of maternal behavior (Benedetto et al. 2014). Also, local delivery of MCH into the ventrolateral preoptic area (VLPO), a key player in the control of sleep, promotes NREM sleep (Benedetto et al. 2013).

In the present chapter, we describe the maternal and sleep behaviors of the lactating rat as well as the interplay between the MCHergic and dopaminergic system in specific brain areas related to the control of both behaviors. As most studies of the neurobiological basis of maternal behavior and sleep were done in rodents, we will focus on these animals.

2 Maternal Behavior

In the rat, newborn pups are altricial, that means that they are highly immature and totally dependent on the maternal care and nutrition for development and survival. They are born with eyes and ears closed (Fig. 1), are unable to regulate their own temperature, and are incapable of urinating and defecating without the anogenital stimulation provided from the mother (Fig. 2b) (Numan and Insel 2003). These characteristics of the pups require a constant maternal attention and care resulting in a highly demanding stage with important metabolic costs for the mother (Thornburg et al. 2006). Thus, the mother rat provides not only the nutritional requirements to the offspring through lactation but also care and protection, contributing to the establishment of the maternal-infant bonding which is crucial for the development and survival of the young (Numan and Insel 2003). As newborn rats grow up, the mother will be continuously accommodating her own physiology and behavior to the pups' requirements.

To fulfill the pups' needs, mother rats develop a wide variety of behaviors termed as maternal behavior that is aimed to provide food, heat, shelter, cleaning, nourishment, and affect to the offspring (Reisbick et al. 1975; Rosenblatt 1975; Pereira 2016). In rats, if the mother is reunited with the pups after a certain time of separation, she will display a sequence of maternal behaviors. She will transport

Fig. 1. Altricial rat pup. Note that the pup rat is born hairless with eyes and ears closed

materials to build a nest, retrieve the pups into it, and lick and rearrange them in the nest (Stern 1989) (Fig. 2a, b). As these behaviors precede and promote mother-infant contact, they are known as pronurturant or active behaviors (Hansen et al. 1991b; Stern 1989). Following the reunion of the pups, the mother stands over them (Fig. 2c), and with sufficient ventral stimulation, she will adopt quiescent nursing postures, referred as nurturant, passive, or nursing behaviors (Hansen et al. 1991a; Stern and Johnson 1990) (Fig. 2d). Nursing is the behavior that defines us as mammals and is present in all mammalian species. However, it presents a wide variability among species, from the rabbit that nurses a few minutes once a day to the rat that nurses during most of the day in the early period of lactation (Gonzalez-Mariscal et al. 2016; Grota and Ader 1969; Zarrow et al. 1965).

In the mother rat, upright crouching is the most typical nursing posture, also known as kyphosis (Stern and Johnson 1990). This posture is elicited by the sensory stimulation of the mother's ventral area provided by a sufficient number (at least four) of pups (Neville 2006; Stern and Johnson 1990; Wakerley 1996). In this posture, the dam stands still over the litter, in high- or low-arched back posture with its legs rigidly splayed (Fig. 2d). Only during quiescent nursing postures milk ejection occurs (Voloschin and Tramezzani 1979).

The postpartum period of the rat lasts approximately three weeks until weaning during which the mother continuously adapts her maternal responses to pups' needs. Thus, during the first days after parturition, mother rats spend 85% of the day with their pups, rarely leaving the nest. While pups grow older, maternal behavior diminishes toward weaning (Numan 1994; Pereira 2016; Rheingold 1963; Rosenblatt et al. 1985). Both active and nursing behaviors decline significantly from the 10th–12th days onward (Grota and Ader 1969; Pereira 2016; Reisbick et al. 1975).

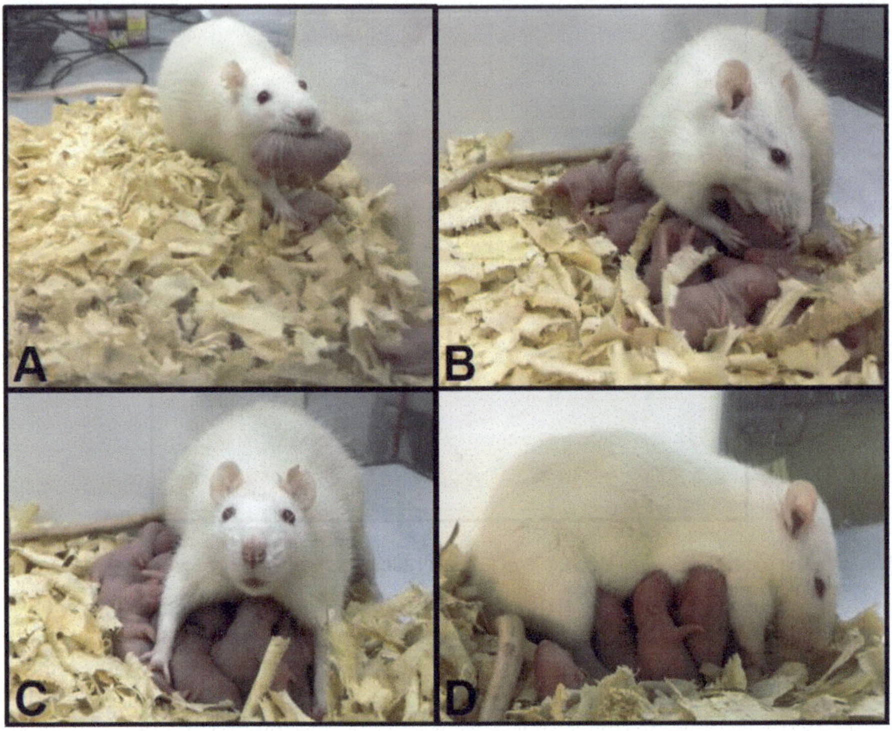

Fig. 2. Active and passive maternal behaviors in the rat. (**a**) Transporting or retrieving a pup. (**b**) Licking a pup. (**c**) Hovering over the pups. (**d**) Nursing the litter

3 Sleep During Postpartum Period

As mentioned, nursing is one of the most energy-consuming stages for the female (Krasnow and Steiner 2006; Thornburg et al. 2006; Zhao et al. 2010). In contrast, sleep is necessary for the conservation and restoration of energy. Although both processes are essential for the survival of the individual or the offspring, the strategies to reconcile both behaviors vary among species. Recently, we demonstrated that the mother rat can nurse and sleep at the same time (Benedetto et al. 2017b) (see Fig. 3). Thus, during the low upright crouching posture (also known as low kyphosis, the most common nursing posture), mother rats mostly sleep, particularly in NREM sleep, both at the light and dark phases of the cycle (see Fig. 4).

Remarkably, while suckling from the pups is sufficient to stimulate the upright crouching posture in the mother rat, this stimulus has to be preceded by a NREM sleep episode for milk ejection occurrence (Lincoln et al. 1980; Sutherland et al. 1987; Voloschin and Tramezzani 1979). In fact, sleep deprivation decreases lactation and impairs pups' weight gain (Voloschin and Tramezzani 1979). This is not the case for other animals, such as rabbits, where suckling is associated with a desynchronized electroencephalogram (Neve et al. 1982). However, as rabbit doe

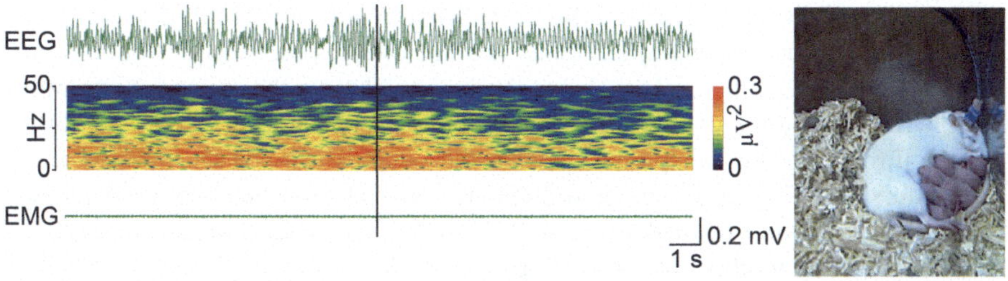

Fig. 3. Mother rat during the first postpartum week. The raw electroencephalogram (EEG), its correspondent spectrogram, and the electromyogram (EMG) of a mother rat during nursing in supine posture. The vertical line indicates the moment when the photograph was taken. Note the transition from NREM to REM sleep

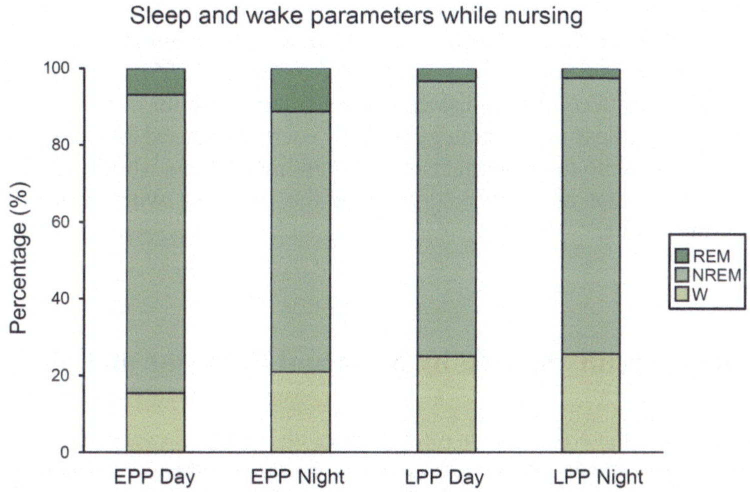

Fig. 4. Graphic charts showing the percentage of time spent in wakefulness (W), NREM, and REM sleep while nursing. Comparisons of the percentage of time of sleep and W during low kyphosis during the first and second postpartum weeks in the light and dark periods. [These data were published in Benedetto et al. (2017b).] *EPP* early postpartum, *LPP* late postpartum

usually nurses once a day (Gonzalez-Mariscal et al. 2016), nursing and pups' attention would not interfere with maternal sleep.

In some species, sleep deprivation and sleep fragmentation are consequences of the early stages of motherhood (Hunter et al. 2009; Lyamin et al. 2007). In the case of the human mother, partial sleep deprivation and sleep fragmentation have been reported; these sleep disturbances are compensated, at least partially, by a deeper sleep (Montgomery-Downs et al. 2010; Nishihara et al. 2004). Sleep deprivation and fragmentation may lead to maternal irritability and possible psychiatric disorders, with negative consequences for the care and welfare of the newborn (Lee 1998; Sharma and Mazmanian 2003). An extreme case is found in dolphins, where both the mother and calf are sleep deprived during the first months after birth (Lyamin et al. 2007). In laboratory rat the antecedents are not conclusive, since Voloschin and Tramezzani (1979) reported no sleep deprivation during the second postpartum week, while Sivadas et al. (2016) showed that mother rats are sleep deprived during the day throughout the postpartum period. Our studies showed that NREM sleep during nursing is highly fragmented compared to that observed when the mother sleeps outside the nest (Benedetto et al. 2017b). In this sense, Sivadas et al. (2016) also showed that mother rats have fragmented sleep compared to non-maternal animals.

It is likely that different species have developed specific and unique strategies to conciliate nursing and sleep according to their habitat, ecology, and the degree of development of the young, among other factors. For instance, in mother dolphins, which do not have a burrow to protect their young from possible predators, it would not be adaptive that sleep has to be a prerequisite for milk ejection, as predation risk would increase dramatically during sleep. Also, for the rabbit doe, which usually nurses once a day and for a few minutes (Zarrow et al. 1965), nursing and sleep would not represent a conflict and sleep is not a prerequisite for milk ejection (Neve et al. 1982). Regardless of the strategy used by each species, adaptations have been developed to reconcile the maternal care and nursing of the newborn with mother's sleep physiology that assure the survival of the offspring without affecting the welfare of the mother.

4 Neural Circuits Shared by Maternal Behavior and Sleep

The rapid onset of maternal responsiveness relies on hormonal events that occur close to parturition. Specifically, the ovarian steroids, estrogen and progesterone, as well as lactogenic hormones and oxytocin initially activate the brain areas that control the maternal responsiveness to the newborn litter (Numan and Insel 2003; Rosenblatt 1980; Rosenblatt et al. 1988). Afterward, maternal behavior is maintained by the continued mother-infant interactions and seems to be relatively independent of endocrine regulation (Numan and Insel 2003; Rosenblatt 1980; Rosenblatt et al. 1988).

Experimental evidence has shown that the mPOA and its connections with the mesolimbic dopamine (DA) system play a predominant role in the control of active aspects of maternal behavior (Stolzenberg and Numan 2011). Interestingly, these brain areas are also key components of neural circuits for sleep generation and maintenance (Benedetto et al. 2017a; Kaushik et al. 2011; Mendelson 1996, 2000, 2001; Monti et al. 2016). Also, there is strong evidence that the MCHergic system is also involved in the regulation of both maternal behavior and sleep. Thus, we will focus on these brain areas and systems in this section.

4.1 The Preoptic Area Controls Maternal and Sleep Behaviors

4.1.1 Preoptic Area and Maternal Behavior

The first evidence that the POA was involved in maternal behavior was in 1956, when Fisher reported that chemical stimulation of this area with testosterone elicits active maternal behavior in male rats (Fisher 1956). Since then, many studies have demonstrated the involvement of mPOA in maternal behavior. For instance, mPOA damage interferes with all maternal behaviors (Numan et al. 1988) or mainly with the active maternal behavior (Numan and Callahan 1980; Terkel et al. 1979), while chemical stimulation with oxytocin (Pedersen et al. 1994), estrogen implants (Rosenblatt and Ceus 1998), dopaminergic agents (Miller and Lonstein 2005), and prolactin (Bridges et al. 1990) promotes full maternal behavior.

Thus, the mPOA is believed to act as a crucial area in the control of maternal behavior, integrating information from diverse modalities related to the pups and adjusting maternal responses to fulfill the pups' needs as they grow older (Pereira 2016; Pereira and Ferreira 2015; Risold et al. 1994; Simerly and Swanson 1986). Also, it is a key neural site where the hormones such as prolactin (Bridges et al. 1990), vasopressin (Bosch et al. 2010), and oxytocin (Pedersen et al. 1994) and a variety of neurotransmitters/neuromodulators such as DA (Miller and Lonstein 2005; Stolzenberg et al. 2007), hypocretin (Rivas et al. 2016), and MCH (Benedetto et al. 2014) modulate maternal behavior.

Particularly, this area is known to present important plasticity both during pregnancy and postpartum periods (Champagne and Curley 2016; Parent et al. 2017; Pereira and Morrell 2009; Rondini et al. 2010; Schrader et al. 2012) allowing the adaptation of maternal responsiveness to the changing needs of the offspring. Several studies point out a significant anatomical and functional reorganization of different brain areas across postpartum period that are crucial for maternal behavior (Driessen et al. 2014; Insel 1990; Pereira 2016; Pereira and Morrell 2009; Rondini et al. 2010). Specifically, the mPOA has been demonstrated to change its role throughout the postpartum period, from a facilitatory role during early lactation period (Jacobson et al. 1980; Lee et al. 2000; Numan et al. 1977; Pereira and Morrell 2009; Rosenblatt and Ceus 1998) to an inhibitory role during mid-lactation (Pereira and Morrell 2009). Interestingly, Schrader et al. (2012) show that the mPOA loses its daily rhythmicity of Fos activation in pregnant rats compared to diestrous females (Schrader et al. 2012).

4.1.2 Preoptic Area and Sleep

Since the early Von Economo's studies of the hypothalamus, the POA has been proposed as a sleep center (Von Economo 1930). Most studies have pointed out the crucial role of the VLPO and the median POA (MnPN) in the regulation of NREM sleep (Benedetto et al. 2012, 2013; Gong et al. 2004; Gvilia et al. 2006; Lu et al.

2000; Torterolo et al. 2009a), while the extended VLPO (eVLPO) has been involved in REM sleep regulation (Lu et al. 2002). Likewise, the mPOA, the same area that is critical for maternal behavior, has been also involved in the control of sleep (Kumar 2004). In fact, microinjections of different neuropeptides and substances, such as glutamate (Kaushik et al. 2011), triazolam (Mendelson and Martin 1992), pentobarbital (Mendelson 1996), and adenosine (Mendelson 2000), are known to promote NREM sleep. Although most studies relate mPOA to NREM sleep, there is also experimental data that spotlight the importance of this area also in the regulation of REM sleep (Asala et al. 1990; Suntsova and Dergacheva 2004). Particularly, most mPOA neurons increase their firing rate during REM sleep compared to W and NREM. In addition, electrical stimulation of the mPOA at low frequency during NREM sleep promotes the entrance to REM sleep (Suntsova and Dergacheva 2004). Moreover, Asala et al. (1990) induced a reduction of NREM sleep and an increase in REM sleep after mPOA lesions.

4.2 Dopaminergic System Modulates Maternal and Sleep Behaviors

4.2.1 Dopaminergic System and Maternal Behavior

The mesocorticolimbic system has been recognized for its central role in several motivated behaviors (Berridge 2004; Salamone and Correa 2012), including maternal behavior (Stolzenberg and Numan 2011). This system is comprised of dopamine (DA)-containing cell bodies in the midbrain ventral tegmental area (VTA) and its major targets, the nucleus accumbens (NAc) and prefrontal cortex (PFC) (Fallon and Moore 1978; Lindvall and Bjorklund 1974; Lindvall et al. 1974; Moore and Bloom 1978). The NAc has been considered as a neural interface between the limbic and motor systems allowing the transition from motivation to motor action (Mogenson et al. 1980). In addition, this nucleus is one of the main projection sites of the MCHergic neurons and has an important MCHergic receptor density (Bittencourt et al. 1992; Hervieu et al. 2000). The interconnection between the mPOA and the mesocorticolimbic DAergic system is known to regulate the motivational aspects of maternal behavior (Numan and Insel 2003; Stolzenberg and Numan 2011).

The central role of the DAergic system in the control of maternal behavior has initially been shown by the actions of systemic DA antagonists that reduced active maternal behavior but enhanced nursing and milk ejection (Stern 1991; Stern and Keer 1999). Interestingly, active maternal behaviors, such as licking the pups, are linked to an elevated DA release and increased DA receptor levels in the NAc (Champagne et al. 2004; Hansen et al. 1993). In accordance, DA receptor antagonism in the NAc inhibits maternal retrieval and licking, but promotes nursing behavior (Keer and Stern 1999). In addition, lesions of DAergic neurons of the VTA also impair pup-directed maternal behavior, but nursing was unaffected (Hansen et al. 1991b).

The administration of DA D1 receptor antagonist (SCH-23390) into mPOA impairs the retrieval and licking of pups but no other components of the maternal behavior (Miller and Lonstein 2005; Numan et al. 2005a), while the infusion of the D2 receptor antagonist raclopride into this area increases nursing, leaving intact active maternal behaviors (Miller and Lonstein 2005). These data suggest that different DA receptors within the mPOA are involved in distinct aspects of maternal behavior, where D1 receptors would control active maternal behavior, while D2 receptors would regulate nursing postures.

4.2.2 Dopaminergic System and Sleep

Interestingly, the DAergic system also plays a critical but complex role in the control of sleep and wakefulness (W). DAergic neurons, mostly present in the VTA and substantia nigra pars compacta (SNc), do not change their mean firing rate across the sleep-wake cycle (Miller et al. 1983; Trulson 1985; Trulson and Preussler 1984; Trulson et al. 1981). However, the temporal pattern of the discharge is strongly modulated during the sleep-wake cycle. Accordingly, during W, DAergic neurons in VTA discharge in burst in response to salient stimuli (Schultz et al. 1993). This increase in bursting activity is accompanied by a substantial increase in DA extracellular levels in striatal regions (Wightman and Robinson 2002). In this regard, microdialysis studies by Lena et al. (2005) have shown that DA release is greater during W in comparison to sleep, both in the prefrontal cortex and the NAc (Lena et al. 2005). In addition, Dahan et al. (2007) demonstrated that there is a prominent burst firing increase in VTA DAergic neurons during REM sleep that resembles the bursting induced by the consumption of palatable food (Dahan et al. 2007). This is in line with previous findings that described an increase in the number of Fos-immunoreactive neurons in the VTA during REM sleep (Maloney et al. 2002). In addition, the release of DA, both in the NAc and prefrontal cortex, increases during REM sleep in comparison to NREM sleep (Lena et al. 2005). In this regard, it has been shown that DAergic VTA neurons participate in the promotion of theta rhythm, a prominent feature of REM sleep (Orzel-Gryglewska et al. 2015).

4.2.3 Dopamine and POA Regulating Maternal Behavior and Sleep Together

In spite of the bunch of evidence that POA modulates both sleep and maternal behavior, there were no studies focused on determining how this brain area acts to promote these two behaviors together.

Based on the findings that (1) nursing and NREM sleep can be co-expressed (Benedetto et al. 2017b), (2) mPOA stimulation increases the time spent in NREM sleep (Kaushik et al. 2011), and (3) the local delivery of a dopaminergic D2 antagonist raclopride increases nursing duration without interfering with active maternal behavior (Miller and Lonstein 2005), we posit that raclopride into mPOA would also be associated with an increase in NREM sleep. Contrary to our expectations, we find neither an increase in nursing nor in NREM sleep after the microinjection of raclopride within the mPOA. Surprisingly, REM sleep and its transitional stage from NREM were significantly reduced in time after this microinjection procedure. Regarding maternal behavior, the latency to reunite the entire litter into the nest was increased, while the time to the start of nursing was reduced, suggesting that some aspects of the maternal sequence were affected (Benedetto et al. 2017a). Thus, we are carrying out more experiments to elucidate how the dopaminergic system acts to integrate sleep and maternal behavior within the mPOA.

4.3 The MCHergic System

The general feature of this system has been described in preceding chapters. Briefly, MCH is an inhibitory neuropeptide synthesized by neurons that are primarily located in the posterolateral hypothalamus and incerto-hypothalamic area (Bittencourt et al. 1992). The MCHergic neurons project widely throughout the central nervous system (CNS).

MCH acts via neurons expressing the MCH receptor 1 (MCHR1) and MCH receptor 2 (called MCHR2) (Macneil 2013; Saito and Nagasaki 2008). The latter has been described in primates (including humans), cats, and dogs but seems to be absent in rodents (Tan et al. 2002).

There is a widespread distribution of the MCHR1 mRNA along with the MCHergic fibers including critical areas for sleep and maternal behavior (Bittencourt et al. 1992; Saito et al. 2001).

4.3.1 MCH and Sleep

The role of MCH as a sleep-promoting factor has been assessed in previous chapters of this book and reviewed in detail (Konadhode et al. 2015; Monti et al. 2013; Torterolo et al. 2011, 2015). Briefly, the intracerebroventricular administration of MCH in the rat produces an increase in REM sleep and a moderate enhancement in the time spent in NREM sleep, while the systemic administration of MCHR1 antagonists both decreases sleep and increases wakefulness (Verret et al. 2003; Ahnaou et al. 2008). Microinjection of MCH into the REM-off neuronal areas, such as the dorsal raphe and the locus coeruleus of the rat, facilitates the generation of REM sleep (Lagos et al. 2009; Monti et al. 2015). MCH also promotes REM sleep when microinjected into either the basal forebrain of the rat or the NPO of the cat (Lagos et al. 2012; Torterolo et al. 2009b). In contrast, the administration of MCH into the VLPO, a NREM sleep-promoting area, induced NREM sleep (Benedetto et al. 2013).

Fos and electrophysiological studies have shown that MCHergic neurons are active mostly during sleep (mainly during REM sleep) in the rat (Verret et al. 2003; Hassani et al. 2009). However, Gonzalez et al. (2016) by means of fiber-optic recordings have recently shown that these neurons are active during novelty exploration.

The concentration of MCH in the CSF of rats increases during the light phase, when the animals are predominantly asleep, and is affected by sleep deprivation or sleep restriction (Pelluru et al. 2013; Dias Abdo Agamme et al. 2015). By means of in vivo microdialysis, Blouin et al. (2013) have shown that the release of MCH in the amygdala of patients reaches a maximum level at sleep onset.

Studies of ppMCH and MCHR1 KO mice indicate that the sleep of these animals is altered (Ahnaou et al. 2008; Willie et al. 2008; Takase et al. 2014). Recent optogenetic and chemogenetic studies have confirmed the role of MCH in sleep generation. Optogenetic stimulation of MCHergic neurons increased both REM and NREM sleep at night, whereas during the day only REM sleep was increased (Konadhode et al. 2013; Blanco-Centurion et al. 2016). In addition, delta power (an indicator of sleep intensity) was also increased (Blanco-Centurion et al. 2016). Optogenetic activation of MCH neurons at the onset of REM sleep extended the duration of REM sleep episodes (Jego et al. 2013). Also, the acute silencing of these neurons reduced hippocampal theta rhythm during REM sleep without affecting the duration of the episodes. Tsunematsu et al. (2014) showed that the acute optogenetic activation of MCH neurons at 10 Hz induced transitions from NREM to REM sleep and increased REM sleep time. Also, Vetrivelan et al. (2016) have shown that the selective chemogenetic activation of the MCHergic neurons causes specific increases in REM sleep without altering wakefulness or NREM sleep.

4.3.2 MCH and Maternal Behavior

A recent review by Diniz and Bittencourt (2017) described that MCH is involved in motivated behaviors such as feeding, drinking, and mating and also active maternal behavior. In fact, not only MCHergic neurons project toward regions involved both in active maternal behavior, such as the mPOA, VTA, and NAc (Bittencourt et al. 1992), but also mRNA for the MCHR1 has been recognized in these regions (Saito et al. 2001).

The importance of this neuropeptide in the postpartum period is evidenced by the fact that in males and cycling females most MCHergic neurons are located within the lateral hypothalamus and incerto-hypothalamic area, while during lactation, neurons of the mPOA express pre-pro-MCH mRNA and MCH, being most of them GABAergic (approximately 95%) (Knollema et al. 1992; Rondini et al. 2010). Remarkably, the number of MCHergic neurons in the mPOA varies along lactation (Alvisi et al. 2016; Knollema et al. 1992; Rondini et al. 2010). Taken together, these interesting results suggest that the number of MCH neurons within the mPOA varies according to a curve in which: their number is very low on PPD5, moderated at PPD12, and maximum at PPD15–16, starts to decrease at PPD17–19 (but is higher than at PPD12), at PPD21, has a similar number as in PPD12, finally disappearing after weaning (Alvisi et al. 2016; Knollema et al. 1992; Rondini et al. 2010). This changing pattern in the number of mPOA MCHergic neurons is likely associated to maternal behavior, such that these cells are maximal when active maternal behaviors begin to decrease, as the pups grow older, become more

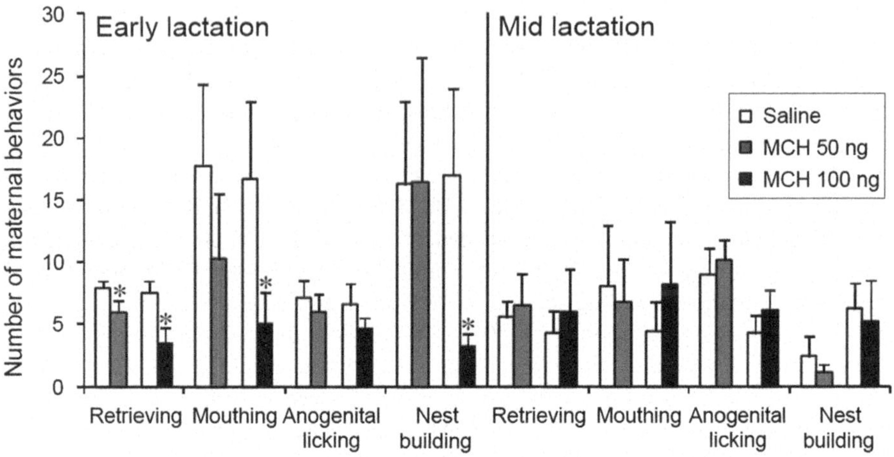

Fig. 5. Effects of MCH microinjections into the mPOA on active components of maternal behavior during early and mid-lactation. Graphic chart showing the number of different active maternal responses after bilateral administration of saline and MCH (50 and 100 ng) in a 30-min maternal test. All values are presented as means ± SEM; significant differences are indicated by asterisks (*$p < 0.05$). Part of the early lactation results was published in Benedetto et al. (2014)

independent, and start to consume food (Hall and Rosenblatt 1978; Reisbick et al. 1975), and decline at the end of lactation. However, these changes could be also related to hormonal variations along lactation (Ferreira et al. 2017).

Benedetto et al. (2014) have shown that bilateral microinjections of MCH into the mPOA decrease active maternal behaviors during the early postpartum period (PPD5–6) to levels characteristic of the late postpartum period (Benedetto et al. 2014). The fact that only active maternal behaviors but not huddling and nursing were reduced after MCH treatment agrees with published evidence showing that interference with mPOA function mainly affects these active components (Jacobson et al. 1980; Numan 1974; Numan and Insel 2003; Pereira and Morrell 2009; Terkel et al. 1979). Interestingly, the same procedure during the PPD14–15, the period in which MCH reaches its maximum levels, did not provoke any change in maternal behavior compared to control saline microinjections (Fig. 5). It could be speculated that during this latter stage, external MCH does not significantly change maternal behavior, such that increasing its levels exogenously does not potentiate the inhibitory effect of the high endogenous levels at this moment. It would be interesting to explore if MCHR1 varies within the mPOA along lactation, promoting these functional changes. These differential results found according to the postpartum stage are in line with the knowledge of the mPOA as a plastic area that changes its functionality during the postpartum period (Champagne and Curley 2016; Pereira 2016; Pereira and Morrell 2009).

It has also been shown that MCHR1 KO mother mice show a decrease in the survival of the pups during PPD1–2 compared to pups from wild-type mothers (Adams et al. 2011; Alachkar et al. 2016). However, beyond the PPD2 pups' mortality did not differ between these two groups. Authors hypothesized that

MCH would be important for the initiation of maternal behavior, but not for its maintenance (Alachkar et al. 2016). Although these results might seem contradictory with Benedetto et al. (2014), it can be speculated that the requirement of MCH may fluctuate at the different stages of the postpartum period to display an appropriate maternal behavior that matches with the changing characteristics and needs of the offspring; when MCH is outside its normal range, the maternal care of the pups could be altered.

The MCHergic system seems to be linked also to an endocrine function during lactation. Recent evidences presented by Ferreira et al. (2017) show that the number of mPOA-MCH neurons is positively correlated with the litter size (Ferreira et al. 2017), likely caused by the suckling stimulus from pups that also promotes the release of prolactin and oxytocin (Wakerley 1996). In this sense, Parkes and Vale (1993) found that MCH directly stimulates the release of oxytocin acting on neurohypophysis cell cultures (Parkes and Vale 1993). On the contrary, MCH was significantly decreased in a model of hyperprolactinemia compared to control rats (Garcia et al. 2003). Thus, prolactin seems to be inversely related to MCH levels. Further experiments are needed to understand the functionality of these mPOA-MCH neurons.

4.4 MCH and DAergic System Interaction in the Control of Sleep and Maternal Behavior

As described before, it can be proposed that the DAergic system promotes motivation and arousal, while the MCHergic system stimulates consummatory behaviors and sleep. Hence, it could be speculated that DAergic and MCHergic systems have opposite functions. Based on this idea, it could be hypothesized that MCH may decrease motivation and arousal, at least in part, by inhibiting the DAergic system.

As mentioned, the NAc is a main target of the mesolimbic DAergic neurons and is an essential area for a proper care of the offspring (Hansen et al. 1993; Mogenson et al. 1980; Numan et al. 2005a, b). This nucleus is also one of the main projection sites of the MCHergic neurons and has an important MCHR1 density (Bittencourt et al. 1992; Hervieu et al. 2000). MCHR1 is co-expressed with DAergic receptors in the shell of the NAc, which raises the possibility that MCH and DA interact in NAc shell during motivated responses (Chung et al. 2009; Georgescu et al. 2005). This idea has been explored using a whole-cell patch-clamp recording from the NAc shell where the administration of MCH alone had no effect on spike firing, but the discharge rate showed an increase when MCH was applied in combination with D1 or D2 agonists (Chung et al. 2009; Hopf et al. 2013). Furthermore, biochemical analysis in NAc shell explants showed that MCH signaling blocks DA-induced phosphorylation of the AMPA glutamate receptor (Georgescu et al. 2005).

In addition, MCHR1 KO mice present hyperactivity that may be explained by interactions between MCHergic and DAergic systems (Lalonde and Qian 2007; Marsh et al. 2002). In fact, MCHR1 KO mice are supersensitive to the locomotor activating effects of d-amphetamine and D1 agonists as compared to wild-type animals. Besides, deletion of MCHR1 results in an upregulation of mesolimbic DA receptors, and the lack of MCH leads to an increase in DA release and in limbic DA transporter levels, indicating that MCH may negatively modulate mesolimbic monoamine function (Mul et al. 2011; Smith et al. 2005).

From these data, it could be expected that during highly aroused motivated behaviors, such as maternal behavior, the DAergic system may inhibit the MCHergic neurons. In fact, DA hyperpolarizes MCHergic neurons by activating G-protein-activated inwardly rectifying K+ (GIRK) channels by means of the activation of alpha-2a noradrenergic receptors (Alberto et al. 2011; Conductier et al. 2011). Conductier et al. (2011) also showed that MCH neurons receive both GABAergic and glutamatergic inputs and that DA modulates these inputs in a complex manner. At low concentrations, DA activates D1-like receptors promoting presynaptic activity, while at higher concentrations, DA activates D2-like receptors that inhibits presynaptic activity. Overall, DA leads to a decrease in the excitability of MCHergic neurons.

Regarding sleep and lactation from these data described above, we hypothesize that the DAergic and MCH system may interact to decrease the activity of the motivational system to promote sleep, thus allowing milk ejection.

5 Conclusions and Future Directions

In the present report, we analyzed data that support the interconnection between sleep and maternal behavior, two crucial behaviors for survival of mammals. We also propose a possible mechanism by which the MCHergic system modulates both behaviors together. We show that MCH induces sleep and reduces active maternal behaviors, thus promoting energy conservation for the mother. As mPOA is a critical area for the modulation of both behaviors, it is possible that MCH, acting in this area, will enhance sleep during nursing, thus promoting milk ejection occurrence. We are currently assessing this possibility.

As the stimulus of sucking from the pups induces not only nursing postures (and thus immobility) but also the synthesis and release of prolactin, hormone known to promote sleep and milk production (Wakerley 1996), mPOA-MCH cells might also have an endocrine function that indirectly promotes sleep and milk ejection, assuring the welfare of the pups and also of the mother (Alachkar et al. 2016). Also it would be important to explore the putative relationship between MCH and oxytocin in the control of both behaviors, as this hormone is stimulated by the sucking from the pups and promotes milk ejection.

Although MCH is expressed in mPOA neurons during the postpartum period, its functionality in lactating rats is still not clear. Also, behavioral experiments during the postpartum period focused on the effect of MCH agents on both maternal behavior and sleep are still needed to fully understand the complex interaction between both behaviors and its relation with the MCHergic system.

References

Adams AC, Domouzoglou EM, Chee MJ, Segal-Lieberman G, Pissios P, Maratos-Flier E (2011) Ablation of the hypothalamic neuropeptide melanin concentrating hormone is associated with behavioral abnormalities that reflect impaired olfactory integration. Behav Brain Res 224(1): 195–200

Ahnaou A, Drinkenburg WH, Bouwknecht JA, Alcazar J, Steckler T, Dautzenberg FM (2008) Blocking melanin-concentrating hormone MCH1 receptor affects rat sleep-wake architecture. Eur J Pharmacol 579(1–3):177–188

Alachkar A, Alhassen L, Wang Z, Wang L, Onouye K, Sanathara N, Civelli O (2016) Inactivation of the melanin concentrating hormone system impairs maternal behavior. Eur Neuropsychopharmacol 26(11):1826–1835

Alberto CO, Trask RB, Hirasawa M (2011) Dopamine acts as a partial agonist for alpha2A adrenoceptor in melanin-concentrating hormone neurons. J Neurosci 31(29):10671–10676

Alvisi RD, Diniz GB, Da-Silva JM, Bittencourt JC, Felicio LF (2016) Suckling-induced Fos activation and melanin-concentrating hormone immunoreactivity during late lactation. Life Sci 148: 241–246

Asala SA, Okano Y, Honda K, Inoue S (1990) Effects of medial preoptic area lesions on sleep and wakefulness in unrestrained rats. Neurosci Lett 114(3):300–304

Benedetto L, Chase MH, Torterolo P (2012) GABAergic processes within the median preoptic nucleus promote NREM sleep. Behav Brain Res 232(1):60–65

Benedetto L, Rodriguez-Servetti Z, Lagos P, D'Almeida V, Monti JM, Torterolo P (2013) Microinjection of melanin concentrating hormone into the lateral preoptic area promotes non-REM sleep in the rat. Peptides 39:11–15

Benedetto L, Pereira M, Ferreira A, Torterolo P (2014) Melanin-concentrating hormone in the medial preoptic area reduces active components of maternal behavior in rats. Peptides 58C: 20–25

Benedetto L, Rivas M, Cavelli M, Pena F, Monti J, Ferreira A, Torterolo P (2017a) Microinjection of the dopamine D2-receptor antagonist Raclopride into the medial preoptic area reduces REM sleep in lactating rats. Neurosci Lett 659:104–109

Benedetto L, Rivas M, Pereira M, Ferreira A, Torterolo P (2017b) A descriptive analysis of sleep and wakefulness states in different maternal behaviors in lactating rats. Arch Ital Biol 155(3):99–109

Berridge KC (2004) Motivation concepts in behavioral neuroscience. Physiol Behav 81(2): 179–209

Bittencourt JC, Presse F, Arias C, Peto C, Vaughan J, Nahon JL, Vale W, Sawchenko PE (1992) The melanin-concentrating hormone system of the rat brain: an immuno- and hybridization histochemical characterization. J Comp Neurol 319(2):218–245

Blanco-Centurion C, Liu M, Konadhode RP, Zhang X, Pelluru D, van den Pol AN, Shiromani PJ (2016) Optogenetic activation of melanin-concentrating hormone neurons increases non-rapid eye movement and rapid eye movement sleep during the night in rats. Eur J Neurosci 44(10): 2846–2857

Blouin AM, Fried I, Wilson CL, Staba RJ, Behnke EJ, Lam HA, Maidment NT, Karlsson KAE, Lapierre JL, Siegel JM (2013) Human hypocretin and melanin-concentrating hormone levels are linked to emotion and social interaction. Nat Commun 4:1547

Bosch OJ, Pfortsch J, Beiderbeck DI, Landgraf R, Neumann ID (2010) Maternal behaviour is associated with vasopressin release in the medial preoptic area and bed nucleus of the stria terminalis in the rat. J Neuroendocrinol 22(5):420–429

Bridges RS, Numan M, Ronsheim PM, Mann PE, Lupini CE (1990) Central prolactin infusions stimulate maternal behavior in steroid-treated, nulliparous female rats. Proc Natl Acad Sci U S A 87(20):8003–8007

Champagne FA, Curley JP (2016) Plasticity of the maternal brain across the lifespan. New Dir Child Adolesc Dev 2016(153):9–21

Champagne FA, Chretien P, Stevenson CW, Zhang TY, Gratton A, Meaney MJ (2004) Variations in nucleus accumbens dopamine associated with individual differences in maternal behavior in the rat. J Neurosci 24(17):4113–4123

Chung S, Hopf FW, Nagasaki H, Li CY, Belluzzi JD, Bonci A, Civelli O (2009) The melanin-concentrating hormone system modulates cocaine reward. Proc Natl Acad Sci U S A 106(16): 6772–6777

Conductier G, Nahon JL, Guyon A (2011) Dopamine depresses melanin concentrating hormone neuronal activity through multiple effects on alpha2-noradrenergic, D1 and D2-like dopaminergic receptors. Neuroscience 178:89–100

Dahan L, Astier B, Vautrelle N, Urbain N, Kocsis B, Chouvet G (2007) Prominent burst firing of dopaminergic neurons in the ventral tegmental area during paradoxical sleep. Neuropsychopharmacology 32(6):1232–1241

Dias Abdo Agamme AL, Aguilar Calegare BF, Fernandes L, Costa A, Lagos P, Torterolo P, D'Almeida V (2015) MCH levels in the CSF, brain preproMCH and MCHR1 gene expression during paradoxical sleep deprivation, sleep rebound and chronic sleep restriction. Peptides 74: 9–15

Diniz GB, Bittencourt JC (2017) The melanin-concentrating hormone as an integrative peptide driving motivated behaviors. Front Syst Neurosci 11:32

Driessen TM, Zhao C, Whittlinger A, Williams H, Gammie SC (2014) Endogenous CNS expression of neurotensin and neurotensin receptors is altered during the postpartum period in outbred mice. PLoS One 9(1):e83098

Fallon JH, Moore RY (1978) Catecholamine innervation of the basal forebrain. IV. Topography of the dopamine projection to the basal forebrain and neostriatum. J Comp Neurol 180(3):545–580

Ferreira JGP, Duarte JCG, Diniz GB, Bittencourt JC (2017) Litter size determines the number of melanin-concentrating hormone neurons in the medial preoptic area of Sprague Dawley lactating dams. Physiol Behav 181:75–79

Fisher AE (1956) Maternal and sexual behavior induced by intracranial chemical stimulation. Science 124(3214):228–229

Garcia MC, Lopez M, Gualillo O, Seoane LM, Dieguez C, Senaris RM (2003) Hypothalamic levels of NPY, MCH, and prepro-orexin mRNA during pregnancy and lactation in the rat: role of prolactin. FASEB J 17(11):1392–1400

Georgescu D, Sears RM, Hommel JD, Barrot M, Bolanos CA, Marsh DJ, Bednarek MA, Bibb JA, Maratos-Flier E, Nestler EJ, DiLeone RJ (2005) The hypothalamic neuropeptide melanin-concentrating hormone acts in the nucleus accumbens to modulate feeding behavior and forced-swim performance. J Neurosci 25(11):2933–2940

Gong H, McGinty D, Guzman-Marin R, Chew KT, Stewart D, Szymusiak R (2004) Activation of c-fos in GABAergic neurones in the preoptic area during sleep and in response to sleep deprivation. J Physiol 556(Pt 3):935–946

Gonzalez JA, Iordanidou P, Strom M, Adamantidis A, Burdakov D (2016) Awake dynamics and brain-wide direct inputs of hypothalamic MCH and orexin networks. Nat Commun 7:11395

Gonzalez-Mariscal G, Caba M, Martinez-Gomez M, Bautista A, Hudson R (2016) Mothers and offspring: the rabbit as a model system in the study of mammalian maternal behavior and sibling interactions. Horm Behav 77:30–41

Grota LJ, Ader R (1969) Continuous recording of maternal behavior in Rattus Norvegicus. Anim Behav 17:722–729

Gvilia I, Xu F, McGinty D, Szymusiak R (2006) Homeostatic regulation of sleep: a role for preoptic area neurons. J Neurosci 26(37):9426–9433

Hall WG, Rosenblatt JS (1978) Development of nutritional control of food intake in suckling rat pups. Behav Biol 24(4):413–427

Hansen S, Harthon C, Wallin E, Lofberg L, Svensson K (1991a) The effects of 6-OHDA-induced dopamine depletions in the ventral or dorsal striatum on maternal and sexual behavior in the female rat. Pharmacol Biochem Behav 39(1):71–77

Hansen S, Harthon C, Wallin E, Lofberg L, Svensson K (1991b) Mesotelencephalic dopamine system and reproductive behavior in the female rat: effects of ventral tegmental 6-hydroxydopamine lesions on maternal and sexual responsiveness. Behav Neurosci 105(4): 588–598

Hansen S, Bergvall AH, Nyiredi S (1993) Interaction with pups enhances dopamine release in the ventral striatum of maternal rats: a microdialysis study. Pharmacol Biochem Behav 45(3): 673–676

Hassani OK, Lee MG, Jones BE (2009) Melanin-concentrating hormone neurons discharge in a reciprocal manner to orexin neurons across the sleep-wake cycle. Proc Natl Acad Sci U S A 106(7):2418–2422

Hervieu GJ, Cluderay JE, Harrison D, Meakin J, Maycox P, Nasir S, Leslie RA (2000) The distribution of the mRNA and protein products of the melanin- concentrating hormone (MCH) receptor gene, slc-1, in the central nervous system of the rat. Eur J Neurosci 12(4):1194–1216

Hopf FW, Seif T, Chung S, Civelli O (2013) MCH and apomorphine in combination enhance action potential firing of nucleus accumbens shell neurons in vitro. PeerJ 1:e61

Hunter LP, Rychnovsky JD, Yount SM (2009) A selective review of maternal sleep characteristics in the postpartum period. J Obstet Gynecol Neonatal Nurs 38(1):60–68

Insel TR (1990) Regional changes in brain oxytocin receptors post-partum: time-course and relationship to maternal behaviour. J Neuroendocrinol 2(4):539–545

Jacobson CD, Terkel J, Gorski RA, Sawyer CH (1980) Effects of small medial preoptic area lesions on maternal behavior: retrieving and nest building in the rat. Brain Res 194(2):471–478

Jego S, Glasgow SD, Herrera CG, Ekstrand M, Reed SJ, Boyce R, Friedman J, Burdakov D, Adamantidis AR (2013) Optogenetic identification of a rapid eye movement sleep modulatory circuit in the hypothalamus. Nat Neurosci 16(11):1637–1643

Kaushik MK, Kumar VM, Mallick HN (2011) Glutamate microinjection at the medial preoptic area enhances slow wave sleep in rats. Behav Brain Res 217(1):240–243

Keer SE, Stern JM (1999) Dopamine receptor blockade in the nucleus accumbens inhibits maternal retrieval and licking, but enhances nursing behavior in lactating rats. Physiol Behav 67(5): 659–669

Knollema S, Brown ER, Vale W, Sawchenko PE (1992) Novel hypothalamic and preoptic sites of prepro-melanin-concentrating hormone messenger ribonucleic acid and peptide expression in lactating rats. J Neuroendocrinol 4(6):709–717

Konadhode RR, Pelluru D, Blanco-Centurion C, Zayachkivsky A, Liu M, Uhde T, Glen WB Jr, van den Pol AN, Mulholland PJ, Shiromani PJ (2013) Optogenetic stimulation of MCH neurons increases sleep. J Neurosci 33(25):10257–10263

Konadhode RR, Pelluru D, Shiromani PJ (2015) Neurons containing orexin or melanin concentrating hormone reciprocally regulate wake and sleep. Front Syst Neurosci 8:244

Krasnow SM, Steiner RA (2006) Physiological mechanisms integrating metabolism and reproduction. In: Neill JD (ed) Knobil and Neill's physiology of reproduction. Elsevier, St. Louis, MO, pp 2553–2625

Kumar VM (2004) Why the medial preoptic area is important for sleep regulation. Indian J Physiol Pharmacol 48(2):137–149

Lagos P, Torterolo P, Jantos H, Chase MH, Monti JM (2009) Effects on sleep of melanin-concentrating hormone (MCH) microinjections into the dorsal raphe nucleus. Brain Res 1265: 103–110

Lagos P, Monti JM, Jantos H, Torterolo P (2012) Microinjection of the melanin-concentrating hormone into the lateral basal forebrain increases REM sleep and reduces wakefulness in the rat. Life Sci 90(23–24):895–899

Lalonde R, Qian S (2007) Exploratory activity, motor coordination, and spatial learning in Mchr1 knockout mice. Behav Brain Res 178(2):293–304

Lee KA (1998) Alterations in sleep during pregnancy and postpartum: a review of 30 years of research. Sleep Med Rev 2(4):231–242

Lee A, Clancy S, Fleming AS (2000) Mother rats bar-press for pups: effects of lesions of the mpoa and limbic sites on maternal behavior and operant responding for pup-reinforcement. Behav Brain Res 108(2):215–231

Lena I, Parrot S, Deschaux O, Muffat-Joly S, Sauvinet V, Renaud B, Suaud-Chagny MF, Gottesmann C (2005) Variations in extracellular levels of dopamine, noradrenaline, glutamate, and aspartate across the sleep–wake cycle in the medial prefrontal cortex and nucleus accumbens of freely moving rats. J Neurosci Res 81(6):891–899

Lincoln DW, Hentzen K, Hin T, van der Schoot P, Clarke G, Summerlee AJ (1980) Sleep: a prerequisite for reflex milk ejection in the rat. Exp Brain Res 38(2):151–162

Lindvall O, Bjorklund A (1974) The organization of the ascending catecholamine neuron systems in the rat brain as revealed by the glyoxylic acid fluorescence method. Acta Physiol Scand Suppl 412:1–48

Lindvall O, Bjorklund A, Moore RY, Stenevi U (1974) Mesencephalic dopamine neurons projecting to neocortex. Brain Res 81(2):325–331

Lu J, Greco MA, Shiromani P, Saper CB (2000) Effect of lesions of the ventrolateral preoptic nucleus on NREM and REM sleep. J Neurosci 20(10):3830–3842

Lu J, Bjorkum AA, Xu M, Gaus SE, Shiromani PJ, Saper CB (2002) Selective activation of the extended ventrolateral preoptic nucleus during rapid eye movement sleep. J Neurosci 22(11): 4568–4576

Lyamin O, Pryaslova J, Kosenko P, Siegel J (2007) Behavioral aspects of sleep in bottlenose dolphin mothers and their calves. Physiol Behav 92(4):725–733

Macneil DJ (2013) The role of melanin-concentrating hormone and its receptors in energy homeostasis. Front Endocrinol (Lausanne) 4:49

Maloney KJ, Mainville L, Jones BE (2002) C-Fos expression in dopaminergic and GABAergic neurons of the ventral mesencephalic tegmentum after paradoxical sleep deprivation and recovery. Eur J Neurosci 15(4):774–778

Marsh DJ, Weingarth DT, Novi DE, Chen HY, Trumbauer ME, Chen AS, Guan XM, Jiang MM, Feng Y, Camacho RE, Shen Z, Frazier EG, Yu H, Metzger JM, Kuca SJ, Shearman LP, Gopal-Truter S, MacNeil DJ, Strack AM, MacIntyre DE, Van der Ploeg LH, Qian S (2002) Melanin-concentrating hormone 1 receptor-deficient mice are lean, hyperactive, and hyperphagic and have altered metabolism. Proc Natl Acad Sci U S A 99(5):3240–3245

Mendelson WB (1996) Sleep induction by microinjection of pentobarbital into the medial preoptic area in rats. Life Sci 59(22):1821–1828

Mendelson WB (2000) Sleep-inducing effects of adenosine microinjections into the medial preoptic area are blocked by flumazenil. Brain Res 852(2):479–481

Mendelson WB (2001) The sleep-inducing effect of ethanol microinjection into the medial preoptic area is blocked by flumazenil. Brain Res 892(1):118–121

Mendelson WB, Martin JV (1992) Characterization of the hypnotic effects of triazolam microinjections into the medial preoptic area. Life Sci 50(15):1117–1128

Miller SM, Lonstein JS (2005) Dopamine d1 and d2 receptor antagonism in the preoptic area produces different effects on maternal behavior in lactating rats. Behav Neurosci 119(4): 1072–1083

Miller JD, Farber J, Gatz P, Roffwarg H, German DC (1983) Activity of mesencephalic dopamine and non-dopamine neurons across stages of sleep and walking in the rat. Brain Res 273(1): 133–141

Mogenson GJ, Jones DL, Yim CY (1980) From motivation to action: functional interface between the limbic system and the motor system. Prog Neurobiol 14(2–3):69–97

Montgomery-Downs HE, Insana SP, Clegg-Kraynok MM, Mancini LM (2010) Normative longitudinal maternal sleep: the first 4 postpartum months. Am J Obstet Gynecol 203(5):465 e461–465 e467

Monti JM, Torterolo P, Lagos P (2013) Melanin-concentrating hormone control of sleep-wake behavior. Sleep Med Rev 17(4):293–298

Monti JM, Lagos P, Jantos H, Torterolo P (2015) Increased REM sleep after intra-locus coeruleus nucleus microinjection of melanin-concentrating hormone (MCH) in the rat. Prog Neuro-Psychopharmacol Biol Psychiatry 56:185–188

Monti JM, Pandi-Perumal SR, Chokroverty S (2016) Dopamine and sleep: molecular, functional, and clinical aspects. Springer, Switzerland

Moore RY, Bloom FE (1978) Central catecholamine neuron systems: anatomy and physiology of the dopamine systems. Annu Rev Neurosci 1:129–169

Mul JD, la Fleur SE, Toonen PW, Afrasiab-Middelman A, Binnekade R, Schetters D, Verheij MM, Sears RM, Homberg JR, Schoffelmeer AN, Adan RA, DiLeone RJ, De Vries TJ, Cuppen E (2011) Chronic loss of melanin-concentrating hormone affects motivational aspects of feeding in the rat. PLoS One 6(5):e19600

Neve HA, Paisley AC, Summerlee AJ (1982) Arousal a prerequisite for suckling in the conscious rabbit? Physiol Behav 28(2):213–217

Neville MC (2006) Lactation and its hormonal control. In: Knobil E, Neill JD (eds) The physiology of reproduction. Elsevier, New York, pp 2993–3054

Nishihara K, Horiuchi S, Eto H, Uchida S, Honda M (2004) Delta and theta power spectra of night sleep EEG are higher in breast-feeding mothers than in non-pregnant women. Neurosci Lett 368(2):216–220

Numan M (1974) Medial preoptic area and maternal behavior in the female rat. J Comp Physiol Psychol 87(4):746–759

Numan M (1994) Maternal behavior. In: Knobil E, Neill JD (eds) The physiology of reproduction. Raven, New York, pp 221–302

Numan M, Callahan EC (1980) The connections of the medial preoptic region and maternal behavior in the rat. Physiol Behav 25(5):653–665

Numan M, Insel TR (2003) The neurobiology of parental behavior. Springer, New York

Numan M, Rosenblatt JS, Komisaruk BR (1977) Medial preoptic area and onset of maternal behavior in the rat. J Comp Physiol Psychol 91(1):146–164

Numan M, Corodimas KP, Numan MJ, Factor EM, Piers WD (1988) Axon-sparing lesions of the preoptic region and substantia innominata disrupt maternal behavior in rats. Behav Neurosci 102 (3):381–396

Numan M, Numan MJ, Pliakou N, Stolzenberg DS, Mullins OJ, Murphy JM, Smith CD (2005a) The effects of D1 or D2 dopamine receptor antagonism in the medial preoptic area, ventral pallidum, or nucleus accumbens on the maternal retrieval response and other aspects of maternal behavior in rats. Behav Neurosci 119(6):1588–1604

Numan M, Numan MJ, Schwarz JM, Neuner CM, Flood TF, Smith CD (2005b) Medial preoptic area interactions with the nucleus accumbens-ventral pallidum circuit and maternal behavior in rats. Behav Brain Res 158(1):53–68

Orzel-Gryglewska J, Matulewicz P, Jurkowlaniec E (2015) Brainstem system of hippocampal theta induction: the role of the ventral tegmental area. Synapse 69(11):553–575

Parent C, Wen X, Dhir SK, Ryan R, Diorio J, Zhang TY (2017) Maternal care associates with differences in morphological complexity in the medial preoptic area. Behav Brain Res 326: 22–32

Parkes DG, Vale WW (1993) Contrasting actions of melanin-concentrating hormone and neuropeptide-E-I on posterior pituitary function. Ann N Y Acad Sci 680:588–590

Pedersen CA, Caldwell JD, Walker C, Ayers G, Mason GA (1994) Oxytocin activates the postpartum onset of rat maternal behavior in the ventral tegmental and medial preoptic areas. Behav Neurosci 108(6):1163–1171

Pelluru D, Konadhode R, Shiromani PJ (2013) MCH neurons are the primary sleep-promoting group. Sleep 36(12):1779–1781

Pereira M (2016) Structural and functional plasticity in the maternal brain circuitry. New Dir Child Adolesc Dev 2016(153):23–46

Pereira M, Ferreira A (2015) Neuroanatomical and neurochemical basis of parenting: dynamic coordination of motivational, affective and cognitive processes. Horm Behav 77:72–85

Pereira M, Morrell JI (2009) The changing role of the medial preoptic area in the regulation of maternal behavior across the postpartum period: facilitation followed by inhibition. Behav Brain Res 205(1):238–248

Reisbick S, Rosenblatt JS, Mayer AD (1975) Decline of maternal behavior in the virgin and lactating rat. J Comp Physiol Psychol 89(7):722–732

Rheingold H (1963) Maternal behavior in mammals. Wiley, New York

Risold PY, Canteras NS, Swanson LW (1994) Organization of projections from the anterior hypothalamic nucleus: a Phaseolus vulgaris-leucoagglutinin study in the rat. J Comp Neurol 348(1): 1–40

Rivas M, Torterolo P, Ferreira A, Benedetto L (2016) Hypocretinergic system in the medial preoptic area promotes maternal behavior in lactating rats. Peptides 81:9–14

Rondini TA, Donato J Jr, Rodrigues Bde C, Bittencourt JC, Elias CF (2010) Chemical identity and connections of medial preoptic area neurons expressing melanin-concentrating hormone during lactation. J Chem Neuroanat 39(1):51–62

Rosenblatt JS (1975) Prepartum and postpartum regulation of maternal behaviour in the rat. Ciba Found Symp 33:17–37

Rosenblatt JS (1980) Hormonal and nonhormonal regulation of maternal behavior: a theoretical survey. Reprod Nutr Dev 20(3B):791–800

Rosenblatt JS, Ceus K (1998) Estrogen implants in the medial preoptic area stimulate maternal behavior in male rats. Horm Behav 33(1):23–30

Rosenblatt J, Mayer A, Siegel H (1985) Maternal behavior among nonprimate mammals. In: Adler N, Pfaff D, Goy RW (eds) Handbook of behavioral neurobiology. Plenum, New York, pp 229–298

Rosenblatt JS, Mayer AD, Giordano AL (1988) Hormonal basis during pregnancy for the onset of maternal behavior in the rat. Psychoneuroendocrinology 13(1–2):29–46

Saito Y, Nagasaki H (2008) The melanin-concentrating hormone system and its physiological functions. Results Probl Cell Differ 46:159–179

Saito Y, Cheng M, Leslie FM, Civelli O (2001) Expression of the melanin-concentrating hormone (MCH) receptor mRNA in the rat brain. J Comp Neurol 435(1):26–40

Salamone JD, Correa M (2012) The mysterious motivational functions of mesolimbic dopamine. Neuron 76(3):470–485

Schrader JA, Smale L, Nunez AA (2012) Pregnancy affects FOS rhythms in brain regions regulating sleep/wake state and body temperature in rats. Brain Res 1480:53–60

Schultz W, Apicella P, Ljungberg T (1993) Responses of monkey dopamine neurons to reward and conditioned stimuli during successive steps of learning a delayed response task. J Neurosci 13(3):900–913

Sharma V, Mazmanian D (2003) Sleep loss and postpartum psychosis. Bipolar Disord 5(2):98–105

Simerly RB, Swanson LW (1986) The organization of neural inputs to the medial preoptic nucleus of the rat. J Comp Neurol 246(3):312–342

Sivadas N, Radhakrishnan A, Aswathy BS, Kumar VM, Gulia KK (2016) Dynamic changes in sleep pattern during post-partum in normal pregnancy in rat model. Behav Brain Res 320: 264–274

Smith DG, Tzavara ET, Shaw J, Luecke S, Wade M, Davis R, Salhoff C, Nomikos GG, Gehlert DR (2005) Mesolimbic dopamine super-sensitivity in melanin-concentrating hormone-1 receptor-deficient mice. J Neurosci 25(4):914–922

Stern JM (1989) Maternal behavior: sensory, hormonal and neural determinants. In: Brush FR, Levine S (eds) Psychoendocrinology. Academic, New York, pp 103–226

Stern JM (1991) Nursing posture is elicited rapidly in maternally naive, haloperidol-treated female and male rats in response to ventral trunk stimulation from active pups. Horm Behav 25(4): 504–517

Stern JM, Johnson SK (1990) Ventral somatosensory determinants of nursing behavior in Norway rats. I. Effects of variations in the quality and quantity of pup stimuli. Physiol Behav 47(5):993–1011

Stern JM, Keer SE (1999) Maternal motivation of lactating rats is disrupted by low dosages of haloperidol. Behav Brain Res 99(2):231–239

Stolzenberg DS, Numan M (2011) Hypothalamic interaction with the mesolimbic DA system in the control of the maternal and sexual behaviors in rats. Neurosci Biobehav Rev 35(3):826–847

Stolzenberg DS, McKenna JB, Keough S, Hancock R, Numan MJ, Numan M (2007) Dopamine D1 receptor stimulation of the nucleus accumbens or the medial preoptic area promotes the onset of maternal behavior in pregnancy-terminated rats. Behav Neurosci 121(5):907–919

Suntsova NV, Dergacheva OY (2004) The role of the medial preoptic area of the hypothalamus in organizing the paradoxical phase of sleep. Neurosci Behav Physiol 34(1):29–35

Sutherland RC, Juss TS, Wakerley JB (1987) Prolonged electrical stimulation of the nipples evokes intermittent milk ejection in the anaesthetised lactating rat. Exp Brain Res 66(1):29–34

Takase K, Kikuchi K, Tsuneoka Y, Oda S, Kuroda M, Funato H (2014) Meta-analysis of melanin-concentrating hormone signaling-deficient mice on behavioral and metabolic phenotypes. PLoS One 9(6):e99961

Tan CP, Sano H, Iwaasa H, Pan J, Sailer AW, Hreniuk DL, Feighner SD, Palyha OC, Pong SS, Figueroa DJ, Austin CP, Jiang MM, Yu H, Ito J, Ito M, Ito M, Guan XM, MacNeil DJ, Kanatani A, Van der Ploeg LH, Howard AD (2002) Melanin-concentrating hormone receptor subtypes 1 and 2: species-specific gene expression. Genomics 79(6):785–792

Terkel J, Bridges RS, Sawyer CH (1979) Effects of transecting lateral neural connections of the medial preoptic area on maternal behavior in the rat: nest building, pup retrieval and prolactin secretion. Brain Res 169(2):369–380

Thornburg KL, Bagby SP, Giraud GD (2006) Maternal adaptation to pregnancy. In: Neill JD (ed) Knobil and Neill's physiology of reproduction. Elsevier, St. Louis, MO, pp 2899–2924

Torterolo P, Benedetto L, Lagos P, Sampogna S, Chase MH (2009a) State-dependent pattern of Fos protein expression in regionally-specific sites within the preoptic area of the cat. Brain Res 1267:44–56

Torterolo P, Sampogna S, Chase MH (2009b) MCHergic projections to the nucleus pontis oralis participate in the control of active (REM) sleep. Brain Res 1268:76–87

Torterolo P, Lagos P, Monti JM (2011) Melanin-concentrating hormone: a new sleep factor? Front Neurol 2:14

Torterolo P, Scorza C, Lagos P, Urbanavicius J, Benedetto L, Pascovich C, Lopez-Hill X, Chase MH, Monti JM (2015) Melanin-concentrating hormone (MCH): role in REM sleep and depression. Front Neurosci 9:475

Trulson ME (1985) Activity of dopamine-containing substantia nigra neurons in freely moving cats. Neurosci Biobehav Rev 9(2):283–297

Trulson ME, Preussler DW (1984) Dopamine-containing ventral tegmental area neurons in freely moving cats: activity during the sleep-waking cycle and effects of stress. Exp Neurol 83(2):367–377

Trulson ME, Preussler DW, Howell GA (1981) Activity of substantia nigra units across the sleep-waking cycle in freely moving cats. Neurosci Lett 26(2):183–188

Tsunematsu T, Ueno T, Tabuchi S, Inutsuka A, Tanaka KF, Hasuwa H, Kilduff TS, Terao A, Yamanaka A (2014) Optogenetic manipulation of activity and temporally controlled cell-

specific ablation reveal a role for MCH neurons in sleep/wake regulation. J Neurosci 34(20): 6896–6909
Verret L, Goutagny R, Fort P, Cagnon L, Salvert D, Leger L, Boissard R, Salin P, Peyron C, Luppi PH (2003) A role of melanin-concentrating hormone producing neurons in the central regulation of paradoxical sleep. BMC Neurosci 4:19
Vetrivelan R, Kong D, Ferrari LL, Arrigoni E, Madara JC, Bandaru SS, Lowell BB, Lu J, Saper CB (2016) Melanin-concentrating hormone neurons specifically promote rapid eye movement sleep in mice. Neuroscience 336:102–113
Voloschin LM, Tramezzani JH (1979) Milk ejection reflex linked to slow wave sleep in nursing rats. Endocrinology 105(5):1202–1207
Von Economo C (1930) Sleep as a problem of localization. J Nerv Ment Dis 71:249–259
Wakerley JB (1996) Milk ejection and its control. In: Knobil E, Neill JD (eds) The physiology of reproduction. Academic, New York, pp 3129–3191
Wightman RM, Robinson DL (2002) Transient changes in mesolimbic dopamine and their association with "reward". J Neurochem 82(4):721–735
Willie JT, Sinton CM, Maratos-Flier E, Yanagisawa M (2008) Abnormal response of melanin-concentrating hormone deficient mice to fasting: hyperactivity and rapid eye movement sleep suppression. Neuroscience 156(4):819–829
Zarrow MX, Denenberg VH, Anderson CO (1965) Rabbit: frequency of suckling in the pup. Science 150(3705):1835–1836
Zhao ZJ, Chi QS, Cao J (2010) Milk energy output during peak lactation in shaved Swiss mice. Physiol Behav 101(1):59–66

Melanin-Concentrating Hormone in Medical Conditions

Alicia Costa, Luciana Benedetto, Patricia Lagos, Jaime M. Monti, and Pablo Torterolo

Abstract In the last decade, there was an impressive advance in the knowledge of the anatomy and physiology of the melanin-concentrating hormone (MCH) neuronal system. However, its role in pathology is still not clear. MCH is a peptidergic neuromodulator synthesized by neurons whose somas are mainly located in the posterolateral hypothalamus and incerto-hypothalamic area. The MCHergic neurons project throughout the central nervous system innervating areas involved in several physiological functions. MCH exerts its biological effects acting through two metabotropic receptors. There are substantive experimental data suggesting that the MCHergic system is involved in the control of energy homeostasis, mood, and sleep. In the present study, we summarize data related to the role of MCH in these functions, as well as preclinical and clinical evidences showing that dysfunction of the MCHergic system might be involved in several medical disorders including obesity, central hypersomnia, mood disorders, and ciliopathies.

1 Melanin-Concentrating Hormone: General Aspects

In the preceding chapters, a detailed description of the anatomy and physiology of the melanin-concentrating hormone (MCH) has been performed. Briefly, MCH is a 19 amino-acid cyclic neuropeptide that was first documented in the pituitary of teleost fish, enabling them to change skin color and blend into their environment (Kawauchi

A. Costa (✉) · L. Benedetto · P. Lagos · P. Torterolo
Department of Physiology, School of Medicine, University of the Republic, Montevideo, Uruguay
e-mail: acosta@fmed.edu.uy

J. M. Monti
Department of Pharmacology and Therapeutics, School of Medicine Clinics Hospital, University of the Republic, Montevideo, Uruguay

S. R. Pandi-Perumal et al. (eds.), *Melanin-Concentrating Hormone and Sleep*,
https://doi.org/10.1007/978-3-319-75765-0_10

et al. 1983). MCH was identified in mammalian brain where it is primarily found in neurons located in the posterolateral hypothalamus and incerto-hypothalamic area; these neurons project to different regions of the central nervous system (CNS) (Table 1) (Bittencourt et al. 1992). Brainstem tanycytes of the cat also show MCH immunoreactivity (Torterolo et al. 2008). MCH is also present in the gastrointestinal tract and pancreas (Kokkotou et al. 2008; Pissios et al. 2007).

Most of the MCHergic neurons co-localize with GABA (Sapin et al. 2010), and additional subpopulations of MCHergic neurons can be differentiated by their co-expression of the peptides nesfatin or cocaine-amphetamine-regulated transcript (CART) (Elias et al. 2001; Foo et al. 2008). Moreover, some MCHergic neurons may also release glutamate (Chee et al. 2015).

MCH acts via neurons expressing the metabotropic MCH receptor-1(MCHR1). Humans, primates, cats, and dogs (but not rodents) also express the MCH receptor-2 (MCHR2). There is a widespread distribution of the MCHR1 mRNA (Hervieu et al. 2000; Saito et al. 2001), and this receptor has also been detected in peripheral organs (Macneil 2013; Presse et al. 2014). The distribution of MCHR2 nearly overlaps that of the MCHR1 (Hill et al. 2001; Mori et al. 2001); however, the function of MCHR2 remains unknown mostly because rodents cannot be used as animal models.

Table 1 Main structures that receive projections from MCHergic neurons

Telencephalon
– Cerebral cortex
– Olfactory regions
Main bulb, anterior olfactory nucleus, olfactory tubercle
– Limbic system
Hippocampal formation, amygdala, lateral septal nucleus
– Basal forebrain
Diagonal band of Broca (horizontal and vertical limbs), medial septal nucleus, bed nucleus of the stria terminalis
– Neostriatum
Caudate-putamen, globus pallidus
Diencephalon
– Thalamus
Midline, intralaminar, anterior, lateral, ventral, and posterior nuclei
– Hypothalamus
Preoptic area, anterior, lateral, and posterior hypothalamic areas, tuberomammillary nucleus
Mesencephalon
Substantia nigra pars compacta, ventral tegmental area, central gray, mesencephalic reticular formation
Rhombencephalon
Pontine reticular formation, pedunculopontine and laterodorsal tegmental nucleus, locus coeruleus, dorsal raphe nucleus, median raphe nucleus, periaqueductal gray
Spinal cord

Data from Bittencourt (2011), Bittencourt and Elias (1998), Bittencourt et al. (1992, 1998), Elias and Bittencourt (1997), Elias et al. (2008), Hong et al. (2011), Torterolo et al. (2009) and Yoon and Lee (2013)

MCH has mainly an inhibitory role, both at the postsynaptic and presynaptic level where it decreases the release of GABA and glutamate (Gao 2009; Macneil 2013; Presse et al. 2014).

2 Physiological Role of the MCHergic System

MCH is involved in a broad spectrum of cerebral functions (Agame et al. 2015; Benedetto et al. 2014; Diniz and Bittencourt 2017; Ferreira et al. 2017; Macneil 2013). In fact, MCHergic neurons receive monosynaptic inputs from widespread regions of the CNS; these neurons seem to orchestrate the adaptation of global physiologic state to metabolic needs (Gonzalez et al. 2016). Overall, by modulation of different neuronal networks, the MCHergic system tends to promote energy storage.

2.1 Energy Homeostasis

The presence of MCHergic neurons in the lateral hypothalamus of mammalian brains, a critical area in the control of energy homeostasis, has generated the proposal that this peptide may regulate appetite and/or energy expenditure. In fact, the role of the MCHergic system in the control of energy balance is now firmly established (Brown et al. 2015; Macneil 2013; Presse et al. 2014; Takase et al. 2014). For instance, central infusion of MCH induces food intake and increases in body weight, and its gene expression is modulated by fasting in normal rats and in animals without leptin (ob/ob mice) (Presse et al. 1996; Qu et al. 1996). Knockout (KO) mice for ppMCH gene are lean due to hypophagia and increase in metabolic rate (Shimada et al. 1998), whereas mice with an overexpression of MCH eat more than controls leading them to obesity (Ludwig et al. 2001). MCHR1-KO mice are lean, are hyperactive, and have altered metabolism (Chen et al. 2002; Marsh et al. 2002). MCH signaling-deficient mice also present an increase in body temperature, oxygen consumption, heart rate, and mean arterial pressure (Takase et al. 2014).

2.2 Sleep

The role of MCH as a sleep-promoting factor has been reviewed in detail (Konadhode et al. 2015; Monti et al. 2013; Torterolo et al. 2011, 2015). Briefly, if conservation of energy is one of the main functions of sleep (Siegel 2005), and MCHergic neurons are critical in the control of energy homeostasis, MCH is expected to be involved in sleep regulation. In fact, MCHergic neurons project toward regions that are critically involved in the control of sleep and wakefulness (Bittencourt et al. 1992; Torterolo et al. 2009) (Table 1).

Intracerebroventricular administration of MCH in the rat produces a marked increase in REM sleep and a moderate enhancement in the time spent in non-REM (NREM) sleep (Verret et al. 2003). Furthermore, the systemic administration of MCHR-1 antagonists decreases both REM and NREM sleep and increases wakefulness (Ahnaou et al. 2008). Microinjection of MCH into REM-off neuronal areas such as the dorsal raphe (DR) and the locus coreuleus (LC) of the rat facilitates the generation of REM sleep (Lagos et al. 2009; Monti et al. 2015). MCH also promotes REM sleep when microinjected into either the basal forebrain of the rat or the NPO of the cat, two areas related to the generation of this behavioral state (Lagos et al. 2012; Torterolo et al. 2009). In contrast, the administration of MCH into a NREM sleep-promoting area such as the ventrolateral preoptic area (VLPO) induced NREM sleep (Benedetto et al. 2013).

Using Fos protein as a marker of neuronal activity, it has been shown that MCHergic neurons are active during REM sleep in the rat (Verret et al. 2003). Furthermore, Hassani et al. (2009) have recorded MCHergic neurons in non-anesthetized animals in restrained conditions. These neurons have a very low frequency of discharge during wakefulness, and their firing rate increases slightly during NREM sleep and reaches the maximum level of activation during REM sleep (Hassani et al. 2009). Utilizing the in vivo microdialysis technique, it has been shown that the release of MCH in the amygdala of patients with treatment-resistant epilepsy is minimal during active wakefulness with social interactions, increases after eating (consummatory behavior), and reaches a maximum level at sleep onset (Blouin et al. 2013). However, by means of fiber-optic recordings of molecularly MCH-defined populations in unrestrained conditions, Gonzalez et al. (2016) have recently shown that these neurons are activated in wakefulness during novelty exploration and inhibited by stress.

Studies of ppMCH and MCHR1-KO mice indicate that the sleep of these animals is altered. A recent meta-analysis study has confirmed that the strongest results of MCH signaling deficiency were the enhancing of wakefulness and suppression of NREM sleep (Takase et al. 2014). Optogenetic and chemogenetic studies have confirmed the role of MCH in sleep generation (Blanco-Centurion et al. 2016; Jego et al. 2013; Tsunematsu et al. 2014; Vetrivelan et al. 2016).

3 Pathophysiology of the MCHergic System

Preclinical experimental evidences in animal models suggest that the MCHergic system is involved in several disorders. However, although there are some data in human beings, more clinical studies are needed to confirm the preclinical findings. Below, we reviewed the possible involvement of MCH in neurological/psychiatric disorders and, briefly, in general medical conditions (Table 2).

Table 2 Main evidences of the involvement of MCH in pathology

Medical condition	Preclinical (*)/clinical (•) evidence
Obesity	* Acute intracerebroventricular (ICV) injections of MCH into the third ventricle of the rat increase food intake (Della-Zuana et al. 2002; Qu et al. 1996; Shearman et al. 2003)
	* Acute intracerebroventricular administrations of MCH increase food intake in sheep (Whitlock et al. 2005)
	* Chronic intracerebroventricular infusions of MCH in rodents induce hyperphagia and body weight gain (Della-Zuana et al. 2002; Gomori et al. 2003; Ito et al. 2003)
	* Microinjections of MCH into the arcuate nucleus, the paraventricular nucleus, or the dorsomedial nucleus increase in food intake (Abbott et al. 2003)
	* Overexpression of MCH in mice is associated with obesity and hyperinsulinemia (Ludwig et al. 2001)
	* Repeated oral administrations of a MCHR1 antagonist result in body weight reduction and in an anorectic effect in diet-induced-obesity mice (Kawata et al. 2017)
	* Administration of MCHR1 antagonists reduced body weight in dogs and diet-induced-obesity in mice (Ploj et al. 2016)
	* MCHR1 deletion significantly impairs the cue-potentiated feeding in mice (Sherwood et al. 2015)
	• Association between severe early-onset obesity and a history of hyperphagia with two variants of the MCHR1 (Gibson et al. 2004)
	• Two MCHR1 single-nucleotide polymorphisms were associated with an obese phenotype (Wermter et al. 2005)
	• Two single-nucleotide polymorphisms in MCHR2 were associated with childhood obesity (Ghoussaini et al. 2007)
Hypersomnia	* A rodent model of the human Africa trypanosomiasis or sleep sickness presents a progressive decrease of hypocretin and MCH-containing neurons (Palomba et al. 2015)
	• MCH CSF levels slightly increase in hypocretin-deficient narcoleptic children (Thebault et al. 2015)
Mood disorders	* Central administration of MCH stimulates the activity of the hypothalamic-pituitary-adrenal axis and produces changes in anxiety-related behavior (Gonzalez et al. 1996; Jezova et al. 1992; Kennedy et al. 2003; Smith et al. 2006)
	* MCHR1 antagonists have an antidepressant-like profile in the forced swim test (Borowsky et al. 2002)
	* Administration of MCH into the dorsal raphe produces a depressive-like behavior (Lagos et al. 2011b)
	• Bipolar affective disorder is associated with an MCHR2 polymorphism (Abou Jamra et al. 2010)
	• MCH serum levels decrease following 4 weeks of antidepressant treatment in depressive patients (Schmidt et al. 2015)
Alzheimer's disease (AD)	• There is an elevated mean level of MCH in CSF of patients suffering from AD, as well as a correlation with Tau and severity of cognitive impairment (Schmidt et al. 2013)

(continued)

Table 2 (continued)

Medical condition	Preclinical (*)/clinical (•) evidence
Parkinson's disease (PD)	• Decrease in number of both hypocretinergic and MCHergic neurons (Thannickal et al. 2007)
	• Loss of MCHergic neurons is lowest in stage I (12%) and maximal in stage V (74%) of PD (Thorpy and Adler 2005)
Schizophrenia	• Five single-nucleotide polymorphisms in the proximal region of the MCHR1 are associated with schizophrenia (Demontis et al. 2012)
	• Association between a single-nucleotide polymorphism in the ppMCH gene and obese schizophrenic patients receiving antipsychotics (Chagnon et al. 2007)
Ciliopathies	* MCH treatment significantly reduces cilia length (Hamamoto et al. 2016)
Postpartum emotional distress	* Microinjections of MCH into the mPOA decrease active maternal behaviors (Benedetto et al. 2014)
	* MCHR1-KO mice display disruption of maternal behavior (Alachkar et al. 2016)
Stroke	* Experimental studies in rats suggest that MCH could be a neuroprotective factor for brain ischemia (Pace et al. 2015)
Seizures	* MCHR1-KO mice present a higher threshold for induced seizures (Parks et al. 2010)
	* Intraventricular microinjection of MCH was shown to prevent PTZ-induced seizures (Knigge and Wagner 1997)
Addiction	* The MCHergic system is involved in mediating alcohol reward and cue-induced alcohol relapse (Cippitelli et al. 2010)
	* MCH in the nucleus accumbens shell plays a critical role in the development but not expression of methamphetamine-induced locomotor sensitization in rats (Sun et al. 2013)
	* MCHR1 antagonism reduces appetite for calories and suppression of addictive-like behaviors in rats (Karlsson et al. 2012)
	* MCHR1-KO mice exhibit psychomotor susceptibility to cocaine (Tyhon et al. 2006)
	* Cannabinoids excite MCHergic neurons (Huang et al. 2007)
Huntington's disease	• Decrease in the number of hypocretinergic neurons while the number of MCHergic neurons remain unchanged (Aziz et al. 2008)
Colitis and intestinal tumors	* Mice genetically deficient in MCH have substantially reduced local inflammatory responses in a model of experimental colitis (Kokkotou et al. 2008; Ziogas et al. 2013)
	* Mice receiving treatments with an anti-MCH antibody develop attenuated colonic inflammation and survive longer (Kokkotou et al. 2008; Ziogas et al. 2013)
	* MCH-deficient mice have a reduction of intestinal tumors (Nagel et al. 2012)
	• Increased colonic expression of MCH and its receptor in patients with inflammatory bowel disease (Kokkotou et al. 2008; Ziogas et al. 2013)

(continued)

Table 2 (continued)

Medical condition	Preclinical (*)/clinical (•) evidence
Skin pathology	• MCHR1 autoantibodies are present in vitiligo patients (Kemp and Weetman 2009)
	• Upregulation of the MCHR1 in the skin lesions of patients with psoriasis (Loite et al. 2013)
	• MCH receptors are present in human melanoma cells (Drozdz and Eberle 1995)
Cachexia	• Increase in the number of MCHR1 in the infundibular nucleus in postmortem brain material of cachectic patients (Unmehopa et al. 2005)
Osteoporosis	* MCHR1-KO mice have osteoporosis caused by a reduction in the cortical bone mass (Bohlooly et al. 2004)
Reproductive pathology	* MCH regulates the release of LHRH in the female rat (Gallardo et al. 2004)

3.1 Obesity

Obesity is one of the most pervasive chronic diseases in need of new strategies for medical treatment and prevention. In western society, it is one of the leading causes of mortality, morbidity, and healthcare costs. Obesity is defined as excess adipose tissue in which a distortion in the energy balance causes overweight.

The hypothalamus is critical for the control of body weight. MCH is involved in the regulation of energy homeostasis, and therefore, it may play a role in the pathogenesis of obesity. Several experimental data support this hypothesis [reviewed by Macneil (2013), Pissios (2009) and Presse et al. (2014)]. For instance, acute intracerebroventricular injections of MCH transiently stimulate food intake in rats, and chronic intracerebroventricular infusion of MCH in rats or mice induces hyperphagia and body weight gain. The development of obesity is accompanied by an increase in liver and adipose tissue. Furthermore, plasma glucose, insulin, and leptin levels are also increased in these animals; all these parameters have been associated with obesity in humans. Overexpression of MCH in mice is also associated with obesity and hyperinsulinemia. Overall, preclinical studies have shown that the MCHergic system has a major role in the control of energy homeostasis; by increasing food intake and mostly regulating energy expenditure, this system tends to accumulate energy.

In humans, Gibson et al. (2004) found an association between severe early-onset obesity and a history of hyperphagia with two variants of the MCHR-1. Association of two MCHR-1 single-nucleotide polymorphisms with an obese phenotype in German children and adolescents was also found by Wermter et al. (2005). Two dysfunctional MCHR1 mutants were identified in markedly underweight individuals, raising the possibility that a lean phenotype may be linked to deficient MCHR1 signaling (Goldstein et al. 2010). Furthermore, two single-nucleotide polymorphisms within the MCHR2 were also associated with childhood obesity

(Ghoussaini et al. 2007). Finally, an association between MCHR2 polymorphisms and body mass index was found in psychiatric patients (Delacretaz et al. 2015).

Because of its important role in energy homeostasis, MCHR antagonists are considered to be a future therapeutic option for treating obesity (Luthin 2007; Rivera et al. 2008). Nowadays, there are substantial efforts in order to obtain an MCHR1 antagonist with the appropriate physical and chemical properties to be used for obesity treatment (Johansson and Lofberg 2015).

3.2 *Hypersomnia*

The American Academy of Sleep Medicine describes hypersomnia of central origin as the group of disorders in which the primary complaint is excessive daytime sleepiness, which is not caused by disturbed nocturnal sleep or misaligned circadian rhythms (American Academy of Sleep Medicine 2005). This group of diseases includes disorders such as narcolepsy with and without cataplexy. Although the MCHergic system is involved in the control of sleep, the relation of this system with hypersomnia is still not clear.

While hypocretin-containing neurons degenerate in narcolepsy with cataplexy, the MCHergic neurons that are intermingled with the hypocretinergic neurons in the lateral hypothalamus do not (Thannickal et al. 2000). Peyron et al. (2011) studied the CSF MCH levels in patients with central hypersomnia. The study included sixteen patients with a diagnosis of narcolepsy with cataplexy, six patients with idiopathic hypersomnia, and two patients with post-traumatic hypersomnia who were compared to neurological patients without hypersomnia. The MCH levels were slightly lower in patients with hypersomnia, but after exclusion of the patients with post-traumatic hypersomnia, these differences became nonsignificant (Peyron et al. 2011). On the contrary, Thebault et al. (2015) observed MCH CSF levels marginally increased ($p = 0.054$) in hypocretin-deficient narcoleptic children following H1N1-AS03 vaccination (Thebault et al. 2015).

One quarter of narcoleptic patients' harbor autoantibodies that, when applied to normal rat and mouse brains, label MCH-containing neurons of the lateral hypothalamus and pro-opiomelanocortine (POMC) neurons of the nucleus arcuatus of the hypothalamus (Bergman et al. 2014). This labeling is blocked by NEI and the melanocyte-stimulating hormone, suggesting that the autoantibody recognizes an epitope present in both neuropeptides. The role of MCH and NEI and its relation with the hypocretinergic neurons in narcolepsy is still unknown.

The rodent model of the human Africa trypanosomiasis or sleep sickness reveals a progressive decrease of hypocretin- and MCH-containing neurons, suggesting that these peptidergic systems could contribute to the pathogenesis of the hypersomnia present in this disease (Palomba et al. 2015).

3.3 Mood Disorders

The enriched distribution of MCHergic fibers and receptors within the brain areas involved in the modulation of mood (Bittencourt et al. 1992; Hervieu et al. 2000) supports recent evidence suggesting the implication of the MCHergic system in the control of mood. In this respect, central administration of MCH stimulated the activity of the hypothalamic-pituitary-adrenal (HPA) axis and produced changes in anxiety-related behaviors (Gonzalez et al. 1996; Jezova et al. 1992; Kennedy et al. 2003; Smith et al. 2006). In addition, a recent study determined that MCHergic neurons are inhibited by stress (Gonzalez et al. 2016).

The relationship between the MCHergic system and major depression (MD) has been recently reviewed (Torterolo et al. 2015). Briefly, MD is a mood disorder that is diagnosed on the basis of symptomatic criteria like sadness or melancholia, guilt, irritability, and anhedonia that can be accompanied by insomnia or hypersomnia and alterations in body weight. MD is one of the most common psychiatric diseases with high prevalence and a strong association with suicide.

Borowsky et al. (2002) demonstrated in rats that MCHR-1 antagonists have an antidepressant-like profile in the forced swim test (FST), a widely used experimental paradigm for screening antidepressant activity. This finding was confirmed by most of the recent studies.

A large number of studies showed that the serotonergic system is involved in the physiopathology of MD (Torterolo et al. 2015). The DR contains the vast majority of serotonergic neurons of the brain (Monti 2010). Administration of MCH into the DR produces a depressive-like behavior in the FST that can be prevented by pretreatment with MCHR1 antagonists, as well as by the antidepressants fluoxetine or nortryptiline (Lagos et al. 2011; Urbanavicius et al. 2014). In addition, anti-MCH antibody infusion into the DR (immunoneutralization) produced an antidepressive effect (Lagos et al. 2011). These results suggest that the MCHergic neurons are involved in the regulation of emotional behaviors through the modulation of the serotonergic neuronal activity within the DR. Recently, we demonstrated that administration of MCH into the DR decreases the DR neuronal firing and the release of serotonin (Devera et al. 2015; Urbanavicius et al. 2016).

In summary, animal studies strongly suggest that MCH is a pro-depressive factor. In this regard, preclinical studies have demonstrated that MCHR1 antagonists have an enormous potential as antidepressant drugs (Chung et al. 2010; Rivera et al. 2008; Shimazaki et al. 2006).

Studies in humans are scarce. However, bipolar affective disorder (BPAD) is associated with an MCHR2 polymorphism (Abou Jamra et al. 2010). This association was most pronounced in BPAD male patients with psychotic symptoms. Furthermore, Schmidt et al. (2015) have shown that MCH serum levels decrease following 4 weeks of antidepressant treatment in MD patients (Schmidt et al. 2015).

Preclinical studies suggest a great potential of the MCHR1 antagonists for the treatment of depression and anxiety disorders, due to the rapid onset of effect and minor adverse CNS side effects (Shimazaki et al. 2006).

3.4 Alzheimer's Disease

Increased life expectancy has produced a dramatic rise in the prevalence and impact of aging-associated diseases, including dementia. Alzheimer's disease (AD) is by far the most common dementia in late life. Although progressive deterioration of memory, language, and intellect are the classic hallmarks of AD, sleep disturbance is a common and highly disruptive symptom associated with AD (Mander et al. 2016; Peter-Derex et al. 2015). Epidemiological studies have reported that up to 45% of patients with AD have sleep disturbances.

AD pathogenesis is widely believed to be driven by the production and deposition of the beta-amyloid peptide (Aβ) (Murphy and LeVine 2010). Aβ accumulation has been hypothesized to result from an imbalance between Aβ production and clearance; indeed, Aβ clearance seems to be impaired in both early and late forms of AD. Tarasoff-Conway et al. (2015) reviewed the clearance systems that act together to drive extracellular Aβ from the brain; any alteration of their function could contribute to AD.

Interestingly, experimental increase in cortical Aβ produce NREM sleep fragmentation (Mander et al. 2016; Peter-Derex et al. 2015). Furthermore, experimental deprivation of NREM sleep accompanied by an extended wakefulness escalates Aβ production and the corresponding cortical deposition. Therefore, NREM sleep represents one crucial pathway through which the brain appears to manage Aβ levels; the absence of sleep contributes to the aggregation of Aβ, while the presence of NREM sleep proactively reduces Aβ burden. Within this proposed framework, disrupted NREM increases Aβ aggregation, which by itself impairs NREM sleep, resulting in a vicious cycle accelerating AD progression. Remarkably, the study by Xie et al. (2013) showed that during sleep, there is a marked increase in convective exchange of CSF with interstitial brain fluid (glymphatic system). So, a disruption of sleep may decrease the removal of potentially neurotoxic waste products that accumulate in the CNS during wakefulness.

Because the MCHergic system is involved in sleep promotion and in learning and memory (Adamantidis and de Lecea 2009), it is likely that a dysfunction of this system may participate in the pathophysiology of AD. In this respect, there is an elevated mean level of MCH in CSF of patients suffering from AD, as well as a correlation with Tau and severity of cognitive impairment; these data point toward an impact of MCH in AD (Schmidt et al. 2013). New data are needed in order to clarify the role of MCH in AD.

3.5 Parkinson's Disease

The raised incidence of Parkinson's disease (PD) becomes a serious issue in an aged society (Pringsheim et al. 2014). The main symptoms of PD are resting tremor and bradykinesia in limbs which are effectively improved by L-dopa, but the disease is usually accompanied by non-motor symptoms (Postuma et al. 2015).

In addition to the well-known degeneration of dopaminergic neurons, PD is accompanied by a decrease in the number of both hypocretinergic and MCHergic neurons (Thannickal et al. 2007). The loss of both neuronal types was correlated with the clinical stage of PD; the loss of MCHergic neurons was lowest in stage I (12%) and maximal in stage V (74%). The tentative relationship between a loss of MCHergic activity and some symptoms of PD has not been studied yet; however, a decrease in the activity of MCHergic cells could partly explain the high incidence of sleep-related symptoms in PD such as REM sleep behavior disorder (Thorpy and Adler 2005). Interestingly, MCH has neuroprotective effects upon dopaminergic neurons in mice PD models (Park et al. 2016). In fact, they revealed that intranasal administration of MCH directly mediates neuronal protection in the brain, which holds great promise for the application of MCH in the therapy of neurodegenerative diseases.

3.6 *Schizophrenia*

Deficits in sensorimotor gating measured by prepulse inhibition (PPI) of the startle response have been known as characteristics of patients with schizophrenia and related neuropsychiatric disorders. PPI disruption is thought to rely on the activity of the mesocorticolimbic dopaminergic system and is inhibited by most antipsychotic drugs. The MCHergic system is a regulator of sensorimotor gating and may be involved in the physiopathology of these psychiatric disorders (Chung et al. 2011).

Relations between MCH and schizophrenia have been also found in humans. Five single-nucleotide polymorphisms in the proximal region of the MCHR1 were associated with schizophrenia (Demontis et al. 2012). The authors suggested that MCHR1 might influence schizophrenia susceptibility, in particular among men and patients responding to conventional treatment. Chagnon et al. (2007) also found an association between a single-nucleotide polymorphism in the ppMCH gene and obese schizophrenic patients receiving the antipsychotic olanzapine.

3.7 *Ciliopathies*

Primary cilia are appendages that are present in most human cell types, and neuronal primary cilia are abundant throughout the rodent brain (Bishop et al. 2007; Niño et al. 2017; Wheatley et al. 1996). MCHR1 is located in the primary cilia of neurons (Berbari et al. 2008). The signaling proteins that are enriched in the membrane of the cilium define its function. Importantly, disruption of the signaling mediated by these receptors can cause disease and altered development (Bisgrove and Yost 2006; Davenport and Yoder 2005; Hildebrandt and Otto 2005; Pan et al. 2005). The importance of these cilia is suggested by the fact that several human ciliary disorders (ciliopathies), including Bardet-Biedl syndrome (BBS), Joubert syndrome, and Meckel syndrome, have prominent functional and structural CNS deficits (Badano et al. 2006). For instance, a recent study in BBS patients showed statistically significant increased CSF volume in both the subarachnoidal space and ventricles (Keppler-Noreuil et al. 2011).

In mice models of BBS, MCHR-1 fails to localize to neuronal primary cilia and accumulates in the cytoplasm in the olfactory tubercle, hypothalamus, and nucleus accumbens (Berbari et al. 2008). These mutant mice exhibit obesity and/or cognitive defects (Hildebrandt et al. 2011). Therefore, MCHR-1 location in neuronal primary cilia might play an important role in the regulation of several physiological aspects. Interestingly, Conductier et al. (2013) showed that MCH modulates ciliary beating of ependymal cells at the third ventricle, suggesting that it could contribute to maintaining CSF homeostasis. In addition, MCH treatment significantly reduced cilia length in hTERT-RPE-1 cell line, suggesting that MCH modulates the sensibility of cells to external environment by controlling cilia length (Hamamoto et al. 2016).

3.8 Other Neurological/Psychiatric Conditions

Postpartum Emotional Distress Postpartum emotional distress is a special kind of mood disorder. Preclinical studies have shown that the medial preoptic area (mPOA) together with components of the mesolimbic dopamine system play a critical role in motivational processes of active maternal behavior (Miller and Lonstein 2005; Numan and Stolzenberg 2009; Stack et al. 2002). Interestingly, neurons of the mPOA expressed MCH only during the postpartum period (Rondini et al. 2010). Notably, whereas MCH expression in the mPOA progressively increases during lactation, maternal behavior naturally declines, suggesting that elevated MCHergic activity in the mPOA inhibits maternal behavior in the late postpartum period. In fact, Benedetto et al. (2014) have shown that microinjections of MCH into the mPOA decrease active maternal behaviors. It has also been shown that MCHR1-KO mice display a disruption of maternal behavior (Alachkar et al. 2016). Hence, it can be speculated that levels of MCH must be within a certain range to display a normal maternal behavior.

Throughout the latter part of human gestation, there is a progressive increase in conjugated estrogens in maternal plasma (Challis et al. 1980), while there is a sharp decrease of them after placental delivery (Cunningham et al. 1989). This rapid decline in estrogen and other reproductive hormone levels, which occurs immediately after delivery, is thought to be the main cause of postpartum depression (Schiller et al. 2015). Interestingly, estrogen and MCH levels have an inverse relationship (Santollo and Eckel 2008, 2013). Thus, taking into account the abovementioned MCH effects on depression, it can be hypothesized that affective alterations after delivery can be at least in part, due to the rise in MCH levels triggered by the estrogen drop. Hence, it would be important to know whether postpartum blues that take place in 50–80% of human mothers or depression in the first days after delivery that has a prevalence of 25% (Gurel and Gurel 2000; Josefsson et al. 2001) is related to a dysfunction of MCH-containing neurons.

Stroke Recent experimental studies in rats have suggested that MCH could be a neuroprotective factor for brain ischemia (Pace et al. 2015).

Seizures In comparison to wild-type (WT) mice, MCHR1-KO mice present a higher threshold for inducing seizures following administration of the convulsing compounds pentylenetetrazol (PTZ) or pilocarpine (Parks et al. 2010). MCHR1-KO mice were also observed to be strongly protected from the development of PTZ kindling. In this regard, intraventricular microinjection of MCH was shown to prevent PTZ-induced seizures (Knigge and Wagner 1997). Thus, although these data suggest that the MCHergic system modulates seizures occurrence, the role of MCH in this function is still unclear.

Addiction The robust anatomical and functional relationship between the MCHergic system and the mesocortical dopaminergic system strongly suggests that MCH is involved in the regulation of reward and addiction (Torterolo et al. 2016). Experimental evidences suggest that the MCHergic system modulates alcohol intake (Duncan et al. 2006). Furthermore, MCHR1-KO mice exhibit an atypical psychomotor susceptibility to cocaine (Tyhon et al. 2006). In addition, cannabinoids excite MCHergic neurons (Huang et al. 2007). Clinical studies are still needed to shed light on the role of MCH in drug addiction.

Huntington's Disease In this hyperkinetic syndrome, there is a decrease in the number of hypocretinergic neurons; however, the number of MCHergic neurons remains intact (Aziz et al. 2008).

3.9 *Other Medical Conditions*

Colitis and Intestinal Tumors Mice genetically deficient in MCH have substantially reduced local inflammatory responses in a model of experimental colitis (Kokkotou et al. 2008; Ziogas et al. 2013). Likewise, mice receiving treatments with an anti-MCH antibody developed attenuated colonic inflammation and survived longer (Kokkotou et al. 2008; Ziogas et al. 2013). Furthermore, there is an increased colonic expression of MCH and its receptor in patients with inflammatory bowel disease, suggesting a role of MCH in inflammatory processes in the intestine (Kokkotou et al. 2008; Ziogas et al. 2013).

Salmonella-induced enterocolitis produces a larger mortality with a higher bacterial load in MCHR1-KO mice in comparison to wild-type mice; this is likely related to the fact that MCH increases the phagocytic capacity of monocytes that have the MCHR1 (Karagiannis et al. 2013).

MCH-deficient mice have a reduction of intestinal tumors which led to the suggestion that blocking MCH could be a therapeutic approach to decrease the risk of colorectal cancer (Nagel et al. 2012). It is important to note that the MCHergic system is considered a modulator of the immune system (Lakaye et al. 2009); therefore, this dysfunction could be involved in different pathological conditions.

Skin Pathology MCH receptor plays a critical role in mammalian skin physiology including pigmentation, inflammation, and cell proliferation (Kemp and Weetman 2009). These important physiological roles have a pathological counterpart. MCHR1 autoantibodies have been recognized in vitiligo patients (Kemp and Weetman 2009). In fact, these autoantibodies inhibit the receptor function when tested in a Chinese hamster ovarian cell line (Gottumukkala et al. 2006).

Patients with psoriasis have an upregulation of the MCHR1 in the skin lesions (Loite et al. 2013); the relation of this finding with the pathogenesis of this condition is still unknown. MCH receptors are also present in human melanoma cells (Drozdz and Eberle 1995).

Cachexia A significant increase in the number of MCHR1 cell body staining was found in the infundibular nucleus in postmortem brain material of cachectic patients compared with matched controls, supporting a role for this receptor in energy homeostasis in humans and in this terminal condition (Unmehopa et al. 2005).

Osteoporosis MCHR1-KO mice have osteoporosis caused by a reduction in the cortical bone mass (Bohlooly et al. 2004). Serum levels of c-telopeptide, a marker of bone resorption, are also increased in these mice. The authors concluded that MCHR1 is involved in a tonic stimulation of bone mass.

Reproductive Pathologies Anatomical data strongly suggest that the MCHergic system may play a critical role in reproduction. MCHergic fibers are in close apposition with approximately 85–90% of gonadotropin-releasing hormone (GnRH) cell bodies throughout the preoptic and anterior hypothalamic area in the rat. In addition, MCHR1 was co-expressed on about 50–55% of GnRH neurons (Williamson-Hughes et al. 2005). These data, as well as the fact that MCH is involved in the regulation of luteinizing hormone-releasing hormone (LHRH) release in the female rat (Gallardo et al. 2004), strongly suggest that MCH may play a role in the regulation of GnRH neuronal function and reproduction. Clinical studies regarding the role of MCH in reproductive pathologies are still lacking.

4 Conclusions

The widespread distribution of the MCHergic fibers and receptors in the CNS strongly suggests that MCH is involved in several and critical physiological functions; the more salient roles are to promote energy conservation and sleep. Mood regulation is thought to be another important function of this system. The involvement of the MCHergic system in pathology is suggested by several preclinical studies. However, although MCH seems to be involved in several human pathologies, clinical research is still needed to shed light on the physiopathological role of this important peptidergic system in medical conditions.

Acknowledgments This work was partially supported by the "Programa de desarrollo de ciencias básicas," PEDECIBA, Uruguay.

Conflict of Interest None to declare.

References

Abbott CR, Kennedy AR, Wren AM, Rossi M, Murphy KG, Seal LJ, Todd JF, Ghatei MA, Small CJ, Bloom SR (2003) Identification of hypothalamic nuclei involved in the orexigenic effect of melanin-concentrating hormone. Endocrinology 144:3943–3949

Abou Jamra R, Schulze TG, Becker T, Brockschmidt FF, Green E, Alblas MA, Wendland JR, Adli M, Grozeva D, Strohmeier J, Georgi A, Craddock N, Propping P, Rietschel M, Nothen MM, Cichon S, Schumacher J (2010) A systematic association mapping on chromosome 6q in bipolar affective disorder—evidence for the melanin-concentrating-hormone-receptor-2 gene as a risk factor for bipolar affective disorder. Am J Med Genet B Neuropsychiatr Genet 153B: 878–884

Adamantidis A, de Lecea L (2009) The hypocretins as sensors for metabolism and arousal. J Physiol 587:33–40

Agame A, Calegare BFA, Fernandes L, Acosta A, Lagos P, Torterolo P, D'Almeida V (2015) MCH levels in the CSF, brain preproMCH and MCHR1 gene expression during sleep deprivation, sleep rebound and sleep restriction. Peptides 74:9–15

Ahnaou A, Drinkenburg WH, Bouwknecht JA, Alcazar J, Steckler T, Dautzenberg FM (2008) Blocking melanin-concentrating hormone MCH1 receptor affects rat sleep-wake architecture. Eur J Pharmacol 579:177–188

Alachkar A, Alhassen L, Wang Z, Wang L, Onouye K, Sanathara N, Civelli O (2016) Inactivation of the melanin concentrating hormone system impairs maternal behavior. Eur Neuropsychopharmacol 26:1826–1835

American Academy of Sleep Medicine (ed) (2005) International classification of sleep disorder, 2nd edn. American Academy of Sleep Medicine, Westchester

Aziz A, Fronczek R, Maat-Schieman M, Unmehopa U, Roelandse F, Overeem S, van Duinen S, Lammers GJ, Swaab D, Roos R (2008) Hypocretin and melanin-concentrating hormone in patients with Huntington disease. Brain Pathol 18:474–483

Badano JL, Leitch CC, Ansley SJ, May-Simera H, Lawson S, Lewis RA, Beales PL, Dietz HC, Fisher S, Katsanis N (2006) Dissection of epistasis in oligogenic Bardet-Biedl syndrome. Nature 439:326–330

Benedetto L, Rodriguez-Servetti Z, Lagos P, D'Almeida V, Monti JM, Torterolo P (2013) Microinjection of melanin concentrating hormone into the lateral preoptic area promotes non-REM sleep in the rat. Peptides 39C:11–15

Benedetto L, Pereira M, Ferreira A, Torterolo P (2014) Melanin-concentrating hormone in the medial preoptic area reduces active components of maternal behavior in rats. Peptides 58:20–25

Berbari NF, Johnson AD, Lewis JS, Askwith CC, Mykytyn K (2008) Identification of ciliary localization sequences within the third intracellular loop of G protein-coupled receptors. Mol Biol Cell 19:1540–1547

Bergman P, Adori C, Vas S, Kai-Larsen Y, Sarkanen T, Cederlund A, Agerberth B, Julkunen I, Horvath B, Kostyalik D, Kalmar L, Bagdy G, Huutoniemi A, Partinen M, Hokfelt T (2014) Narcolepsy patients have antibodies that stain distinct cell populations in rat brain and influence sleep patterns. Proc Natl Acad Sci USA 111:E3735–E3744

Bisgrove BW, Yost HJ (2006) The roles of cilia in developmental disorders and disease. Development 133:4131–4143

Bishop GA, Berbari NF, Lewis J, Mykytyn K (2007) Type III adenylyl cyclase localizes to primary cilia throughout the adult mouse brain. J Comp Neurol 505:562–571

Bittencourt JC (2011) Anatomical organization of the melanin-concentrating hormone peptide family in the mammalian brain. Gen Comp Endocrinol 172:185–197

Bittencourt JC, Elias CF (1998) Melanin-concentrating hormone and neuropeptide EI projections from the lateral hypothalamic area and zona incerta to the medial septal nucleus and spinal cord: a study using multiple neuronal tracers. Brain Res 805:1–19

Bittencourt JC, Presse F, Arias C, Peto C, Vaughan J, Nahon JL, Vale W, Sawchenko PE (1992) The melanin-concentrating hormone system of the rat brain: an immuno- and hybridization histochemical characterization. J Comp Neurol 319:218–245

Bittencourt JC, Frigo L, Rissman RA, Casatti CA, Nahon JL, Bauer JA (1998) The distribution of melanin-concentrating hormone in the monkey brain (Cebus apella). Brain Res 804:140–143

Blanco-Centurion C, Liu M, Konadhode RP, Zhang X, Pelluru D, van den Pol AN, Shiromani PJ (2016) Optogenetic activation of melanin-concentrating hormone neurons increases non-rapid eye movement and rapid eye movement sleep during the night in rats. Eur J Neurosci 44:2846–2857

Blouin AM, Fried I, Wilson CL, Staba RJ, Behnke EJ, Lam HA, Maidment NT, Karlsson KAE, Lapierre JL, Siegel JM (2013) Human hypocretin and melanin-concentrating hormone levels are linked to emotion and social interaction. Nat Commun 4:1547

Bohlooly YM, Mahlapuu M, Andersen H, Astrand A, Hjorth S, Svensson L, Tornell J, Snaith MR, Morgan DG, Ohlsson C (2004) Osteoporosis in MCHR1-deficient mice. Biochem Biophys Res Commun 318:964–969

Borowsky B, Durkin MM, Ogozalek K, Marzabadi MR, DeLeon J, Lagu B, Heurich R, Lichtblau H, Shaposhnik Z, Daniewska I, Blackburn TP, Branchek TA, Gerald C, Vaysse PJ, Forray C (2002) Antidepressant, anxiolytic and anorectic effects of a melanin-concentrating hormone-1 receptor antagonist. Nat Med 8:825–830

Brown JA, Woodworth HL, Leinninger GM (2015) To ingest or rest? Specialized roles of lateral hypothalamic area neurons in coordinating energy balance. Front Syst Neurosci 9:9

Chagnon YC, Bureau A, Gendron D, Bouchard RH, Merette C, Roy MA, Maziade M (2007) Possible association of the pro-melanin-concentrating hormone gene with a greater body mass index as a side effect of the antipsychotic olanzapine. Am J Med Genet B Neuropsychiatr Genet 144B:1063–1069

Challis JR, Patrick JE, Campbell K, Natale R, Richardson B (1980) Diurnal changes in maternal plasma oestrone and oestradiol at 30 to 31, 34 to 35 and 38 to 39 weeks gestational age. Br J Obstet Gynaecol 87:983–988

Chee MJ, Arrigoni E, Maratos-Flier E (2015) Melanin-concentrating hormone neurons release glutamate for feedforward inhibition of the lateral septum. J Neurosci 35:3644–3651

Chen Y, Hu C, Hsu CK, Zhang Q, Bi C, Asnicar M, Hsiung HM, Fox N, Slieker LJ, Yang DD, Heiman ML, Shi Y (2002) Targeted disruption of the melanin-concentrating hormone receptor-1 results in hyperphagia and resistance to diet-induced obesity. Endocrinology 143:2469–2477

Chung S, Parks GS, Lee C, Civelli O (2010) Recent updates on the melanin-concentrating hormone (MCH) and its receptor system: lessons from MCH1R antagonists. J Mol Neurosci 43:115–121

Chung S, Verheij MM, Hesseling P, van Vugt RW, Buell M, Belluzzi JD, Geyer MA, Martens GJ, Civelli O (2011) The melanin-concentrating hormone (MCH) system modulates behaviors associated with psychiatric disorders. PLoS One 6:e19286

Cippitelli A, Karlsson C, Shaw JL, Thorsell A, Gehlert DR, Heilig M (2010) Suppression of alcohol self-administration and reinstatement of alcohol seeking by melanin-concentrating hormone receptor 1 (MCH1-R) antagonism in Wistar rats. Psychopharmacology 211:367–375

Conductier G, Martin AO, Risold PY, Jego S, Lavoie R, Lafont C, Mollard P, Adamantidis A, Nahon JL (2013) Control of ventricular ciliary beating by the melanin concentrating hormone-expressing neurons of the lateral hypothalamus: a functional imaging survey. Front Endocrinol 4:182

Cunningham FG, MacDonald PC, Gant NF (1989) Williams obstetrics, 19th edn. Appleton and Lange, Norwalk

Davenport JR, Yoder BK (2005) An incredible decade for the primary cilium: a look at a once-forgotten organelle. Am J Physiol 289:F1159–F1169

Delacretaz A, Preisig M, Vandenberghe F, Saigi Morgui N, Quteineh L, Choong E, Gholam-Rezaee M, Kutalik Z, Magistretti P, Aubry JM, von Gunten A, Castelao E, Vollenweider P, Waeber G, Conus P, Eap CB (2015) Influence of MCHR2 and MCHR2-AS1 genetic polymorphisms on body mass index in psychiatric patients and in population-based subjects with present or past atypical depression. PLoS One 10:e0139155

Della-Zuana O, Presse F, Ortola C, Duhault J, Nahon JL, Levens N (2002) Acute and chronic administration of melanin-concentrating hormone enhances food intake and body weight in Wistar and Sprague-Dawley rats. Int J Obes Relat Metab Disord 26:1289–1295

Demontis D, Nyegaard M, Christensen JH, Severinsen J, Hedemand A, Hansen T, Werge T, Mors O, Borglum AD (2012) The gene encoding the melanin-concentrating hormone receptor 1 is associated with schizophrenia in a Danish case-control sample. Psychiatr Genet 22:62–69

Devera A, Pascovich C, Lagos P, Falconi A, Sampogna S, Chase MH, Torterolo P (2015) Melanin-concentrating hormone (MCH) modulates the activity of dorsal raphe neurons. Brain Res 1598:114–128

Diniz GB, Bittencourt JC (2017) The melanin-concentrating hormone as an integrative peptide driving motivated behaviors. Front Syst Neurosci 11:32

Drozdz R, Eberle AN (1995) Binding sites for melanin-concentrating hormone (MCH) in brain synaptosomes and membranes from peripheral tissues identified with highly tritiated MCH. J Recept Signal Transduct Res 15:487–502

Duncan EA, Rider TR, Jandacek RJ, Clegg DJ, Benoit SC, Tso P, Woods SC (2006) The regulation of alcohol intake by melanin-concentrating hormone in rats. Pharmacol Biochem Behav 85:728–735

Elias CF, Bittencourt JC (1997) Study of the origins of melanin-concentrating hormone and neuropeptide EI immunoreactive projections to the periaqueductal gray matter. Brain Res 755:255–271

Elias CF, Lee CE, Kelly JF, Ahima RS, Kuhar M, Saper CB, Elmquist JK (2001) Characterization of CART neurons in the rat and human hypothalamus. J Comp Neurol 432:1–19

Elias CF, Sita LV, Zambon BK, Oliveira ER, Vasconcelos LA, Bittencourt JC (2008) Melanin-concentrating hormone projections to areas involved in somatomotor responses. J Chem Neuroanat 35:188–201

Ferreira JGP, Bittencourt JC, Adamantidis A (2017) Melanin-concentrating hormone and sleep. Curr Opin Neurobiol 44:152–158

Foo KS, Brismar H, Broberger C (2008) Distribution and neuropeptide coexistence of nucleobindin-2 mRNA/nesfatin-like immunoreactivity in the rat CNS. Neuroscience 156:563–579

Gallardo MG, Chiocchio SR, Tramezzani JH (2004) Changes of melanin-concentrating hormone related to LHRH release in the median eminence of rats. Brain Res 1030:152–158

Gao XB (2009) Electrophysiological effects of MCH on neurons in the hypothalamus. Peptides 30:2025–2030

Ghoussaini M, Vatin V, Lecoeur C, Abkevich V, Younus A, Samson C, Wachter C, Heude B, Tauber M, Tounian P, Hercberg S, Weill J, Levy-Marchal C, Le Stunff C, Bougneres P, Froguel P, Meyre D (2007) Genetic study of the melanin-concentrating hormone receptor 2 in childhood and adulthood severe obesity. J Clin Endocrinol Metab 92:4403–4409

Gibson WT, Pissios P, Trombly DJ, Luan J, Keogh J, Wareham NJ, Maratos-Flier E, O'Rahilly S, Farooqi IS (2004) Melanin-concentrating hormone receptor mutations and human obesity: functional analysis. Obes Res 12:743–749

Goldstein C, Schroeder JC, Fortin JP, Goss JM, Schaus SE, Beinborn M, Kopin AS (2010) Two naturally occurring mutations in the type 1 melanin-concentrating hormone receptor abolish agonist-induced signaling. J Pharmacol Exp Ther 335:799–806

Gomori A, Ishihara A, Ito M, Mashiko S, Matsushita H, Yumoto M, Ito M, Tanaka T, Tokita S, Moriya M, Iwaasa H, Kanatani A (2003) Chronic intracerebroventricular infusion of MCH causes obesity in mice. Melanin-concentrating hormone. Am J Physiol Endocrinol Metab 284: E583–E588

Gonzalez MI, Vaziri S, Wilson CA (1996) Behavioral effects of alpha-MSH and MCH after central administration in the female rat. Peptides 17:171–177

Gonzalez JA, Iordanidou P, Strom M, Adamantidis A, Burdakov D (2016) Awake dynamics and brain-wide direct inputs of hypothalamic MCH and orexin networks. Nat Commun 7:11395

Gottumukkala RV, Gavalas NG, Akhtar S, Metcalfe RA, Gawkrodger DJ, Haycock JW, Watson PF, Weetman AP, Kemp EH (2006) Function-blocking autoantibodies to the melanin-concentrating hormone receptor in vitiligo patients. Lab Invest 86:781–789

Gurel S, Gurel H (2000) The evaluation of determinants of early postpartum low mood: the importance of parity and inter-pregnancy interval. Eur J Obstet Gynecol Reprod Biol 91:21–24

Hamamoto A, Yamato S, Katoh Y, Nakayama K, Yoshimura K, Takeda S, Kobayashi Y, Saito Y (2016) Modulation of primary cilia length by melanin-concentrating hormone receptor 1. Cell Signal 28:572–584

Hassani OK, Lee MG, Jones BE (2009) Melanin-concentrating hormone neurons discharge in a reciprocal manner to orexin neurons across the sleep-wake cycle. Proc Natl Acad Sci U S A 106:2418–2422

Hervieu GJ, Cluderay JE, Harrison D, Meakin J, Maycox P, Nasir S, Leslie RA (2000) The distribution of the mRNA and protein products of the melanin- concentrating hormone (MCH) receptor gene, slc-1, in the central nervous system of the rat. Eur J Neurosci 12:1194–1216

Hildebrandt F, Otto E (2005) Cilia and centrosomes: a unifying pathogenic concept for cystic kidney disease? Nat Rev 6:928–940

Hildebrandt F, Benzing T, Katsanis N (2011) Ciliopathies. N Engl J Med 364:1533–1543

Hill J, Duckworth M, Murdock P, Rennie G, Sabido-David C, Ames RS, Szekeres P, Wilson S, Bergsma DJ, Gloger IS, Levy DS, Chambers JK, Muir AI (2001) Molecular cloning and functional characterization of MCH2, a novel human MCH receptor. J Biol Chem 276:20125–20129

Hong EY, Yoon YS, Lee HS (2011) Differential distribution of melanin-concentrating hormone (MCH)- and hypocretin (Hcrt)-immunoreactive neurons projecting to the mesopontine cholinergic complex in the rat. Brain Res 1424:20–31

Huang H, Acuna-Goycolea C, Li Y, Cheng HM, Obrietan K, van den Pol AN (2007) Cannabinoids excite hypothalamic melanin-concentrating hormone but inhibit hypocretin/orexin neurons: implications for cannabinoid actions on food intake and cognitive arousal. J Neurosci 27:4870–4881

Ito M, Gomori A, Ishihara A, Oda Z, Mashiko S, Matsushita H, Yumoto M, Ito M, Sano H, Tokita S, Moriya M, Iwaasa H, Kanatani A (2003) Characterization of MCH-mediated obesity in mice. Am J Physiol Endocrinol Metab 284:E940–E945

Jego S, Glasgow SD, Herrera CG, Ekstrand M, Reed SJ, Boyce R, Friedman J, Burdakov D, Adamantidis AR (2013) Optogenetic identification of a rapid eye movement sleep modulatory circuit in the hypothalamus. Nat Neurosci 16:1637–1643

Jezova D, Bartanusz V, Westergren I, Johansson BB, Rivier J, Vale W, Rivier C (1992) Rat melanin-concentrating hormone stimulates adrenocorticotropin secretion: evidence for a site of action in brain regions protected by the blood-brain barrier. Endocrinology 130:1024–1029

Johansson A, Lofberg C (2015) Novel MCH1 receptor antagonists: a patent review. Expert Opin Ther Pat 25:193–207

Josefsson A, Berg G, Nordin C, Sydsjo G (2001) Prevalence of depressive symptoms in late pregnancy and postpartum. Acta Obstet Gynecol Scand 80:251–255

Karagiannis AK, Ziogas DC, Gras-Miralles B, Geiger BM, Nagel J, Trebicka E, Najarian R, Cherayil BJ, Kokkotou E (2013) Increased susceptibility of melanin-concentrating hormone-deficient mice to infection with Salmonella enterica serovar Typhimurium. Infect Immun 81:166–172

Karlsson C, Zook M, Ciccocioppo R, Gehlert DR, Thorsell A, Heilig M, Cippitelli A (2012) Melanin-concentrating hormone receptor 1 (MCH1-R) antagonism: reduced appetite for calories and suppression of addictive-like behaviors. Pharmacol Biochem Behav 102:400–406

Kawata Y, Okuda S, Hotta N, Igawa H, Takahashi M, Ikoma M, Kasai S, Ando A, Satomi Y, Nishida M, Nakayama M, Yamamoto S, Nagisa Y, Takekawa S (2017) A novel and selective melanin-concentrating hormone receptor 1 antagonist ameliorates obesity and hepatic steatosis in diet-induced obese rodent models. Eur J Pharmacol 796:45–53

Kawauchi H, Kawazoe I, Tsubokawa M, Kishida M, Baker BI (1983) Characterization of melanin-concentrating hormone in chum salmon pituitaries. Nature 305:321–323

Kemp EH, Weetman AP (2009) Melanin-concentrating hormone and melanin-concentrating hormone receptors in mammalian skin physiopathology. Peptides 30:2071–2075

Kennedy AR, Todd JF, Dhillo WS, Seal LJ, Ghatei MA, O'Toole CP, Jones M, Witty D, Winborne K, Riley G, Hervieu G, Wilson S, Bloom SR (2003) Effect of direct injection of melanin-concentrating hormone into the paraventricular nucleus: further evidence for a stimulatory role in the adrenal axis via SLC-1. J Neuroendocrinol 15:268–272

Keppler-Noreuil KM, Blumhorst C, Sapp JC, Brinckman D, Johnston J, Nopoulos PC, Biesecker LG (2011) Brain tissue- and region-specific abnormalities on volumetric MRI scans in 21 patients with Bardet-Biedl syndrome (BBS). BMC Med Genet 12:101

Knigge KM, Wagner JE (1997) Melanin-concentrating hormone (MCH) involvement in pentylenetetrazole (PTZ)-induced seizure in rat and guinea pig. Peptides 18:1095–1097

Kokkotou E, Moss AC, Torres D, Karagiannides I, Cheifetz A, Liu S, O'Brien M, Maratos-Flier E, Pothoulakis C (2008) Melanin-concentrating hormone as a mediator of intestinal inflammation. Proc Natl Acad Sci U S A 105:10613–10618

Konadhode RR, Pelluru D, Shiromani PJ (2015) Neurons containing orexin or melanin concentrating hormone reciprocally regulate wake and sleep. Front Syst Neurosci 8:244

Lagos P, Torterolo P, Jantos H, Chase MH, Monti JM (2009) Effects on sleep of melanin-concentrating hormone (MCH) microinjections into the dorsal raphe nucleus. Brain Res 1265:103–110

Lagos P, Urbanavicius J, Scorza C, Miraballes R, Torterolo P (2011) Depressive-like profile induced by MCH microinjections into the dorsal raphe nucleus evaluated in the forced swim test. Behav Brain Res 218:259–266

Lagos P, Monti JM, Jantos H, Torterolo P (2012) Microinjection of the melanin-concentrating hormone into the lateral basal forebrain increases REM sleep and reduces wakefulness in the rat. Life Sci 90:895–899

Lakaye B, Coumans B, Harray S, Grisar T (2009) Melanin-concentrating hormone and immune function. Peptides 30:2076–2080

Loite U, Kingo K, Reimann E, Reemann P, Vasar E, Silm H, Koks S (2013) Gene expression analysis of the corticotrophin-releasing hormone-proopiomelanocortin system in psoriasis skin biopsies. Acta Derm Venereol 93:400–405

Ludwig DS, Tritos NA, Mastaitis JW, Kulkarni R, Kokkotou E, Elmquist J, Lowell B, Flier JS, Maratos-Flier E (2001) Melanin-concentrating hormone overexpression in transgenic mice leads to obesity and insulin resistance. J Clin Invest 107:379–386

Luthin DR (2007) Anti-obesity effects of small molecule melanin-concentrating hormone receptor 1 (MCHR1) antagonists. Life Sci 81:423–440

Macneil DJ (2013) The role of melanin-concentrating hormone and its receptors in energy homeostasis. Front Endocrinol (Lausanne) 4:49

Mander BA, Winer JR, Jagust WJ, Walker MP (2016) Sleep: a novel mechanistic pathway, biomarker, and treatment target in the pathology of Alzheimer's disease? Trends Neurosci 39:552–566

Marsh DJ, Weingarth DT, Novi DE, Chen HY, Trumbauer ME, Chen AS, Guan XM, Jiang MM, Feng Y, Camacho RE, Shen Z, Frazier EG, Yu H, Metzger JM, Kuca SJ, Shearman LP, Gopal-Truter S, MacNeil DJ, Strack AM, MacIntyre DE, Van der Ploeg LH, Qian S (2002) Melanin-concentrating hormone 1 receptor-deficient mice are lean, hyperactive, and hyperphagic and have altered metabolism. Proc Natl Acad Sci U S A 99:3240–3245

Miller SM, Lonstein JS (2005) Dopamine d1 and d2 receptor antagonism in the preoptic area produces different effects on maternal behavior in lactating rats. Behav Neurosci 119:1072–1083

Monti JM (2010) The structure of the dorsal raphe nucleus and its relevance to the regulation of sleep and wakefulness. Sleep Med Rev 14:307–317

Monti JM, Torterolo P, Lagos P (2013) Melanin-concentrating hormone control of sleep-wake behavior. Sleep Med Rev 17:293–298

Monti JM, Lagos P, Jantos H, Torterolo P (2015) Increased REM sleep after intra-locus coeruleus nucleus microinjection of melanin-concentrating hormone (MCH) in the rat. Prog Neuro-Psychopharmacol Biol Psychiatry 56:185–188

Mori M, Harada M, Terao Y, Sugo T, Watanabe T, Shimomura Y, Abe M, Shintani Y, Onda H, Nishimura O, Fujino M (2001) Cloning of a novel G protein-coupled receptor, SLT, a subtype of the melanin-concentrating hormone receptor. Biochem Biophys Res Commun 283:1013–1018

Murphy MP, LeVine H 3rd (2010) Alzheimer's disease and the amyloid-beta peptide. J Alzheimers Dis 19:311–323

Nagel JM, Geiger BM, Karagiannis AK, Gras-Miralles B, Horst D, Najarian RM, Ziogas DC, Chen X, Kokkotou E (2012) Reduced intestinal tumorigenesis in APCmin mice lacking melanin-concentrating hormone. PLoS One 7:e41914

Niño S, Torterolo P, Lagos P (2017) Identificación del receptor de la hormona concentradora de melanina (MCHR-1) en las cilias primarias de neuronas del SNC de la rata. Congreso Nacional de Biociencias 2017:237

Numan M, Stolzenberg DS (2009) Medial preoptic area interactions with dopamine neural systems in the control of the onset and maintenance of maternal behavior in rats. Front Neuroendocrinol 30:46–64

Pace M, Baracchi F, Gao B, Bassetti C (2015) Identification of sleep-modulated pathways involved in Neuroprotection from stroke. Sleep 38:1707–1718

Palomba M, Seke-Etet PF, Laperchia C, Tiberio L, Xu YZ, Colavito V, Grassi-Zucconi G, Bentivoglio M (2015) Alterations of orexinergic and melanin-concentrating hormone neurons in experimental sleeping sickness. Neuroscience 290:185–195

Pan J, Wang Q, Snell WJ (2005) Cilium-generated signaling and cilia-related disorders. Lab Invest 85:452–463

Park JY, Kim SN, Yoo J, Jang J, Lee A, Oh JY, Kim H, Oh ST, Park SU, Kim J, Park HJ, Jeon S (2016) Novel neuroprotective effects of melanin-concentrating hormone in Parkinson's disease. Mol Neurobiol 54(10):7706–7721

Parks GS, Okumura SM, Gohil K, Civelli O (2010) Mice lacking melanin concentrating hormone 1 receptor are resistant to seizures. Neurosci Lett 484:104–107

Peter-Derex L, Yammine P, Bastuji H, Croisile B (2015) Sleep and Alzheimer's disease. Sleep Med Rev 19:29–38

Peyron C, Valentin F, Bayard S, Hanriot L, Bedetti C, Rousset B, Luppi PH, Dauvilliers Y (2011) Melanin concentrating hormone in central hypersomnia. Sleep Med 12:768–772

Pissios P (2009) Animals models of MCH function and what they can tell us about its role in energy balance. Peptides 30:2040–2044

Pissios P, Ozcan U, Kokkotou E, Okada T, Liew CW, Liu S, Peters JN, Dahlgren G, Karamchandani J, Kudva YC, Kurpad AJ, Kennedy RT, Maratos-Flier E, Kulkarni RN (2007) Melanin concentrating hormone is a novel regulator of islet function and growth. Diabetes 56:311–319

Ploj K, Benthem L, Kakol-Palm D, Gennemark P, Andersson L, Bjursell M, Borjesson J, Karrberg L, Mansson M, Antonsson M, Johansson A, Iverson S, Carlsson B, Turnbull A, Linden D (2016) Effects of a novel potent melanin-concentrating hormone receptor 1 antagonist, AZD1979, on body weight homeostasis in mice and dogs. Br J Pharmacol 173:2739–2751

Postuma RB, Berg D, Stern M, Poewe W, Olanow CW, Oertel W, Obeso J, Marek K, Litvan I, Lang AE, Halliday G, Goetz CG, Gasser T, Dubois B, Chan P, Bloem BR, Adler CH, Deuschl G (2015) MDS clinical diagnostic criteria for Parkinson's disease. Mov Disord 30:1591–1601

Presse F, Sorokovsky I, Max JP, Nicolaidis S, Nahon JL (1996) Melanin-concentrating hormone is a potent anorectic peptide regulated by food-deprivation and glucopenia in the rat. Neuroscience 71:735–745

Presse F, Conductier G, Rovere C, Nahon JL (2014) The melanin-concentrating hormone receptors: neuronal and non-neuronal functions. Int J Obes Suppl 4:S31–S36

Pringsheim T, Jette N, Frolkis A, Steeves TD (2014) The prevalence of Parkinson's disease: a systematic review and meta-analysis. Mov Disord 29:1583–1590

Qu D, Ludwig DS, Gammeltoft S, Piper M, Pelleymounter MA, Cullen MJ, Mathes WF, Przypek R, Kanarek R, Maratos-Flier E (1996) A role for melanin-concentrating hormone in the central regulation of feeding behaviour. Nature 380:243–247

Rivera G, Bocanegra-Garcia V, Galiano S, Cirauqui N, Ceras J, Perez S, Aldana I, Monge A (2008) Melanin-concentrating hormone receptor 1 antagonists: a new perspective for the pharmacologic treatment of obesity. Curr Med Chem 15:1025–1043

Rondini TA, Donato J Jr, Rodrigues Bde C, Bittencourt JC, Elias CF, (2010) Chemical identity and connections of medial preoptic area neurons expressing melanin-concentrating hormone during lactation. J Chem Neuroanat 39:51–62

Saito Y, Cheng M, Leslie FM, Civelli O (2001) Expression of the melanin-concentrating hormone (MCH) receptor mRNA in the rat brain. J Comp Neurol 435:26–40

Santollo J, Eckel LA (2008) Estradiol decreases the orexigenic effect of neuropeptide Y, but not agouti-related protein, in ovariectomized rats. Behav Brain Res 191:173–177

Santollo J, Eckel LA (2013) Oestradiol decreases melanin-concentrating hormone (MCH) and MCH receptor expression in the hypothalamus of female rats. J Neuroendocrinol 25:570–579

Sapin E, Berod A, Leger L, Herman PA, Luppi PH, Peyron C (2010) A very large number of GABAergic neurons are activated in the tuberal hypothalamus during paradoxical (REM) sleep hypersomnia. PLoS One 5:e11766

Schiller CE, Meltzer-Brody S, Rubinow DR (2015) The role of reproductive hormones in postpartum depression. CNS Spectr 20:48–59

Schmidt FM, Kratzsch J, Gertz HJ, Tittmann M, Jahn I, Pietsch UC, Kaisers UX, Thiery J, Hegerl U, Schonknecht P (2013) Cerebrospinal fluid melanin-concentrating hormone (MCH) and hypocretin-1 (HCRT-1, orexin-a) in Alzheimer's disease. PLoS One 8:e63136

Schmidt FM, Nowak C, Kratzsch J, Sander C, Hegerl U, Schonknecht P (2015) Dynamics of melanin-concentrating hormone (MCH) serum levels in major depressive disorder during antidepressant treatment. J Affect Disord 180:207–213

Shearman LP, Camacho RE, Sloan Stribling D, Zhou D, Bednarek MA, Hreniuk DL, Feighner SD, Tan CP, Howard AD, Van der Ploeg LH, MacIntyre DE, Hickey GJ, Strack AM (2003) Chronic MCH-1 receptor modulation alters appetite, body weight and adiposity in rats. Eur J Pharmacol 475:37–47

Sherwood A, Holland PC, Adamantidis A, Johnson AW (2015) Deletion of melanin concentrating hormone receptor-1 disrupts overeating in the presence of food cues. Physiol Behav 152:402–407

Shimada M, Tritos NA, Lowell BB, Flier JS, Maratos-Flier E (1998) Mice lacking melanin-concentrating hormone are hypophagic and lean. Nature 396:670–674

Shimazaki T, Yoshimizu T, Chaki S (2006) Melanin-concentrating hormone MCH(1) receptor antagonists: a potential new approach to the treatment of depression and anxiety disorders. CNS Drugs 20:801–811

Siegel JM (2005) REM sleep. In: Kryger MH, Roth T, Dement W (eds) Principles and practices of sleep medicine, 5th edn. Elsevier-Saunders, Philadelphia, pp 92–111

Smith DG, Davis RJ, Rorick-Kehn L, Morin M, Witkin JM, McKinzie DL, Nomikos GG, Gehlert DR (2006) Melanin-concentrating hormone-1 receptor modulates neuroendocrine, behavioral, and corticolimbic neurochemical stress responses in mice. Neuropsychopharmacology 31:1135–1145

Stack EC, Balakrishnan R, Numan MJ, Numan M (2002) A functional neuroanatomical investigation of the role of the medial preoptic area in neural circuits regulating maternal behavior. Behav Brain Res 131:17–36

Sun LL, Zhang Y, Liu JF, Wang J, Zhu WL, Zhao LY, Xue YX, Lu L, Shi J (2013) Role of melanin-concentrating hormone in the nucleus accumbens shell in rats behaviourally sensitized to methamphetamine. Int J Neuropsychopharmacol 16:1767–1780

Takase K, Kikuchi K, Tsuneoka Y, Oda S, Kuroda M, Funato H (2014) Meta-analysis of melanin-concentrating hormone signaling-deficient mice on behavioral and metabolic phenotypes. PLoS One 9:e99961

Tarasoff-Conway JM, Carare RO, Osorio RS, Glodzik L, Butler T, Fieremans E, Axel L, Rusinek H, Nicholson C, Zlokovic BV, Frangione B, Blennow K, Menard J, Zetterberg H, Wisniewski T, de Leon MJ (2015) Clearance systems in the brain-implications for Alzheimer disease. Nat Rev Neurol 11:457–470

Thannickal TC, Moore RY, Nienhuis R, Ramanathan L, Gulyani S, Aldrich M, Cornford M, Siegel JM (2000) Reduced number of hypocretin neurons in human narcolepsy. Neuron 27:469–474

Thannickal TC, Lai YY, Siegel JM (2007) Hypocretin (orexin) cell loss in Parkinson's disease. Brain 130:1586–1595

Thebault S, Waters P, Snape MD, Cottrell D, Darin N, Hallbook T, Huutoniemi A, Partinen M, Pollard AJ, Vincent A (2015) Neuronal antibodies in children with or without narcolepsy following H1N1-AS03 vaccination. PLoS One 10:e0129555

Thorpy MJ, Adler CH (2005) Parkinson's disease and sleep. Neurol Clin 23:1187–1208

Torterolo P, Lagos P, Sampogna S, Chase MH (2008) Melanin-concentrating hormone (MCH) immunoreactivity in non-neuronal cells within the raphe nuclei and subventricular region of the brainstem of the cat. Brain Res 1210:163–178

Torterolo P, Sampogna S, Chase MH (2009) MCHergic projections to the nucleus pontis oralis participate in the control of active (REM) sleep. Brain Res 1268:76–87

Torterolo P, Lagos P, Monti JM (2011) Melanin-concentrating hormone: a new sleep factor? Front Neurol 2:14

Torterolo P, Scorza C, Lagos P, Urbanavicius J, Benedetto L, Pascovich C, Lopez Hill X, Chase M, Monti j (2015) Melanin-concentrating hormone (MCH): role in REM sleep and depression. Front Neurosci 9:1–10

Torterolo P, Benedetto L, Monti JM (2016) Functional interactions between MCHergic and dopaminergic neurons: role in the control of wakefulness and sleep. In: Monti JM, Pandi-Perumal SR, Chokroverty S (eds) Dopamine and sleep: molecular, functional, and clinical aspects. Springer, Switzerland

Tsunematsu T, Ueno T, Tabuchi S, Inutsuka A, Tanaka KF, Hasuwa H, Kilduff TS, Terao A, Yamanaka A (2014) Optogenetic manipulation of activity and temporally controlled cell-specific ablation reveal a role for MCH neurons in sleep/wake regulation. J Neurosci 34:6896–6909

Tyhon A, Adamantidis A, Foidart A, Grisar T, Lakaye B, Tirelli E (2006) Mice lacking the melanin-concentrating hormone receptor-1 exhibit an atypical psychomotor susceptibility to cocaine and no conditioned cocaine response. Behav Brain Res 173:94–103

Unmehopa UA, van Heerikhuize JJ, Spijkstra W, Woods JW, Howard AD, Zycband E, Feighner SD, Hreniuk DL, Palyha OC, Guan XM, Macneil DJ, Van der Ploeg LH, Swaab DF (2005) Increased melanin concentrating hormone receptor type I in the human hypothalamic infundibular nucleus in cachexia. J Clin Endocrinol Metab 90:2412–2419

Urbanavicius J, Lagos P, Torterolo P, Scorza C (2014) Pro-depressive effect induced by microinjections of MCH into the dorsal raphe: time-course, dose-dependence, effects on anxiety-related behaviors and reversion by nortriptyline. Behav Pharmacol 25:316–324

Urbanavicius J, Lagos P, Torterolo P, Abin-Carriquiry JA, Scorza C (2016) Melanin-concentrating hormone projections to the dorsal raphe nucleus: an immunofluorescence and in vivo microdialysis study. J Chem Neuroanat 72:16–24

Verret L, Goutagny R, Fort P, Cagnon L, Salvert D, Leger L, Boissard R, Salin P, Peyron C, Luppi PH (2003) A role of melanin-concentrating hormone producing neurons in the central regulation of paradoxical sleep. BMC Neurosci 4:19

Vetrivelan R, Kong D, Ferrari LL, Arrigoni E, Madara JC, Bandaru SS, Lowell BB, Lu J, Saper CB (2016) Melanin-concentrating hormone neurons specifically promote rapid eye movement sleep in mice. Neuroscience 336:102–113

Wermter AK, Reichwald K, Buch T, Geller F, Platzer C, Huse K, Hess C, Remschmidt H, Gudermann T, Preibisch G, Siegfried W, Goldschmidt HP, Li WD, Price RA, Biebermann H, Krude H, Vollmert C, Wichmann HE, Illig T, Sorensen TI, Astrup A, Larsen LH, Pedersen O, Eberle D, Clement K, Blundell J, Wabitsch M, Schafer H, Platzer M, Hinney A, Hebebrand J (2005) Mutation analysis of the MCHR1 gene in human obesity. Eur J Endocrinol 152:851–862

Wheatley DN, Wang AM, Strugnell GE (1996) Expression of primary cilia in mammalian cells. Cell Biol Int 20:73–81

Whitlock BK, Daniel JA, McMahon CD, Buonomo FC, Wagner CG, Steele B, Sartin JL (2005) Intracerebroventricular melanin-concentrating hormone stimulates food intake in sheep. Domest Anim Endocrinol 28:224–232

Williamson-Hughes PS, Grove KL, Smith MS (2005) Melanin concentrating hormone (MCH): a novel neural pathway for regulation of GnRH neurons. Brain Res 1041:117–124

Xie L, Kang H, Xu Q, Chen MJ, Liao Y, Thiyagarajan M, O'Donnell J, Christensen DJ, Nicholson C, Iliff JJ, Takano T, Deane R, Nedergaard M (2013) Sleep drives metabolite clearance from the adult brain. Science 342:373–377

Yoon YS, Lee HS (2013) Projections from melanin-concentrating hormone (MCH) neurons to the dorsal raphe or the nuclear core of the locus coeruleus in the rat. Brain Res 1490:72–82

Ziogas DC, Gras-Miralles B, Mustafa S, Geiger BM, Najarian RM, Nagel JM, Flier SN, Popov Y, Tseng YH, Kokkotou E (2013) Anti-melanin-concentrating hormone treatment attenuates chronic experimental colitis and fibrosis. Am J Physiol Gastrointest Liver Physiol 304:G876–G884

MCH and Depression

Jessika Urbanavicius, Patricia Lagos, Ximena López, Pablo Torterolo, and Cecilia Scorza

Abstract Depression is a mood disorder affecting emotional, somatic, and cognitive domains. The efficacy of the antidepressant therapies is generally acceptable; however, unpleasant or adverse effects, delayed onset of action, and unresponsive patients continue to be the most common problem in the psychiatric practice. Under these circumstances, the recent implication of the hypothalamic neuropeptide melanin-concentrating hormone (MCH) in the regulation of emotion and mood has offered a great opportunity to study this neuropeptidergic system in the neural bases of depression and its treatment.

1 Introduction

Progress in the understanding of the anatomical, biochemical, and molecular bases of depression has allowed the identification of various neurotransmitter systems involved in the physiopathology and therapy of this mood disorder. The relevance of the monoaminergic systems (e.g., serotoninergic and noradrenergic) in depression and the therapeutic use of the selective serotonin (5-HT) and noradrenaline (NA) reuptake inhibitor (SSRI and SNRI, respectively) antidepressants are widely known. However, a significant amount of evidence supports the role of the MCHergic system in mood and emotion regulation. To assess the basis of this role, it is imperative to be aware of the complete bibliographic data that support

J. Urbanavicius · X. López · C. Scorza (✉)
Departamento de Neurofarmacología Experimental, Instituto de Investigaciones Biológicas Clemente Estable, Montevideo, Uruguay
e-mail: cscorza@iibce.edu.uy

P. Lagos · P. Torterolo
Department of Physiology, School of Medicine, University of the Republic, Montevideo, Uruguay

S. R. Pandi-Perumal et al. (eds.), *Melanin-Concentrating Hormone and Sleep*,
https://doi.org/10.1007/978-3-319-75765-0_11

the hypothesis that a dysfunction in the MCHergic system underlies a pathological process as severe as depression. In the present chapter, we will introduce general aspects of the major depressive disorder followed by a brief description of the MCHergic system in the central nervous system (CNS). A summary of the anatomy and physiology of the MCHergic system and preclinical evidence regarding its role in the modulation of emotional-related behaviors will be reviewed. Considering that MCH can modulate the functionality of monoaminergic systems such as the serotoninergic, we will provide data about the interaction between both systems and its relevance for the pathophysiology of major depression.

2 Major Depressive Disorder (MDD) and Therapeutics

MDD is a severe and chronic psychiatric syndrome with a high prevalence and socioeconomic impact (Greenberg et al. 2003; Andlin-Sobocki and Wittchen 2005; López et al. 2006; WHO 2017). Social, occupational, or other important areas of daily life are affected in patients with MDD (Cuijpers et al. 2012; Murray et al. 2012). MDD is characterized by the presence of a depressed mood and loss of interest or pleasure (e.g., anhedonia) in activities that are normally enjoyable (DSM-V). Also, loss or gain of weight, sleep disturbances, fatigue, and feelings of worthlessness or guilt, impairment in attention and concentration, and recurrent thoughts of death or suicidal ideation are some other symptoms which MDD patients suffer (DSM-V). Although it is pretty clear that a brain dysfunction in connectivity and functional impairments may occur in depression, the underlying pathophysiology of MDD has not yet been clearly defined so far. To elucidate the neurobiological basis of depression and to find new effective therapeutic strategies are the most crucial challenges in mental health research. Recently, an integrated view of the neurobiology of depression has been published (Dean and Keshavan 2017). It was proposed that MDD results of an interactive matrix of several pathophysiological mechanisms. An alteration in any site of the matrix can trigger several biological effects, which could lead to depression (Dean and Keshavan 2017). Changes in monoaminergic and glutamatergic neurotransmission (Caddy et al. 2014; Browne and Lucki 2013), hypothalamic adrenal axis dysregulation and the stress response (Kendler et al. 1999; Pruessner et al. 2003), increased levels of inflammatory markers, and reduced neurogenesis and neuroplasticity capacity (Castrén and Rantamäki 2010; Castrén 2013; Felger and Lotrich 2013) are, among others, pathological conditions which are strongly associated with depression. Moreover, neuroimaging studies of depressed patients have evidenced changes in functional connectivity in the neural circuits involved in mood regulation (Sheline 2003; Drevets et al. 2008), e.g., limbic areas such as the amygdala (Am), hippocampus (Hc), nucleus accumbens (NAcc), and prefrontal cortex (PFC). It is very interesting to highlight that almost all symptoms observed in MDD patients are closely related to MCH functions. Moreover, the widespread innervation of MCH-containing fibers and also the expression of MCHergic receptors in brain areas conforming the limbic/

emotional circuit suggest a potential involvement of MCH in the control of emotion and cognition. However, only a few reports have directly addressed this issue so far.

Many strategies are used to treat depression including pharmacotherapy, electroconvulsive therapy, and deep brain stimulation, among others (Shan et al. 2016; Rosenthal et al. 2016). Specifically, the pharmacological treatments are mainly based on the enhancement of serotonergic and noradrenergic neurotransmission throughout the use of SSRI and SNRI, respectively (Willner et al. 2013). Although these antidepressants represent a therapeutic improvement compared to the traditional tricyclic antidepressants (TCAs) and the monoamine oxidase inhibitors (IMAOs), their efficacy is still limited. Low clinical efficacy, severe side effects (nausea, tremor, insomnia/activation or sedation, dry mouth, and sexual dysfunction), and delayed onset of action are the most common disadvantages (Nestler et al. 2002). Currently, all available antidepressant medications have essentially the same mechanism of action than older ones. Moreover, the antidepressant mechanisms of mood-elevating effects only occur after prolonged administration. This suggests that the neurochemical enhancement of 5-HT or NA neurotransmission could not be per se responsible for the clinical improvement. Several studies have shown that clinical efficacy of the antidepressant drugs depends on presynaptic serotonergic and noradrenergic functions (Romero et al. 1996, 2003; Scorza et al. 2012; Artigas 2013). MCH is one of the several neuromodulators that influence presynaptic monoaminergic neurotransmission (Torterolo et al. 2015; Urbanavicius et al. 2016). Therefore, an alteration in the modulation of the 5-HTergic and NAergic systems through a dysregulation of the MCHergic system could be in the base of depression. Additionally, the interaction between these three neurotransmission systems opens new insights in the treatment of depression.

3 MCH: Anatomy and Function in Rodents

MCH is a neuropeptide of 19-amino acids synthetized from the prepro-MCH (i.e., synthesis precursor) that was described in various species including rodents, nonhuman primates, and humans. Neurons that synthesize MCH are mainly located in the incerto-hypothalamic area, the lateral sector of the posterior hypothalamus, and the dorsomedial hypothalamus (Bittencourt et al. 1992; Torterolo et al. 2006, 2009). Although to a lesser extent, an additional population of MCHergic neurons was described in the olfactory tubercle and the pontine reticular formation (Bittencourt et al. 1992). MCH sends projections to almost all brain regions of the CNS indicating that it is involved in a broad range of functions (Saito et al. 2001; Chung et al. 2011). Functional actions of MCH are mediated through the activation of two G-protein-coupled receptors classified as types 1 and 2, MCHR-1 and MCHR-2, respectively (Saito et al. 1999; Sailer et al. 2001). Almost all studies of MCH have been performed in rodents, where only the MCH-R1 is functional. MCHR-2 is expressed in humans and nonhuman primates and a few other species, but not in rodents (Hill et al. 2001; Sailer et al. 2001). Electrophysiological studies have demonstrated that MCH acts mainly as an inhibitory neuropeptide decreasing the release of different neurotransmitters (Rao et al. 2008; Sears et al. 2010; MacNeil 2013; Devera et al. 2015; Urbanavicius et al. 2016).

Several reports have shown that MCH is implicated in the control of feeding behavior, energy balance, and sleep and wakefulness behaviors. Among these, the classical functions assigned to MCH are the control of feeding and energy balance. Acute intracerebroventricular (i.c.v.) MCH administration in rodents was able to increase feeding, whereas its chronic administration increased weight gain as well as the susceptibility to diet-induced obesity (Qu et al. 1996; Gomori et al. 2003). Moreover, transgenic mice that overexpress MCH displayed an increase in body weight and circulating levels of leptin and were insulin-resistant (Ludwig et al. 2001). On the other hand, prepro-MCH or MCHR-1 knockout (KO) mice (preventing the MCHergic transmission) were both hypophagic and lean (Shimada et al. 1998; Chen et al. 2002; Marsh et al. 2002), while the systemic administration of MCHR-1 antagonists decreased feeding and body weight gain (Borowsky et al. 2002; Eric Hu et al. 2008; Gehlert et al. 2009; Ito et al. 2010; MacNeil 2013). The control on the energy balance suggests that MCH would be linked to motivated behaviors (Diniz and Bittencourt 2017). It has been described that MCHR-1 is highly expressed in the mesolimbic dopaminergic system (Saito et al. 2001), particularly in the NAcc shell (Chung et al. 2009), a region highly related to the regulation of reward and motivation. Interestingly, some works have demonstrated that MCH modulates rewarding behaviors including food reward (Chung et al. 2009; Nair et al. 2009; Mul et al. 2009).

Numerous studies have proposed that MCH may act as a sleep promoter, especially REM sleep, through the regulation of the activating and somnogenic systems (Torterolo et al. 2011, 2015; Monti et al. 2013; Konadhode et al. 2015). On line with this evidence, a decrease in the latency to the first episode of REM sleep is a trait of major depression and is considered to be one of the most robust and specific biological markers of this condition (Adrien 2002; Palagini et al. 2013). Interestingly, the 5-HTergic neurons are involved both in the generation of REM sleep and in the pathophysiology of MDD (Portas et al. 2000; Monti 2010). MCH can inhibit the firing of 5-HTergic neurons at the dorsal raphe nucleus (DRN; Devera et al. 2015), so it is possible to consider that an alteration of the MCHergic system could be part of the matrix in the neurobiology of depression. In fact, MCH microinjections into the DRN increased the time spent in REM sleep, accordingly with an inhibitory action of MCH over the 5-HTergic DRN neurons (Lagos et al. 2011a). In the same direction, MCH also increased REM sleep when administered into the *locus coeruleus* (LC), a wake-promoter nucleus, suggesting an inhibitory action on NAergic LC neurons (Monti et al. 2015).

4 MCH and Mood-Related Behaviors: Preclinical Evidence

Among the brain regions containing MCHergic fibers are those belonging to the limbic and emotional circuits including PFC, NAcc, Hc, Am, and mesencephalic or brain stem nuclei such as ventral tegmental area (VTA), median raphe (MnR), DRN, and LC. Mostly based on this neuroanatomical distribution, a role of MCH in the regulation of emotional or mood-related behaviors has been proposed (Saito et al. 2001; Bittencourt 2011). In fact, the first functional evidence of the participation of MCH in the regulation of emotional-related behaviors came from studies that had explored the effects of MCH on endocrine and behavioral responses to stress. It has been reported that the i.c.v. MCH administration induced the activation of the hypothalamic-pituitary-adrenocortical axis (HPA) as was evidenced by the elevation of plasma levels of the adrenocorticotropic and corticosterone hormones (Jezová et al. 1992; Kennedy et al. 2003; Smith et al. 2006). Moreover, and regarding experimental anxiety, it is still unclear if MCH is able to induce anxiogenic or anxiolytic effects. It has been demonstrated that MCH i.c.v. or locally administered into the medial preoptic area reduced the time spent in the open arms and the number of entries to the open arm in the elevated plus maze (EPM), indicating a clear anxiogenic effect (González et al. 1996; Smith et al. 2006). On the other hand, an anxiolytic response to novelty evaluated both in the EPM and the Vogel's punished drinking test was described after i.c.v. MCH microinjection (Monzón and De Barioglio 1999; Kela et al. 2003). Taking into account these data, we have studied the effect of MCH on anxiety-related behaviors evaluated in the EPM after the local MCH microinjection into the DRN, a region highly related with the anxiety response. In these experiments, anxiety-related behaviors were unaltered by MCH (Urbanavicius et al. 2014).

Studies performed by Borowsky and colleagues have demonstrated that SNAP-7941, the high-affinity MCHR-1 antagonist, elicited an anxiolytic-like effect in different behavioral models in rodents. This anxiolytic action has been shown in successive preclinical studies (Chaki et al. 2005; David et al. 2007; Lee et al. 2010). Taken together, these data were crucial to propose an MCH facilitating but not blocking role in anxiety-related behaviors. Interestingly, in the same study, Borowsky and colleagues also reported an antidepressant effect of SNAP-7941 evaluated in the forced swimming test (FST) in rats, and this effect was similar to that induced by fluoxetine, an SSRI antidepressant. Moreover, the pattern of distribution of [^{3}H]-SNAP-7941 binding sites was similar to the distribution of mRNA MCHR-1 (Chambers et al. 1999; Saito et al. 1999; Hervieu et al. 2000) suggesting that both DRN and LC are the brain sites of action of SNAP-7941 to exert its antidepressive action.

In another study, it has been reported that SNAP-94847, another MCHR-1 antagonist, prevented the decrease of sucrose intake provoked by a chronic mild stress, an ethological animal paradigm of depression used to measure anhedonia (Smith et al. 2009). In agreement with all these findings, several other studies have described antidepressive actions of a new series of MCHR-1 antagonists after its acute or chronic administration using different depression paradigms in rodents (Chaki et al. 2005; Georgescu et al. 2005; David et al. 2007; Gehlert et al. 2009). Albeit these data lead to propose the use of MCHR-1 antagonists in the treatment of depression, further studies are needed to address the specific role of endogenous MCH in the regulation of emotional-related behaviors.

5 Interaction Between MCHergic and 5-HTergic Systems

DRN and MnR are the origins of the great majority of 5-HTergic fibers that innervate most forebrain structures of the CNS. Specifically, the DRN of the rat has been extensively studied, and 5-HT neurons represent one-third of the total number of cells of this nucleus. This region has been highly related with markers of human depression (Arango et al. 2002; Bach-Mizrachi et al. 2006; Nestler et al. 2002), while much less attention has been paid to the MnR, probably because only a minor proportion of its cells is 5-HTergic (Adell et al. 2002). It has been reported that MCHergic fibers reach the rat DRN (Bittencourt et al. 1992; Lagos et al. 2011b; Yoon and Lee 2013). A further description of the distribution and density of MCHergic fibers reaching the rat DRN subregions was performed. Additionally, the identification of appositions between MCH fibers and 5-HT- and GABA-containing neurons, the most abundant cell subtypes within this nucleus (Calizo et al. 2011; Jacobs and Azmitia 1992; Stamp and Semba 1995), has been recently reported (Urbanavicius et al. 2016). It can be suggested that the depressive-like behavior induced by the acute MCH microinjection into the rat DRN evaluated in the FST (identified by an increment and a decrement in immobility and climbing time, respectively) relies on that morphological evidence. This pro-depressive effect was partially sensitive to fluoxetine (Lagos et al. 2011a) and entirely prevented by nortriptyline (a NAergic antidepressant) pretreatment. Also, this response was blocked by ATC-0175, a specific MCHR-1 antagonist (Urbanavicius et al. 2014). Accordingly, it was proposed that the MCHergic system plays a main role in the emotional-related behaviors through the regulation of the neuronal activity in the DRN (Lagos et al. 2011a, b). In fact, MCH can modify the 5-HTergic release in DRN suggesting the direct participation of local 5-HTergic and GABAergic neurons in this effect (Urbanavicius et al. 2016). Considering that MCH elicited a stronger and long-lasting effect on 5-HTergic neurons, it is possible that the depressive-like behavior induced by MCH is mediated by the decrease of 5-HT output from the DRN (Urbanavicius et al. 2016). A pro-depressive action of MCH was also observed after its administration into the MnR, although eliciting a different pattern (i.e., increased immobility and decreased swimming behavior) and a different sensitivity to MCH concentration (López Hill et al. 2013).

Additional evidence provided by studies based on the use of genetically modified animals with a functional inactivation of MCH transmission suggests a potential role for MCHergic system in the modulation of mood-related behavior and its association with 5-HT neurotransmission. Prepro-MCH KO mice developed an antidepressive profile in the FST (decreased immobility and increased swimming time) compared with wild-type mice (Georgescu et al. 2005). MCHR-1 KO mice displayed reduced anxiety-like responses in different behavioral paradigms (Roy et al. 2006). However, only female MCHR-1 KO mice exhibited an antidepressant-like behavior in the tail suspension test and FST evidenced by a decreased immobility and increased climbing time (Roy et al. 2007). Using in vivo brain microdialysis in freely moving mice, basal extracellular levels of 5-HT were significantly lower in the PFC of MCHR-1 KO mice. Moreover, acute forced swim stress significantly increased 5-HT levels in the PFC of wild-type but not in MCHR-1 KO mice (Roy et al. 2006). All these data show that a deficiency in the MCH transmission results in an anxiolytic or antidepressant-like phenotypes and changes in prefrontal cortical 5-HT transmission, a region of the brain critical for the regulation of emotion and cognition. This result also suggests that the development of MCHR-1 antagonists may constitute a novel alternative strategy for the treatment of depressive and anxiety disorders (Roy et al. 2006). Evidence provided by García-Fuster et al. (2012) reinforced this suggestion and led to propose the MCHergic system as a putative biomarker of the severity of depressive-like behaviors and as a predisposing factor for the development of affective disorders (García-Fuster et al. 2012). In agreement with this result, Kim et al. (2015) reported that mice subjected to a chronic restraining stress (during 14 days) developed a depressive-like profile evidenced by an increase in the immobility time in the FST and tail suspension test and a decrease in sucrose consumption. Interestingly, this behavioral response was associated with an increase in the MCH immunoreactivity and mRNA expression in the Am, while siRNA-MCH blocked the appearance of this profile. Local administration of MCH into the amygdala induced a depressive effect in the FST mimicking a stress-induced depressive-like behavior.

It is interesting to note that fluoxetine treatment (following the schedule used to prevent MCH pro-depressive effect in the FST) induced a reduction in MCH cerebrospinal fluid levels, without changes in the prepro-MCH mRNA expression, suggesting that a decrease in the release of MCH may underlie fluoxetine antidepressant action (Calegare et al. 2016). Similarly, Schmidt et al. (2015) demonstrated that MCH cerebrospinal fluid levels decrease after 4 weeks of mirtazapine antidepressant treatment in depressive patients.

All the collected information encourages future work based on the hypothesis that an abnormal increase in the activity of MCHergic neurons is involved in the pathophysiology of depression. However, future research on MCH in human affective disorders is imperative warranted.

References

Adell A, Celada P, Abellan MT, Artigas F (2002) Origin and functional role of the extracellular serotonin in the midbrain raphe nuclei. Brain Res Rev 39:154–180

Adrien J (2002) Neurobiological bases for the relation between sleep and depression. Sleep Med Rev 6:341–351

Andlin-Sobocki P, Wittchen HU (2005) Cost of affective disorders in Europe. Eur J Neurol 1:34–38

Arango V, Underwood MD, Mann JJ (2002) Serotonin brain circuits involved in major depression and suicide. Prog Brain Res 136:443–453

Artigas F (2013) Serotonin receptors involved in antidepressant effects. Pharmacol Ther 137 (1):119–131

Bach-Mizrachi H, Underwood MD, Kassir SA, Bakalian MJ, Sibille E, Tamir H, Mann JJ, Arango V (2006) Neuronal tryptophan hydroxylase mRNA expression in the human dorsal and median raphe nuclei: major depression and suicide. Neuropsychopharmacology 31:814–824

Bittencourt JC (2011) Anatomical organization of the melanin-concentrating hormone peptide family in the mammalian brain. Gen Comp Endocrinol 172(2):185–197

Bittencourt JC, Presse F, Arias C, Peto C, Vaughan J, Nahon JL, Vale W, Sawchenko PE (1992) The melanin-concentrating hormone system of the rat brain: an immuno- and hybridization histochemical characterization. J Comp Neurol 319:218–245

Borowsky B, Durkin MM, Ogozalek K, Marzabadi MR, DeLeon J, Lagu B, Heurich R, Lichtblau H, Shaposhnik Z, Daniewska I, Blackburn TP, Branchek TA, Gerald C, Vaysse PJ, Forray C (2002) Antidepressant, anxiolytic and anorectic effects of a melanin-concentrating hormone-1 receptor antagonist. Nat Med 8(8):825–830

Browne CA, Lucki I (2013) Antidepressant effects of ketamine: mechanisms under lying fast-acting novel antidepressants. Front Pharmacol 4:161

Caddy C, Giaroli G, White TP, Shergill SS, Tracy DK (2014) Ketamine as the prototype glutamatérgica antidepressant: pharmacodynamic actions, and a systematic review and meta-analysis of efficacy. Ther Adv Psychopharmacol 4(2):75–99

Calegare BF, Costa A, Fernandes L, Dias AL, Torterolo P, Almeida VD (2016) Subchronical treatment with Fluoxetine modifies the activity of the MCHergic and hypocretinergic systems. Evidences from peptide CSF concentration and gene expression. Sleep Sci 9(2):89–93

Calizo LH, Akanwa A, Ma X, Pan YZ, Lemos JC, Craige C, Heemstra LA, Beck SG (2011) Raphe serotonin neurons are not homogenous: electrophysiological, morphological and neurochemical evidence. Neuropharmacology 61(3):524–543

Castrén E (2013) Neuronal network plasticity and recovery from depression. JAMA Psychiatry 70 (9):983–989

Castrén E, Rantamäki T (2010) The role of BDNF and its receptors in depression and antidepressant drug action: reactivation of developmental plasticity. Dev Neurobiol 70(5):289–297

Chaki S, Funakoshi T, Hirota-Okuno S, Nishiguchi M, Shimazaki T, Iijima M, Grottick AJ, Kanuma K, Omodera K, Sekiguchi Y, Okuyama S, Tran TA, Semple G, Thomsen W (2005) Anxiolytic- and antidepressant-like profile of ATC0065 and ATC0175: nonpeptidic and orally active melanin-concentrating hormone receptor 1 antagonists. J Pharmacol Exp Ther 313:831–839

Chambers J, Ames RS, Bergsma D, Muir A, Fitzgerald LR, Hervieu G, Dytko GM, Foley JJ, Martin J, Liu WS, Park J, Ellis C, Ganguly S, Konchar S, Cluderay J, Leslie R, Wilson S, Sarau HM (1999) Melanin-concentrating hormone is the cognate ligand for the orphan G-protein-coupled receptor SLC-1. Nature 400(6741):261–265

Chen Y, Hu C, Hsu CK, Zhang Q, Bi C, Asnicar M, Hsiung HM, Fox N, Slieker LJ, Yang DD, Heiman ML, Shi Y (2002) Targeted disruption of the melanin-concentrating hormone receptor-1 results in hyperphagia and resistance to diet-induced obesity. Endocrinology 143(7):2469–2477

Chung S, Hopf FW, Nagasaki H, Li CY, Belluzzi JD, Bonci A, Civelli O (2009) The melanin-concentrating hormone system modulates cocaine reward. Proc Natl Acad Sci 106(16):6772–6777

Chung S, Parks GS, Lee C, Civelli O (2011) Recent updates on the melanin-concentrating hormone (MCH) and its receptor system: lessons from MCH1R antagonists. J Mol Neurosci 43:115–121

Cuijpers P, Beekman ATF, Reynolds CF (2012) Preventing depression. A global priority. JAMA 307(10):1033

David DJ, Klemenhagen KC, Holick KA, Saxe MD, Mendez I, Santarelli L, Craig DA, Zhong H, Swanson CJ, Hegde LG, Ping XI, Dong D, Marzabadi MR, Gerald CP, Hen R (2007) Efficacy of the MCHR1 antagonist N-[3-(1-{[4-(3,4-difluorophenoxy)phenyl]methyl}(4-piperidyl))-4-methylphenyl]-2-methylpropanamide (SNAP 94847) in mouse models of anxiety and depression following acute and chronic administration is independent of hippocampal neurogenesis. J Pharmacol Exp Ther 321:237–248

Dean J, Keshavan M (2017) The neurobiology of depression: an integrated view. Asian J Psychiatr 27:101–111

Devera A, Paskovich C, Lagos P, Falconi A, Sampogna S, Chase MH, Torterolo P (2015) Melanin-concentrating hormone (MCH) modulates the activity of dorsal raphe neurons. Brain Res 1598:114–128

Diagnostic and statistical manual of mental disorders (DSM-5) (2013) 5th ed. American Psychiatric Association

Diniz GB, Bittencourt JC (2017) The melanin-concentrating hormone as an integrative peptide driving motivated behaviors. Front Syst Neurosci 11:32

Drevets WC, Price JL, Furey ML (2008) Brain structural and functional abnormalities in mood disorders: implications for neurocircuitry models of depression. Brain Struct Funct 213 (1–2):93–118

Eric Hu X, Wos JA, Dowty ME, Suchanek PM, Ji W, Chambers JB, Benoit SC, Clegg DJ, Reizes O (2008) Small-molecule melanin-concentrating hormone-1 receptor antagonists require brain penetration for inhibition of food intake and reduction in body weight. J Pharmacol Exp Ther 324(1):206–213

Felger JC, Lotrich FE (2013) Inflammatory cytokines in depression: neurobiological mechanisms and therapeutic implications. Neuroscience 246:199–229

Garcia-Fuster MJ, Parks GS, Clinton SM, Watson SJ, Akil H, Civelli O (2012) The melanin-concentrating hormone (MCH) system in an animal model of depression-like behavior. Eur Neuropsychopharmacol 22(8):607–613

Gehlert DR, Rasmussen K, Shaw J, Li X, Ardayfio P, Craft L, Coskun T, Zhang HY, Chen Y, Witkin JM (2009) Preclinical evaluation of melanin-concentrating hormone receptor 1 antagonism for the treatment of obesity and depression. J Pharmacol Exp Ther 329(2):429–438

Georgescu D, Sears RM, Hommel JD, Barrot M, Bolanos CA, Marsh DJ, Bednarek MA, Bibb JA, Maratos-Flier E, Nestler EJ, DiLeone RJ (2005) The hypothalamic neuropeptide melanin-concentrating hormone acts in the nucleus accumbens to modulate feeding behavior and forced-swim performance. J Neurosci 25:2933–2940

Gomori A, Ishihara A, Ito M, Mashiko S, Matsushita H, Yumoto M, Ito M, Tanaka T, Tokita S, Moriya M, Iwaasa H, Kanatani A (2003) Chronic intracerebroventricular infusion of MCH causes obesity in mice. Melanin-concentrating hormone. Am J Physiol Endocrinol Metab 284 (3):E583–E588

González MI, Vaziri S, Wilson CA (1996) Behavioral effects of alpha-MSH and MCH after central administration in the female rat. Peptides 17(1):171–177

Greenberg PE, Kessler RC, Birnbaum HG, Leong SA, Lowe SW, Berglund PA, Corey-Lisle PK (2003) The economic burden of depression in the United States: how did it change between 1990 and 2000? J Clin Psychiatry 64(12):1465–1475

Hervieu GJ, Cluderay JE, Harrison D, Meakin J, Maycox P, Nasir S, Leslie RA (2000) The distribution of the mRNA and protein products of the melanin-concentrating hormone (MCH) receptor gene, slc-1, in the central nervous system of the rat. Eur J Neurosci 12:1194–1216

Hill J, Duckworth M, Murdock P, Rennie G, Sabido-David C, Ames RS, Szekeres P, Wilson S, Bergsma DJ, Gloger IS, Levy DS, Chambers JK, Muir AI (2001) Molecular cloning and functional characterization of MCH2, a novel human MCH receptor. J Biol Chem 276:20125–20129

Ito M, Ishihara A, Gomori A, Matsushita H, Ito M, Metzger JM, Marsh DJ, Haga Y, Iwaasa H, Tokita S, Takenaga N, Sato N, MacNeil DJ, Moriya M, Kanatani A (2010) Mechanism of the anti-obesity effects induced by a novel melanin-concentrating hormone 1-receptor antagonist in mice. Br J Pharmacol 159(2):374–383

Jacobs BL, Azmitia EC (1992) Structure and function of the brain serotonin system. Physiol Rev 72:165–229

Jezová D, Bartanusz V, Westergren I, Johansson BB, Rivier J, Vale W, Rivier C (1992) Rat melanin-concentrating hormone stimulates adrenocorticotropin secretion: evidence for a site of action in brain regions protected by the blood-brain barrier. Endocrinology 130 (2):1024–1029

Kela J, Salmi P, Rimondini-Giorgini R, Heilig M, Wahlestedt C (2003) Behavioural analysis of melanin-concentrating hormone in rats: evidence for orexigenic and anxiolytic properties. Regul Pept 114:109–114

Kendler KS, Karkowski LM, Prescott CA (1999) Causal relationship between stressful life events and the onset of major depression. Am J Psychiatry 156(6):837–841

Kennedy AR, Todd JF, Dhillo WS, Seal LJ, Ghatei MA, O'Toole CP, Jones M, Witty D, Winborne K, Riley G, Hervieu G, Wilson S, Bloom SR (2003) Effect of direct injection of melanin-concentrating hormone into the paraventricular nucleus: further evidence for a stimulatory role in the adrenal axis via SLC-1. J Neuroendocrinol 15(3):268–272

Kim TK, Kim JE, Park JY, Lee JE, Choi J, Kim H, Lee EH, Kim SW, Lee JK, Kang HS, Han PL (2015) Antidepressant effects of exercise are produced via suppression of hypocretin/orexin and melanin-concentrating hormone in the basolateral amygdala. Neurobiol Dis 79:59–69

Konadhode RR, Pelluru D, Shiromani PJ (2015) Neurons containing orexin or melanin concentrating hormone reciprocally regulate wake and sleep. Front Syst Neurosci 8:244

Lagos P, Torterolo P, Jantos H, Monti J (2011a) Effects on sleep of melanin concentrating hormone into the dorsal raphe nucleus. Brain Res 1265:103–110

Lagos P, Urbanavicius J, Scorza MC, Miraballes R, Torterolo P (2011b) Depressive-like profile induced by MCH microinjections into the dorsal raphe nucleus evaluated in the forced swim test. Behav Brain Res 218(2):259–266

Lee C, Parks CS, Civelli O (2010) Anxiolytic effects of the MCH1R antagonist TPI 1361-17. J Mol Neurosci 43(2):132–137

López Hill X, Pascovich C, Urbanavicius J, Torterolo P, Scorza MC (2013) The median raphe nucleus participates in the depressive-like behavior induced by MCH: differences with the dorsal raphe nucleus. Peptides 50:96–99

Lopez AD, Mathers CD, Ezzati M, Jamison DT, Murray CJ (2006) Global and regional burden of disease and risk factors, 2001: systematic analysis of population health data. Lancet 367 (9524):1747–1757

Ludwig DS, Tritos NA, Mastaitis JW, Kulkarni R, Kokkotou E, Elmquist J, Lowell B, Flier JS, Maratos-Flier E (2001) Melanin-concentrating hormone overexpression in transgenic mice leads to obesity and insulin resistance. J Clin Invest 107(3):379–386

MacNeil DJ (2013) The role of melanin-concentrating-hormone and its receptors in energy homeostasis. Front Endocrinol 4(49):1–14

Marsh DJ, Weingarth DT, Novi DE, Chen HY, Trumbauer ME, Chen AS, Guan XM, Jiang MM, Feng Y, Camacho RE, Shen Z, Frazier EG, Yu H, Metzger JM, Kuca SJ, Shearman LP, Gopal-Truter S, MacNeil DJ, Strack AM, MacIntyre DE, Van der Ploeg LH, Qian S (2002) Melanin-concentrating hormone 1 receptor-deficient mice are lean, hyperactive, and hyperphagic and have altered metabolism. Proc Natl Acad Sci USA 99(5):3240–3245

Monti JM (2010) The role of the dorsal raphe nucleus serotonergic and non-serotonergic neurons, and their receptors in the regulating waking and rapid eye movement (REM) sleep. Sleep Med Rev 14:319–327

Monti JM, Torterolo P, Lagos P (2013) Melanin-concentrating hormone control of sleep-wake behavior. Sleep Med Rev 17(4):293–298

Monti JM, Lagos P, Jantos H, Torterolo P (2015) Increased REM sleep after intra-locus coeruleus nucleus microinjection of melanin-concentrating hormone (MCH) in the rat. Prog Neuro-Psychopharmacol Biol Psychiatry 56:185–188

Monzon ME, De Barioglio SR (1999) Response to novelty after i.c.v. injection of melanin-concentrating hormone (MCH) in rats. Physiol Behav 67(5):813–817

Mul JD, la Fleur SE, Toonen PW, Afrasiab-Middelman A, Binnekade R, Schetters D, Verheij MM, Sears RM, Homberg JR, Schoffelmeer AN, Adan RA, Nair SG, Adams-Deutsch T, Pickens CL, Smith DG, Shaham Y (2009) Effects of the MCH1 receptor antagonist SNAP 94847 on high-fat food-reinforced operant responding and reinstatement of food seeking in rats. Psychopharmacology 205(1):129–140

Murray CJ, Vos T, Lozano R, Naghavi M, Flaxman AD, Michaud C et al (2012) Disability-adjusted life years (DALYs) for 291 diseases and injuries in 21 regions, 1990–2010: a systematic analysis for the Global Burden of Disease Study 2010. Lancet 380(9859):2197–2223

Nair SG, Adams-Deutsch T, Pickens CL, Smith DG, Shaham Y (2009) Effects of the MCH1 receptor antagonist SNAP 94847 on high-fat food-reinforced operant responding and reinstatement of food seeking in rats. Psychopharmacology 205(1):129–140

Nestler EJ, Barrot M, DiLeone RJ, Eisch AJ, Gold SJ, Monteggia LM (2002) Neurobiology of depression. Neuron 34:13–25

Palagini L, Baglioni C, Ciapparelli A, Gemignani A, Riemann D (2013) REM sleep dysregulation in depression: state of the art. Sleep Med Rev 17:377–390

Portas CM, Bjorvatn B, Ursin R (2000) Serotonin and the sleep/wake cycle: special emphasis on microdialysis studies. Prog Neurobiol 60(1):13–35

Pruessner M, Hellhammer DH, Pruessner JC, Lupien SJ (2003) Self-reported depressive symptoms and stress levels in healthy young men: associations with the cortisol response to awakening. Psychosom Med 65(1):92–99

Qu D, Ludwig DS, Gammeltoft S, Piper M, Pelleymounter MA, Cullen MJ, Mathes WF, Przypek R, Kanarek R, Maratos-Flier E (1996) A role for melanin-concentrating hormone in the central regulation of feeding behaviour. Nature 380(6571):243–247

Rao Y, Lu M, Ge F, Marsh DJ, Qian S, Wang AH, Picciotto MR, Gao XB (2008) Regulation of synaptic efficacy in hypocretin/orexin-containing neurons by melanin concentrating hormone in the lateral hypothalamus. J Neurosci 28(37):9101–9110

Romero L, Bel N, Artigas F, De Montigny C, Blier P (1996) Effect of pindolol on the function of pre- and postsynaptic 5-HT1A receptors: in vivo microdialysis and electrophysiological studies in the rat brain. Neuropsychopharmacology 15:349–360

Romero L, Celada P, Martin-Ruiz R, Diaz-Mataix L, Mourelle M, Delgadillo J, Hervás I, Artigas F (2003) Modulation of serotonergic function in rat brain by VN2222, a serotonin reuptake inhibitor and 5-HT1A receptor agonist. Neuropsychopharmacology 28:445–456

Rosenthal DL, Leibu E, Aloysi AS, Kopell BH, Goodman WK, Kellner CH (2016) Safety and efficacy of electroconvulsive therapy for depression in the presence of deep brain stimulation in obsessive-compulsive disorder. J Clin Psychiatry 77(5):689–690

Roy M, David NK, Danao JV, Baribault H, Tian H, Giorgetti M (2006) Genetic inactivation of melanin-concentrating hormone receptor subtype 1 (MCHR1) in mice exerts anxiolytic-like behavioral effects. Neuropsychopharmacology 31(1):112–120

Roy M, David N, Cueva M, Giorgetti M (2007) A study of the involvement of Melanin-Concentrating Hormone Receptor 1 (MCHR1) in murine models of depression. Biol Psychiatry 61(2):174–180

Sailer AW, Sano H, Zeng Z, McDonald TP, Pan J, Pong SS, Feighner SD et al (2001) Identification and characterization of a second melanin-concentrating hormone receptor, MCH-2R. Proc Natl Acad Sci USA 98(13):7564–7569

Saito Y, Nothacker HP, Wang Z, Lin SH, Leslie F, Civelli O (1999) Molecular characterization of the melanin-concentrating-hormone receptor. Nature 400:265–269

Saito Y, Cheng M, Leslie FM, Civelli O (2001) Expression of the melanin-concentrating hormone (MCH) receptor mRNA in the rat brain. J Comp Neurol 435(1):26–40

Schmidt FM, Nowak C, Kratzsch J, Sander C, Hegerl U et al (2015) Dynamics of melanin-concentrating hormone (MCH) serum levels in major depressive disorder during antidepressant treatment. J Affect Disord 180:207–2013

Scorza MC, Llado-Pelfort L, Oller S, Cortes R, Puigdemont D, Portella MJ, Pérez-Egea R, Alvarez E, Celada P, Pérez V, Artigas F (2012) Preclinical and clinical characterization of the selective serotonin-1A receptor antagonist DU-125530 for antidepressant treatment. Br J Pharmacol 167(5):1021–1034

Sears RM, Liu RJ, Narayanan NS, Sharf R, Yeckel MF, Laubach M, Aghajanian GK, DiLeone RJ (2010) Regulation of nucleus accumbens activity by the hypothalamic neuropeptide melanin-concentrating hormone. J Neurosci 30(24):8263–8273

Shan GW, Makmor-Bakry M, Omar MS (2016) Long term use of lithium and factors associated with treatment response among patients with bipolar disorder. Psychiatr Danub 28:146–153

Sheline YI (2003) Neuroimaging studies of mood disorder effects on the brain. Biol Psychiatry 54 (3):338–352

Shimada M, Tritos NA, Lowell BB, Flier JS, Maratos-Flier E (1998) Mice lacking melanin-concentrating hormone are hypophagic and lean. Nature 396(6712):670–674

Smith DG, Davis RJ, Rorick-Kehn L, Morin M, Witkin JM, McKinzie DL, Nomikos GG, Gehlert DR (2006) Melanin-concentrating hormone-1 receptor modulates neuroendocrine, behavioral, and corticolimbic neurochemical stress responses in mice. Neuropsychopharmacology 31 (6):1135–1145

Smith DG, Hegde LG, Wolinsky TD, Miller S, Papp M, Ping X, Edwards T, Gerald CP, Craig DA (2009) The effects of stressful stimuli and hypothalamic-pituitary-adrenal axis activation are reversed by the melanin-concentrating hormone 1 receptor antagonist SNAP 94847 in rodents. Behav Brain Res 197:284–291

Stamp JA, Semba K (1995) Extent of colocalization of serotonin and GABA in the neurons of the rat raphe nuclei. Brain Res 677(1):39–49

Torterolo P, Sampogna S, Morales FR, Chase MH (2006) MCH-containing neurons in the hypothalamus of the cat: searching for a role in the control of sleep and wakefulness. Brain Res 1119:101–114

Torterolo P, Sampogna S, Chase MH (2009) MCHergic projections to the nucleus pontis oralis participate in the control of active (REM) sleep. Brain Res 1268:76–87

Torterolo P, Lagos P, Monti JM (2011) Melanin-concentrating hormone: a new sleep factor? Front Neurol 2:14

Torterolo P, Scorza C, Lagos P, Urbanavicius J, Benedetto L, Pascovich C, López-Hill X, Chase MH, Monti JM (2015) Melanin-Concentrating Hormone (MCH): role in REM sleep and depression. Front Neurosci 9:475

Urbanavicius J, Lagos P, Torterolo P, Scorza C (2014) Prodepressive effect induced by microinjections of MCH into the dorsal raphe: time course, dose dependence, effects on anxiety-related behaviors, and reversion by nortriptyline. Behav Pharmacol 25(4):316–324

Urbanavicius J, Lagos P, Torterolo P, Scorza C (2016) Melanin-concentrating hormone projections to the dorsal raphe nucleus: an immunofluorescence and in vivo microdialysis study. J Chem Neuroanat 72:16–24

Willner P, Scheel-Kruger J, Belzung C (2013) The neurobiology of depression and antidepressant action. Neurosci Biobehav Rev 37(10):2331–2371

World Health Organization (2017) Depression and other common mental disorders. Global health estimates. http://www.who.int/mental_health/management/depression/prevalence_global_health_estimates/en/

Yoon YS, Lee HS (2013) Projections from melanin-concentrating hormone (MCH) neurons to the dorsal raphe or the nuclear core of the locus coeruleus in the rat. Brain Res 1490:72–82

MCH Receptor 1 Antagonists: Antidepressant/Anxiolytic Potential in Animal Models

Shigeyuki Chaki

Abstract Melanin-concentrating hormone (MCH), a cyclic 19-amino acid neuropeptide, has been involved in a variety of physiological events, including the regulation of stress responses and mood. Two subtypes of MCH receptor, MCH1 and MCH2, have been identified, and MCH1 mediates most of the physiological functions of MCH.

To date, numerous non-peptidic MCH1 antagonists have been developed, and studies using these MCH1 antagonists and genetically manipulated mice lacking MCH1 have revealed that the blockade of MCH1 produces antidepressant and anxiolytic effects in a variety of rodent models. In addition, the mechanisms underlying the antidepressant/anxiolytic effects of MCH1 antagonists have been investigated, and the regulation of the hypothalamus-pituitary-adrenal axis activity, mesolimbic dopaminergic system, and serotonergic system may be responsible for the stress-coping and mood-modulating effects of MCH1 antagonists. Importantly, MCH1 antagonists showed a faster onset of action, compared with currently available medications, and a smaller number of side effects, compared with conventional antidepressants and anxiolytic treatments. Therefore, MCH1 antagonists may enable improved treatment for depression and anxiety disorders and deserve further investigation, particularly in clinical trials, as potential new treatments for these disorders.

1 Introduction: Role of MCH in Depression and Anxiety

1.1 MCH System

Melanin-concentrating hormone (MCH) is a cyclic neuropeptide originally isolated from chum salmon pituitaries (Kawauchi et al. 1983). Subsequently, MCH was identified in the mammalian hypothalamus as a cyclic 19-amino acid neuropeptide (Vaughan et al. 1989). MCH is mainly produced in the lateral hypothalamus and the

S. Chaki (✉)
Research Headquarters, Taisho Pharmaceutical Co., Ltd., Saitama, Saitama, Japan
e-mail: s-chaki@taisho.co.jp

S. R. Pandi-Perumal et al. (eds.), *Melanin-Concentrating Hormone and Sleep*,
https://doi.org/10.1007/978-3-319-75765-0_12

adjacent zona incerta, and MCH neurons project broadly throughout the brain (Bittencourt et al. 1992).

MCH is implicated in a variety of physiological functions, including feeding behavior, energy balance, sleep–wake regulation, reproduction function, memory and learning, and the regulation of stress responses and emotion (see review, Shimazaki et al. 2006b). These physiological functions of MCH are mediated through its specific receptors: MCH receptor 1 (MCH1) (Bächner et al. 1999; Chambers et al. 1999; Lembo et al. 1999; Saito et al. 1999; Shimomura et al. 1999) and MCH receptor 2 (MCH2) (Sailer et al. 2001), both of which are G-protein-coupled receptors. Among these MCH receptor subtypes, MCH1 is thought to mediate most of the physiological functions of MCH. Reportedly, MCH2 is not expressed in a functional form in rodents and rabbits, while it is expressed in ferrets, dogs, rhesus monkeys, and humans (Fried et al. 2002; Tan et al. 2002), and the physiological functions of MCH2 are not fully understood. MCH1 mRNA is detected at high levels in the brain and at lower levels in some peripheral tissues in humans (Kolakowski et al. 1996) and rodents (Lembo et al. 1999; Saito et al. 1999). Within the central nervous system (CNS), MCH1 mRNA and protein are expressed in various hypothalamic nuclei including the paraventricular nucleus (PVN), in several limbic structures, and in monoaminergic nuclei including the dorsal raphe nucleus (DRN) and locus coeruleus (Saito et al. 2001). In addition, MCH1 is densely distributed in the nucleus accumbens (NAc) shell (Borowsky et al. 2002; Saito et al. 2001), which has been implicated in reward and motivation (Shirayama and Chaki 2006). Therefore, based on the MCH-containing fiber projections and distributions of MCH1, the MCH/MCH1 system likely plays important roles in the regulation of stress responses and emotional states.

1.2 *Role of MCH/MCH1 System in Depressive-Like/Anxiety-Like Behaviors*

Several lines of evidence have indicated that MCH induces depressive-like behavior in rodents. As described above, MCH1 is densely expressed in the NAc shell (Borowsky et al. 2002; Hervieu et al. 2000; Saito et al. 2001). The injection of MCH into the NAc shell produces depressive-like behavior (increased immobility) in the forced swimming test, while the injection of an MCH1 antagonist in this region exhibited antidepressant-like behavior in the same test (Georgescu et al. 2005). Regarding the nuclei involved in depression, MCH1 is also localized in the DRN, where serotonin neurons are mainly localized (Saito et al. 2001). The injection of MCH into the DRN produced depressive-like behavior (increased immobility) in the forced swimming test (Lagos et al. 2011; Urbanavicius et al. 2014), while the injection of anti-MCH antibody (Lagos et al. 2011) or an MCH1 antagonist (ATC0175) (Urbanavicius et al. 2014) into the same nuclei exhibited antidepressant-like behavior. Therefore, MCH1 stimulation induces depressive-like behavior in rodents, and the NAc shell and DRN may be the brain regions responsible for the regulation of the depressive-like behavior induced by MCH. Consistent with these results, knockout mice lacking MCH (Georgescu et al. 2005) or MCH1 (Roy et al. 2006, 2007) displayed antidepressant-like behavior in the forced swimming and tail suspension tests. In contrast, Gehlert et al. (2009) reported that there is no difference in immobility during the forced swimming test between MCH1 null mice and wild-type mice. Moreover, in a chronic mild stress model that displayed depressive-like behavior, increased MCH1 expression was observed in the hippocampus but not in the prefrontal cortex (Roy et al. 2007). Increased MCH1 expression was reversed by repeated treatment with fluoxetine, which coincided with the reversal of chronic mild stress-induced depressive-like behavior in the tail suspension test (Roy et al. 2007). An increased level of MCH mRNA in the hypothalamus and a decreased level of MCH1 mRNA (possibly because of down-regulation) in the CA1 of the hippocampus have been reported in a genetic model of depression (García-Fuster et al. 2012). Interestingly, a higher MCH mRNA level and consequently the downregulation of the receptor were correlated with the severity of depressive behavior (immobility during the forced swimming test) (García-Fuster et al. 2012). These findings provide additional evidence that the MCH/MCH1 system is involved in depressive-like behavior.

In contrast to the effects of MCH on depressive-like behavior, the effects of MCH on anxiety-like behavior are more complicated. MCH induces not only anxiety-like behavior but also anxiolytic-like behavior in rodents, depending on the brain regions that are injected and the experimental tasks that are used. The local injection of MCH into the medial preoptic area resulted in anxiety-like behaviors in the elevated plus maze in female rats (Gonzalez et al. 1996). Likewise, the intracerebroventricular (i.c. v.) injection of MCH resulted in anxiogenic-like behavior in the elevated plus maze in mice, and this behavior was abrogated by an MCH1 antagonist (Smith et al. 2006). In contrast, the i.c.v. injection of MCH produced anxiolytic effects in the Vogel's punished drinking test (Kela et al. 2003) and in an elevated plus maze (Monzón and De Barioglio 1999) in male rats. Of note, in contrast to the depressive-like behavior, the injection of MCH into the DRN did not affect anxiety-like behavior in the elevated plus maze task (Urbanavicius et al. 2014). The role of the MCH system in anxiety-like behavior has been demonstrated using knockout mice lacking MCH1. MCH1 null mice displayed anxiolytic-like phenotypes across different behavioral paradigms commonly used to assess fear and anxiety responses, including increased social investigation toward a novel intruder in the social interaction test, reduced stress-induced hyperthermia, and an increased investigation time in open areas in the open field and elevated plus maze task (Roy et al. 2006, 2007). Likewise, MCH1 null mice showed a smaller increase in the stress-induced hyperthermia response, which is indicative of an anxiolytic-like phenotype (Smith et al. 2006). In contrast, in the same report, it was demonstrated that the basal anxiety-like behaviors during the elevated plus maze task were not significantly different between the MCH1 null mice and wild-type mice (Smith et al. 2006).

1.3 Role of MCH/MCH1 System in Stress Responses

MCH1 is expressed in the PVN (Hervieu et al. 2000; Saito et al. 2001) and is linked to the regulation of the hypothalamus-pituitary-adrenal (HPA) axis, the dysfunction of which has been observed in patients with major depressive disorder as well as those with anorexia nervosa and post-traumatic stress disorder (Licinio et al. 1996; Rasmusson et al. 2003; Tichomirowa et al. 2005). The i.c.v. injection of MCH and the local injection of MCH into the PVN increased the plasma adrenocorticotropic hormone (ACTH) level (Kennedy et al. 2003; Smith et al. 2006, 2009), and the i.c.v. injection of MCH increased the plasma level of corticosterone. The increase in plasma ACTH and corticosterone levels induced by MCH was blocked by an MCH1 antagonist (Smith et al. 2006), indicating that the stimulation of MCH1 in the PVN activated the HPA axis activity, that is, increases in the ACTH and corticosterone levels. In addition, MCH increases the release of corticotropin-releasing factor (CRF) from hypothalamic explants by stimulating MCH1 (Kennedy et al. 2003), and the MCH-induced increase in the plasma ACTH level was attenuated by an anti-CRF antibody (Jezová et al. 1992). Therefore, MCH1 stimulation activates the HPA axis activity via the stimulation of CRF secretion in the PVN. CRF plays a pivotal role in the regulation of HPA axis activity through CRF receptor 1 (CRF1) on the PVN (Vale et al. 1981) and has been implicated in a variety of stress-related disorders such as depression and anxiety (Nemeroff 1996). Indeed, several CRF1 antagonists have been reported to exert antidepressant and anxiolytic effects in several animal models (Chaki et al. 2004; Griebel and Holsboer 2012; Okuyama et al. 1999). Therefore, the regulation of CRF secretion in the PVN by the MCH/MCH1 system may have important roles in the regulation of mood and anxiety by the MCH/MCH1 system. Moreover, stress-induced acetylcholine release within the prefrontal cortex was no longer observed in knockout mice lacking MCH1, and similar to observations with chlordiazepoxide, the MCH antagonist GW3430 blocked stress-induced acetylcholine release (Smith et al. 2006). Since the prefrontal cortex has been proposed as an area that regulates the neuroendocrine responses to stress (Diorio et al. 1993), this effect could be involved in the anxiogenic/anxiolytic effect mediated through MCH1.

1.4 Role of MCH/MCH1 System in Regulation of Dopaminergic and Serotonergic Transmissions

Mesolimbic dopaminergic pathways are known to be involved in reward and mood and to play important roles in depression (Berton and Nestler 2006; Shirayama and Chaki 2006). This has also been demonstrated by recent findings that the optogenetic inhibition of dopaminergic neurons in the ventral tegmental area (VTA) induced depressive- and anhedonic-like behavior, while optogenetic stimulation of the VTA dopaminergic neurons reversed the depressive- and anhedonic-like behavior induced by chronic mild stress (Tye et al. 2013). MCH-expressing axons are expressed in the NAc shell, and MCH1 is highly expressed in the medium spiny neurons (MSNs) of the NAc shell (Georgescu et al. 2005; Saito et al. 2001). Consistent with its localization, the MCH/MCH1 system has been reported to regulate the activity of the mesolimbic dopaminergic system. MCH inhibits the phosphorylation of the α-amino-3-hydroxy-5-methyl-4-isoxazolepropionic acid (AMPA) receptor subunit GluR1 at Ser845 that is induced by a dopamine D1 receptor agonist in NAc shell slices (Georgescu et al. 2005). Given that the phosphorylation of GluR1 at Ser845 increases the AMPA receptor channel current by increasing the surface expression of GluR1-containing AMPA receptors (Roche et al. 1996), MCH negatively regulates the dopamine D1 receptor-mediated stimulation of AMPA receptor transmission in the NAc shell. Indeed, MCH treatment decreased the AMPA receptor-mediated mEPSCs of MSNs in the NAc shell (Sears et al. 2010). This regulation of AMPA receptor transmission by MCH is particularly interesting, because AMPA receptor potentiators exhibit antidepressant effects in animal models (Alt et al. 2006; Knapp et al. 2002), and the injection of MCH into the NAc shell induces depressive-like behavior (Georgescu et al. 2005). Of note, a reduction in the phosphorylation of GluR1 at Ser845 was observed after the application of MCH alone in the NAc shell slices, and the injection of MCH into the NAc shell also reduced basal GluR1 Ser845 phosphorylation, both of which were blocked by an MCH1 antagonist (Sears et al. 2010). Contrary to the negative regulation of the dopamine D1 receptor by MCH, knockout mice lacking MCH1 are supersensitive to the locomotor-activating effects of a dopamine D1 receptor agonist (Smith et al. 2005). The expression levels of dopamine D1-like and D2-like receptors were increased in the NAc of MCH1 null mice (Smith et al. 2005). Moreover, Pissios et al. (2008) reported that the expression of the dopamine transporter is increased in the NAc of MCH1 null mice, which is likely to compensate for the increased evoked dopamine release in the region. Thus, MCH1 null mice are more sensitive to the locomotor-activating activity of a dopamine transporter inhibitor (Pissios et al. 2008). Therefore, MCH1 blockade increases dopaminergic activity in the NAc. MCH may regulate dopamine release in the NAc shell via a dopaminergic terminal. MCH1 has been reported to co-localize with dynorphin-positive MSNs in the NAc shell (Georgescu et al. 2005). Dynorphin serves as a negative feedback mechanism by acting on presynaptic k opioid receptor to inhibit dopamine release in the NAc shell, while k opioid receptor antagonists increase dopamine release and exert antidepressant effects (Rorick-Kehn et al. 2014; Shirayama et al. 2004). Therefore, it is interesting to hypothesize that MCH1 may regulate dopamine release in the NAc, presumably through the dynorphin/k opioid receptor system. All these findings clearly indicate that the MCH/MCH1 system negatively regulates mesolimbic dopaminergic activity and subsequent AMPA receptor transmission on the MSNs of the NAc shell, while the blockade of MCH1 increases mesolimbic dopaminergic activity.

Serotonergic transmission has long been known to be involved in the effects of antidepressant drugs. The MCH/MCH1 system reportedly regulates the activity of the serotonergic system. Immunostaining and immunolabeling studies have revealed that MCH fibers densely project to the DRN (Lagos et al. 2011) and appear in close apposition to 5-HT and GABA neurons (Urbanavicius et al. 2016). Moreover, the i.c.v. injection of rhodamine-labeled MCH is internalized in both 5-HT and non-5-HT neurons in the DRN, indicating that the i.c.v injection of MCH projects to the DRN, where it binds to MCH1 expressed on both 5-HT and non-5-HT neurons (Devera et al. 2015). Moreover, the i.c.v. injection of MCH reduced 5-HT neuron activity in the DRN (Devera et al. 2015). Therefore, MCH reduces 5-HT neuron activity in the DRN through MCH1 on 5-HT neurons. Consistent with this result, the local injection of MCH into the DRN affected 5-HT release in this region, as evaluated by microdialysis; a lower concentration of MCH resulted in a smaller release of 5-HT, while a higher concentration of MCH resulted in a larger release of 5-HT (Urbanavicius et al. 2016). The authors speculated that a lower concentration of MCH directly inhibits the activity of 5-HT neurons, while a higher concentration of MCH inhibits the activity of GABA neurons, leading to the disinhibition of 5-HT neuron activity. These findings indicate that MCH fibers project to the DRN 5-HT neurons to inhibit their activity, and this mechanism may be involved in MCH-induced depressive-like behavior.

The roles of the MCH/MCH1 system in stress responses, the dopaminergic system, and the serotonergic system are depicted in Fig. 1.

2 Antidepressant Effects of MCH1 Antagonists in Animal Models

To date, numerous non-peptidic MCH1 antagonists with different scaffolds have been developed, and the antidepressant potentials of MCH1 antagonists have been demonstrated in several animal models using these non-peptidic MCH1 antagonists as pharmacological tools. The structures of these compounds used as tools in animal models are shown in Fig. 2.

A study by Borowsky et al. (2002) was the first to demonstrate the antidepressant effects of an MCH1 antagonist using SNAP-7941 in the forced swimming test. Subsequently, the present author's group demonstrated further evidence of antidepressant effects using ATC0065 and ATC0175 (Chaki et al. 2005). Interestingly, both ATC0065 and ATC0175 increased swimming behavior without changing climbing behavior when the results of a behavior sampling method were analyzed, consistent with the results reported by Borowsky et al. (2002). Compounds acting on serotonergic transmission have been reported to increase swimming behavior, while those acting on noradrenergic transmission have been reported to increase climbing behavior (Detke et al. 1995). In this regard, MCH1 antagonists exert their antidepressant effects by acting on the serotonergic system in the forced swimming test, consistent with the results that the injection of ATC0175 into the DRN blocked MCH-induced depressive-like behavior during the forced swimming test (Urbanavicius et al. 2014). It should be noted that two other MCH1 antagonists, TASP0382650 and TASP0489838, exerted antidepressant effects during the forced swimming test not only during acute treatment but also during

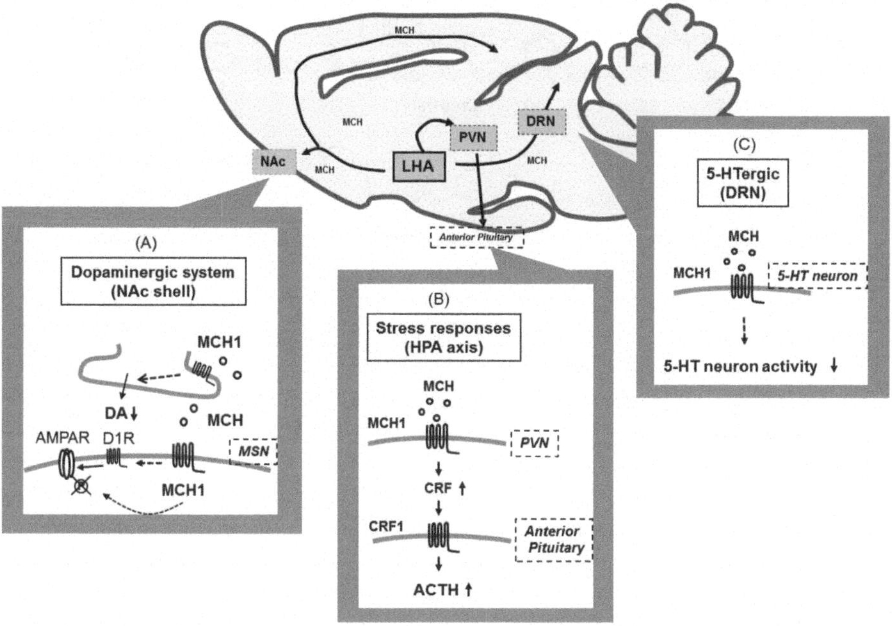

Fig. 1 Proposed mechanisms for the involvement of the MCH/MCH1 system in depression and anxiety. MCH is mainly produced in the LHA, and MCH fibers project throughout the brain. Abundant MCH fibers are detected in the PVN, NAc, and DRN. (**a**) MCH stimulates MCH1 on medium spiny neurons in the NAc shell and inhibits the dopamine D1 receptor-stimulated phosphorylation of GluR1, leading to a decrease in AMPA receptor transmission. MCH1 also directly reduces AMPA receptor transmission on medium spiny neurons. MCH may stimulate MCH1 at dopaminergic terminals in the NAc shell, thereby reducing dopamine release. (**b**) MCH stimulates MCH1 in the PVN and presumably stimulates MCH1 on CRF-producing cells, thereby increasing CRF secretion. The increase in CRF subsequently causes an increase in ACTH release from the anterior pituitary, leading to the activation of HPA axis activity. (**c**) MCH stimulates MCH1 on 5-HT neurons in the DRN, reducing 5-HT neuron activity. *LHA* lateral hypothalamus, *PVN* paraventricular nucleus, *NAc* nucleus accumbens, *DRN* dorsal raphe nucleus, *HPA axis* hypothalamus-pituitary-adrenal axis, *CRF* corticotropin-releasing factor, *CRF1* CRF receptor 1, *ACTH* adrenocorticotropic hormone, *DA* dopamine, *D1R* dopamine D1 receptor, *AMPAR* AMPA receptor, *MSN* medium spiny neuron

chronic treatment for 14 days, indicating an absence of tolerance to the antidepressant effects (Chaki et al. 2015). This result is supported by another report in which an MCH1 antagonist, GW803430, exerted antidepressant effects in the mouse forced swimming test after both acute and subchronic treatment for 5 days (Gehlert et al. 2009). Moreover, combining subeffective doses of GW803430 with a subeffective dose of imipramine produced significant antidepressant effects in the forced swimming test (Gehlert et al. 2009), suggesting that MCHR1 antagonists might enhance the antidepressant actions of currently prescribed antidepressants.

One of the shortcomings of currently prescribed antidepressants is that it takes several weeks for their antidepressant effects to become apparent. A faster onset of

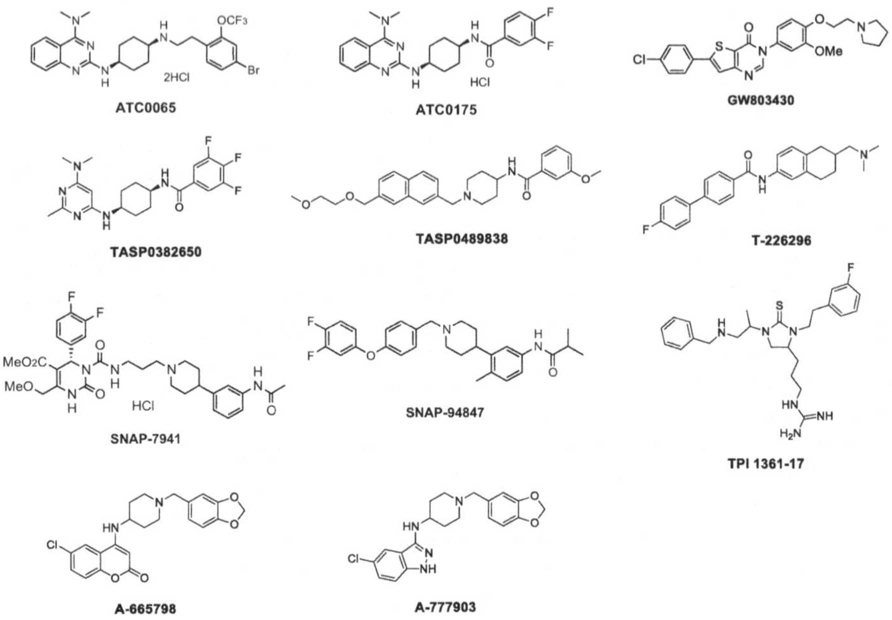

Fig. 2 Chemical structures of non-peptidic MCH1 antagonists used as pharmacological tools in animal models

antidepressant actions has been demonstrated for MCH1 antagonists in some animal models. First, in the olfactory bulbectomy model, in which traditional antidepressants require repeated administration (2 weeks) for their antidepressant effects to become apparent, the MCH1 antagonist TASP0382650 exerted significant antidepressant effects even after a single administration, and another MCH1 antagonist, TASP0489838, tended to reduce hyperemotionality after a single administration, although the reduction was not statistically significant (Chaki et al. 2015). Second, in a chronic mild stress model in which traditional antidepressants require several weeks to manifest their antidepressant effects, SNAP 94847 significantly attenuated the reduction in sucrose intake (anhedonic effects) within 1 week of treatment (Smith et al. 2009). Moreover, in the social dominant test, GW803430 showed faster onset of antidepressant actions than imipramine and fluoxetine (Gehlert et al. 2009). Importantly, SNAP 94847 and GW803430 exerted these antidepressant effects at doses that occupied brain MCH1 (Gehlert et al. 2009; Smith et al. 2009) and, in the case of SNAP 94847, at doses that also blocked drinking behavior induced by the i.c.v. injection of MCH (Smith et al. 2009). Therefore, these effects are mediated through the blockade of brain MCH1. In contrast, there is a report showing a lack of antidepressant activities with MCH1 antagonists (Basso et al. 2006). In the study, significant antidepressant effects were not observed with some MCH1 antagonists including SNAP-7941 (racemic SNAP-7941 and active enantiomer) and other orally bioavailable MCH1 antagonists [T-226296 (Takekawa et al. 2002), A-665798 (Kym et al. 2005), A-777903 (Vasudevan et al. 2005)] in the forced swimming test and the tail suspension test. Although the reason for this discrepancy is not known, it is noteworthy to mention that MCH1 antagonists with diverse structures have demonstrated antidepressant effects in several animal models in several laboratories, as described above. The antidepressant effects of MH1 antagonists in animal models are summarized in Table 1.

3 Anxiolytic Effects of MCH1 Antagonists in Animal Models

Contrary to the inconsistent results obtained in injection studies examining the MCH peptide, MCH1 antagonists have consistently shown anxiolytic effects in several animal models.

In the mouse elevated plus maze task, GW803430 increased the time in the open arms, which is indicative of anxiolytic effects (Smith et al. 2006). Likewise, the i.c.v. injection of a selective MCH1 antagonist, TPI 1361-17 (Nagasaki et al. 2009), exhibited anxiolytic effects in the mouse elevated plus maze test (Lee et al. 2011). In contrast, TASP0382650 did not increase the time in the open arms in the rat elevated plus maze task (Chaki et al. 2015). Interestingly, some MCH1 antagonists such as ATC0065, ATC0175, and TASP0382650 reversed stress-induced anxiety-like behavior (reduction in exploratory behavior in the open arms) in the elevated plus maze (Chaki et al. 2005, 2015). Therefore, MCH1 antagonists may be more effective under higher stressful conditions than under less stressful conditions. This hypothesis is consistent with the observation that the MCH/MCH1 system regulates HPA axis activity, presumably by increasing CRF secretion in the PVN. The abovementioned hypothesis is underpinned by the results that CRF1 antagonists (Chaki et al. 2004; Okuyama et al. 1999) and a vasopressin V1b antagonist (Iijima et al. 2014) (the vasopressin/V1b system is a functional regulator or potentiator of CRF actions on HPA axis activity) reversed anxiety-like behavior under stressful conditions without showing anxiolytic effects under less stressful conditions in the elevated plus maze task (the anxiolytic activity of the V1b antagonist in non-stressed rats was not tested using the same paradigm). Moreover, MCH1 antagonists exerted anxiolytic effects in a social interaction test and a stress-induced hyperthermia test in which an increase in blood ACTH levels was observed (File and Seth 2003; Spooren et al. 2002), indicating that these tests examine stressful conditions. Of note, agents that block stress-related neuropeptide receptors, such as CRF1 (Chaki et al. 2004) and V1b (Iijima et al. 2014; Shimazaki et al. 2006a), also showed anxiolytic effects in these tests. In addition, the anxiolytic effects of TASP0382650 in the social interaction test were not altered after subchronic administration for 7 days (Chaki et al. 2015), indicating that MCH1 antagonists may not cause tolerance, unlike benzodiazepine anxiolytics (Fernandes and File 1999). The absence of tolerance to the anxiolytic effects of MCH1 antagonists was also confirmed using the light/dark paradigm, in which the anxiolytic effects of SNAP 94847 were preserved after treatment for 28 days (David et al. 2007).

Table 1 Effects of MCH1 antagonists in animal models of depression

Model	Species	ATC0065	ATC0175	SNAP-7941	SNAP 98487	GW803430	TASP0382650	TASP0489838	T-226296	A-665798/ A-777903
Forced swimming test										
Acute	Rat	Antidepressant	Antidepressant	Antidepressant/ No effect	n.t.	n.t.	Antidepressant	Antidepressant	Antidepressant (highest dose)	n.t.
Chronic	Rat	n.t.	n.t.	n.t.	n.t.	n.t.	Antidepressant	Antidepressant	n.t.	n.t.
Acute	Mouse	n.t.	n.t.	n.t.	Antidepressant/ No effect	Antidepressant	n.t.	n.t.	n.t.	n.t.
Chronic	Mouse	n.t.	n.t.	n.t.	Antidepressant/ No effect	Antidepressant	n.t.	n.t.	n.t.	n.t.
Tail suspension test	Mouse	n.t.	n.t.	No effect	n.t.	Antidepressant	n.t.	n.t.	No effect	No effect
Olfactory bulbectomy										
Acute	Rat	n.t.	n.t.	n.t.	n.t.	n.t.	Antidepressant	Antidepressant (trend)	n.t.	n.t.
Chronic	Rat	n.t.	n.t.	n.t.	n.t.	n.t.	Antidepressant	Antidepressant	n.t.	n.t.
Chronic mild stress	Rat	n.t.	n.t.	n.t.	Antidepressant (<1 week)	n.t.	n.t.	n.t.	n.t.	n.t.
Social dominant test	Rat	n.t.	n.t.	n.t.	n.t.	Antidepressant (<2 weeks)	n.t.	n.t.	n.t.	n.t.

n.t.: not tested

MCH1 antagonists reduced maternal separation-induced vocalization in guinea pig pups (Borowsky et al. 2002; Chaki et al. 2005, 2015) and freezing behavior in a conditioned fear stress model in rats (Chaki et al. 2015), both of which are indicative of anxiolytic effects. In these tests, agents acting on serotonergic transmission, including fluvoxamine, were also effective (Borsini et al. 2002; Chaki et al. 2005, 2015). Therefore, it is conceivable that serotonergic transmission may be involved in the anxiolytic effects of MCH1 antagonists in some animal models. Presently, antidepressants such as selective serotonin reuptake inhibitors (SSRIs) are the first line of treatment for most anxiety disorders. However, one of the clinical drawbacks of the use of SSRIs is that, as for the treatment of depression, several weeks are required before a therapeutic benefit is observed (Rocca et al. 1997). In this regard, the effects of MCH1 antagonists in a novelty-suppressed feeding test are noteworthy, since MCH1 antagonists exerted anxiolytic effects after a single administration (David et al. 2007) while antidepressants required 4 weeks of treatment before any effects were observed (David et al. 2007; Santarelli et al. 2003). In the novelty-suppressed feeding test, the effects of SNAP 98489 were observed after 4 weeks of treatment as well, with no sign of tolerance to the anxiolytic effects of the MCH1 antagonists (David et al. 2007). The anxiolytic effects of MCH1 antagonists were also observed in the marble burying test (Gehlert et al. 2009), which is proposed in an animal model for obsessive-compulsive disorder. Gehlert et al. (2009) reported that GW803430 potently reduced the marble burying behavior at doses which did not affect motor behavior, while a benzodiazepine anxiolytic reduced the behavior at doses that also produced motor-impairing effects. In contrast, other MCH1 antagonists (ATC0065 and ATC0175) only exhibited a tendency to reduce marble burying behavior (Chaki et al. 2005). Therefore, efficacy in this model should be confirmed using other MCH1 antagonists. Moreover, other MCH1 antagonists, such as A-665798 and A-777903, did not produce any effects in a Vogel conflict test (Basso et al. 2006).

The anxiolytic effects of MCH1 antagonists in animal models are summarized in Table 2.

4 Adverse Side Effects of MCH1 Antagonists in Animal Models

Because MCH/MCH1 system is involved in a variety of physiological functions, the blockade of MCH1 may cause unwanted side effects. The MCH system reportedly plays an important role in memory and learning. Indeed, the injection of MCH into the hippocampus or amygdala modulates memory retention in the one-trial step-down inhibitory avoidance test (Monzon et al. 1999), and the injection of MCH into the hippocampus increases susceptibility to the induction of long-term potentiation in the dentate gyrus and increases the mRNA expression of the NMDA receptor subunit in the dentate gyrus (Varas et al. 2003). These reports raise concerns that

Table 2 Effects of MCH1 antagonists in animal models of anxiety

Model	Species	ATC0065	ATC0175	SNAP-7941	SNAP 98487	GW803430	TASP0382650	TASP0489838	TPI 1361-17	A-665798/A-777903
Elevated plus maze										
Stress	Rat	Anxiolytic	Anxiolytic	n.t.	n.t.	n.t.	Anxiolytic	n.t.	n.t.	n.t.
Non-stress	Rat	n.t.	n.t.	n.t.	n.t.	n.t.	No effect	n.t.	n.t.	n.t.
	Mouse	n.t.	n.t.	n.t.	n.t.	Anxiolytic	n.t.	n.t.	Anxiolytic	n.t.
Light/dark										
Acute	Mouse	n.t.	n.t.	n.t.	Anxiolytic	n.t.	n.t.	n.t.	Anxiolytic	n.t.
Chronic	Mouse	n.t.	n.t.	n.t.	Anxiolytic	n.t.	n.t.	n.t.	n.t.	n.t.
Social interaction										
Acute	Rat	n.t.	Anxiolytic	Anxiolytic	n.t.	n.t.	Anxiolytic	Anxiolytic	n.t.	n.t.
Chronic	Rat	n.t.	n.t.	n.t.	n.t.	n.t.	Anxiolytic	n.t.	n.t.	n.t.
Stress-induced hyperthermia	Mouse	Anxiolytic	Anxiolytic	n.t.	Anxiolytic	Anxiolytic	Anxiolytic	Anxiolytic	n.t.	n.t.
Maternal separation-induced vocalization	Guinea pig pups	n.t.	Anxiolytic	Anxiolytic	n.t.	n.t.	Anxiolytic	Anxiolytic	n.t.	n.t.
Marble burying	Mouse	No effect	No effect	n.t.	n.t.	Anxiolytic	n.t.	n.t.	n.t.	n.t.
Conditioned fear stress	Rat	n.t.	n.t.	n.t.	n.t.	n.t.	Anxiolytic	n.t.	n.t.	n.t.
Novelty-suppressed feeding										
Acute	Mouse	n.t.	n.t.	n.t.	Anxiolytic	n.t.	n.t.	n.t.	n.t.	n.t.
Chronic	Mouse	n.t.	n.t.	n.t.	Anxiolytic	n.t.	n.t.	n.t.	n.t.	n.t.
Vogel conflict test	Rat	n.t.	n.t.	n.t.	n.t.	n.t.	n.t.	n.t.	n.t.	No effect

n.t.: not tested

MCH1 blockade may impair memory and learning. However, both TASP0382650 and TASP0489838 did not affect hippocampus-dependent spatial memory in a Morris water maze task (Chaki et al. 2015). Rather, MCH1 antagonists, such as SNAP-7941 and GW3430, have been reported to improve social memory impairment induced by scopolamine and to enhance social memory in a social recognition test (Millan et al. 2008). Contrary to the pro-cognitive effects of MCH1 antagonists on social memory, both the anxiolytic diazepam and the antidepressant imipramine impaired social recognition in the same paradigm (Millan et al. 2008), indicating a beneficial profile of MCH1 antagonists compared with conventional anxiolytics and antidepressants. Currently prescribed anxiolytics cause sedation and impaired motor function, which are associated with global CNS depression. In contrast, MCH1 antagonists did not affect spontaneous locomotor activity, hexobarbital-induced sleeping, or the impairment of motor function (rotarod performance, induction of catalepsy) in both rats and mice (Chaki et al. 2005, 2015; David et al. 2007; Gehlert et al. 2009), exhibiting a smaller likelihood of causing sedation or impaired motor function than anxiolytics. Moreover, MCH is an endogenous modulator of seizure activity, based on the report that MCH inhibits pentylenetetrazole (PTZ)-induced seizures (Knigge and Wagner 1997). Therefore, there are concerns that MCH1 antagonists may have a pro-convulsive profile. However, TASP0382650 did not affect susceptibility to PTZ-induced convulsion at a dose 10 times higher than that exhibiting antidepressant/anxiolytic effects (Chaki et al. 2015). Rather, MCH1 null mice are resistant to seizures induced by PTZ and pilocarpine (Parks et al. 2010). All these findings indicate that MCH1 antagonists may not produce the adverse CNS side effects that sometimes occur with currently available anxiolytics and antidepressants and that can be anticipated based on the disruption of the MCH/MCH1 system.

The effects of MCH1 antagonists on adverse side effects in animal models are summarized in Table 3.

5 Concluding Remarks

To date, numerous MCH1 antagonists have been developed as pharmacological tools, and by using prototype MCH1 antagonists with different scaffolds, the antidepressant and anxiolytic potential of MCH1 antagonists has been demonstrated in a variety of animal models. These effects are supported by results obtained in studies using MCH or MCH1 null mice. Moreover, behavioral, neurochemical, and histochemical studies have revealed that MCH1 antagonists exert their effects by modulating the HPA axis and dopaminergic and serotonergic activities.

The beneficial effects of MCH1 antagonists over currently available antidepressants and anxiolytics include a faster onset of action and a reduction in the side effects associated with current medications. Indeed, in some animal models, the onset of the antidepressant and anxiolytic actions of MCH1 antagonists was faster than those of currently available antidepressants, which are used as first-line treatments for both disorders. In addition, MCH1 antagonists do not cause the unwanted

Table 3 Side effect liabilities of MCH1 antagonists in animal models

Model	Species	ATC0065	ATC0175	SNAP-7941	SNAP 98487	GW803430	TASP0382650	TASP0489838
Spontaneous locomotor activity	Rat	No effect	No effect	n.t.	No effect	n.t.	No effect	No effect
	Mouse	n.t.	n.t.	n.t.	No effect	No effect	No effect	No effect
Rotarod performance	Rat	No effect	No effect	n.t.	n.t.	n.t.	No effect	No effect
	Mouse	n.t.	n.t.	n.t.	n.t.	n.t.	No effect	No effect
Hexobarbital-induced sleeping	Rat	n.t.	n.t.	n.t.	n.t.	n.t.	No effect	No effect
	Mouse	n.t.	n.t.	n.t.	n.t.	n.t.	No effect	No effect
Catalepsy induction	Rat	n.t.	n.t.	n.t.	n.t.	n.t.	No effect	No effect
Morris water maze	Rat	n.t.	n.t.	n.t.	n.t.	n.t.	No effect	No effect
PTZ-induced convulsion	Mouse	n.t.	n.t.	n.t.	n.t.	n.t.	No effect	n.t.

n.t.: not tested

side effects sometimes observed with currently available antidepressants and anxiolytic treatment. Although these preclinical results are very promising, none of the MCH1 antagonists have been tested their efficacy in clinical trials. Therefore, whether the data obtained in rodents can be extrapolated to humans remains to be determined. Nevertheless, MCH1 antagonists may provide a novel approach for the treatment of depression and anxiety disorders, which could potentially help to overcome the shortcomings of currently available medications.

References

Alt A, Nisenbaum ES, Bleakman D, Witkin JM (2006) A role for AMPA receptors in mood disorders. Biochem Pharmacol 71:1273–1288

Bächner D, Kreienkamp H, Weise C, Buck F, Richter D (1999) Identification of melanin concentrating hormone (MCH) as the natural ligand for the orphan somatostatin-like receptor 1 (SLC-1). FEBS Lett 457:522–524

Basso AM, Bratcher NA, Gallagher KB, Cowart MD, Zhao C, Sun M, Esbenshade TA, Brune ME, Fox GB, Schmidt M, Collins CA, Souers AJ, Iyengar R, Vasudevan A, Kym PR, Hancock AA, Rueter LE (2006) Lack of efficacy of melanin-concentrating hormone-1 receptor antagonists in models of depression and anxiety. Eur J Pharmacol 540:115–120

Berton O, Nestler EJ (2006) New approaches to antidepressant drug discovery: beyond monoamines. Nat Rev Neurosci 7:137–151

Bittencourt JC, Presse F, Arias C, Peto C, Vaughan J, Nahon JL, Vale W, Sawchenko PE (1992) The melanin-concentrating hormone system of the rat brain: an immuno- and hybridization histochemical characterization. J Comp Neurol 319:218–245

Borowsky B, Durkin MM, Ogozalek K, Marzabadi MR, DeLeon J, Lagu B, Heurich R, Lichtblau H, Shaposhnik Z, Daniewska I, Blackburn TP, Branchek TA, Gerald C, Vaysse PJ, Forray C (2002) Antidepressant, anxiolytic and anorectic effects of a melanin-concentrating hormone-1 receptor antagonist. Nat Med 8:825–830

Borsini F, Podhorna J, Marazziti D (2002) Do animal models of anxiety predict anxiolytic-like effects of antidepressants? Psychopharmacology 163:121–141

Chaki S, Nakazato A, Kennis L, Nakamura M, Mackie C, Sugiura M, Vinken P, Ashton D, Langlois X, Steckler T (2004) Anxiolytic- and antidepressant-like profile of a new CRF1 receptor antagonist, R278995/CRA0450. Eur J Pharmacol 485:145–158

Chaki S, Funakoshi T, Hirota-Okuno S, Nishiguchi M, Shimazaki T, Iijima M, Grottick AJ, Kanuma K, Omodera K, Sekiguchi Y, Okuyama S, Tran TA, Semple G, Thomsen W (2005) Anxiolytic- and antidepressant-like profile of ATC0065 and ATC0175: nonpeptidic and orally active melanin-concentrating hormone receptor 1 antagonists. J Pharmacol Exp Ther 313:831–839

Chaki S, Shimazaki T, Nishiguchi M, Funakoshi T, Iijima M, Ito A, Kanuma K, Sekiguchi Y (2015) Antidepressant/anxiolytic potential and adverse effect liabilities of melanin-concentrating hormone receptor 1 antagonists in animal models. Pharmacol Biochem Behav 135:154–168

Chambers J, Ames RS, Bergsma D, Muir A, Fitzgerald LR, Hervieu G, Dytko GM, Foley JJ, Martin J, Liu WS, Park J, Ellis C, Ganguly S, Konchar S, Cluderay J, Leslie R, Wilson S, Sarau HM (1999) Melanin-concentrating hormone is the cognate ligand for the orphan G-protein-coupled receptor SLC-1. Nature 400:261–265

David DJ, Klemenhagen KC, Holick KA, Saxe MD, Mendez I, Santarelli L, Craig DA, Zhong H, Swanson CJ, Hegde LG, Ping XI, Dong D, Marzabadi MR, Gerald CP, Hen R (2007) Efficacy of the MCHR1 antagonist N-[3-(1-{[4-(3,4-difluorophenoxy)phenyl]methyl}(4-piperidyl))-4-methylphenyl]-2-methylpropanamide (SNAP 94847) in mouse models of anxiety and depression following acute and chronic administration is independent of hippocampal neurogenesis. J Pharmacol Exp Ther 321:237–248

Detke MJ, Rickels M, Lucki I (1995) Active behaviors in the rat forced swimming test differentially produced by serotonergic and noradrenergic antidepressants. Psychopharmacology 121:66–72

Devera A, Pascovich C, Lagos P, Falconi A, Sampogna S, Chase MH, Torterolo P (2015) Melanin-concentrating hormone (MCH) modulates the activity of dorsal raphe neurons. Brain Res 1598:114–128

Diorio D, Viau V, Meaney MJ (1993) The role of the medial prefrontal cortex (cingulate gyrus) in the regulation of hypothalamic-pituitary-adrenal responses to stress. J Neurosci 13:3839–3847

Fernandes C, File SE (1999) Dizocilpine does not prevent the development of tolerance to the anxiolytic effects of diazepam in rats. Brain Res 815:431–434

File SE, Seth P (2003) A review of 25 years of the social interaction test. Eur J Pharmacol 463:35–53

Fried S, O'Neill K, Hawes BE (2002) Cloning and characterization of rhesus monkey MCH-R1 and MCH-R2. Peptides 23:1401–1408

García-Fuster MJ, Parks GS, Clinton SM, Watson SJ, Akil H, Civelli O (2012) The melanin-concentrating hormone (MCH) system in an animal model of depression-like behavior. Eur Neuropsychopharmacol 22:607–613

Gehlert DR, Rasmussen K, Shaw J, Li X, Ardayfio P, Craft L, Coskun T, Zhang HY, Chen Y, Witkin JM (2009) Preclinical evaluation of melanin-concentrating hormone receptor 1 antagonism for the treatment of obesity and depression. J Pharmacol Exp Ther 329:429–438

Georgescu D, Sears RM, Hommel JD, Barrot M, Bolaños CA, Marsh DJ, Bednarek MA, Bibb JA, Maratos-Flier E, Nestler EJ, DiLeone RJ (2005) The hypothalamic neuropeptide melanin-concentrating hormone acts in the nucleus accumbens to modulate feeding behavior and forced-swim performance. J Neurosci 25:2933–2940

Gonzalez MI, Vaziri S, Wilson CA (1996) Behavioral effects of alpha-MSH and MCH after central administration in the female rat. Peptides 17:171–177

Griebel G, Holsboer F (2012) Neuropeptide receptor ligands as drugs for psychiatric diseases: the end of the beginning? Nat Rev Drug Discov 11:462–478

Hervieu GJ, Cluderay JE, Harrison D, Meakin J, Maycox P, Nasir S, Leslie RA (2000) The distribution of the mRNA and protein products of the melanin-concentrating hormone (MCH) receptor gene, slc-1, in the central nervous system of the rat. Eur J Neurosci 12:1194–1216

Iijima M, Yoshimizu T, Shimazaki T, Tokugawa K, Fukumoto K, Kurosu S, Kuwada T, Sekiguchi Y, Chaki S (2014) Antidepressant and anxiolytic profiles of newly synthesized arginine vasopressin V1B receptor antagonists: TASP0233278 and TASP0390325. Br J Pharmacol 171:3511–3525

Jezová D, Bartanusz V, Westergren I, Johansson BB, Rivier J, Vale W, Rivier C (1992) Rat melanin-concentrating hormone stimulates adrenocorticotropin secretion: evidence for a site of action in brain regions protected by the blood-brain barrier. Endocrinology 130:1024–1029

Kawauchi H, Kawazoe I, Tsubokawa M, Kishida M, Baker BI (1983) Characterization of melanin-concentrating hormone in chum salmon pituitaries. Nature 305:321–323

Kela J, Salmi P, Rimondini-Giorgini R, Heilig M, Wahlestedt C (2003) Behavioural analysis of melanin-concentrating hormone in rats: evidence for orexigenic and anxiolytic properties. Regul Pept 114:109–114

Kennedy AR, Todd JF, Dhillo WS, Seal LJ, Ghatei MA, O'Toole CP, Jones M, Witty D, Winborne K, Riley G, Hervieu G, Wilson S, Bloom SR (2003) Effect of direct injection of melanin-concentrating hormone into the paraventricular nucleus: further evidence for a stimulatory role in the adrenal axis via SLC-1. J Neuroendocrinol 15:268–272

Knapp RJ, Goldenberg R, Shuck C, Cecil A, Watkins J, Miller C, Crites G, Malatynska E (2002) Antidepressant activity of memory-enhancing drugs in the reduction of submissive behavior model. Eur J Pharmacol 440:27–35

Knigge KM, Wagner JE (1997) Melanin-concentrating hormone (MCH) involvement in pentylenetetrazole (PTZ)-induced seizure in rat and guinea pig. Peptides 18:1095–1097

Kolakowski LF Jr, Jung BP, Nguyen T, Johnson MP, Lynch KR, Cheng R, Heng HH, George SR, O'Dowd BF (1996) Characterization of a human gene related to genes encoding somatostatin receptors. FEBS Lett 398:253–258

Kym PR, Iyengar R, Souers AJ, Lynch JK, Judd AS, Gao J, Freeman J, Mulhern M, Zhao G, Vasudevan A, Wodka D, Blackburn C, Brown J, Che JL, Cullis C, Lai SJ, LaMarche MJ, Marsilje T, Roses J, Sells T, Geddes B, Govek E, Patane M, Fry D, Dayton BD, Brodjian S, Falls D, Brune M, Bush E, Shapiro R, Knourek-Segel V, Fey T, McDowell C, Reinhart GA, Preusser LC, Marsh K, Hernandez L, Sham HL, Collins CA (2005) Discovery and characterization of aminopiperidinecoumarin melanin concentrating hormone receptor 1 antagonists. J Med Chem 48:5888–5891

Lagos P, Urbanavicius J, Scorza MC, Miraballes R, Torterolo P (2011) Depressive-like profile induced by MCH microinjections into the dorsal raphe nucleus evaluated in the forced swim test. Behav Brain Res 218:259–266

Lee C, Parks GS, Civelli O (2011) Anxiolytic effects of the MCH1R antagonist TPI 1361-17. J Mol Neurosci 43:132–137

Lembo PM, Grazzini E, Cao J, Hubatsch DA, Pelletier M, Hoffert C, St-Onge S, Pou C, Labrecque J, Groblewski T, O'Donnell D, Payza K, Ahmad S, Walker P (1999) The receptor for the orexigenic peptide melanin-concentrating hormone is a G-protein-coupled receptor. Nat Cell Biol 1:267–271

Licinio J, Wong ML, Gold PW (1996) The hypothalamic-pituitary-adrenal axis in anorexia nervosa. Psychiatry Res 62:75–83

Millan MJ, Gobert A, Panayi F, Rivet JM, Dekeyne A, Brocco M, Ortuno JC, Di Cara B (2008) The melanin-concentrating hormone1 receptor antagonists, SNAP-7941 and GW3430, enhance social recognition and dialysate levels of acetylcholine in the frontal cortex of rats. Int J Neuropsychopharmacol 11:1105–1122

Monzón ME, De Barioglio SR (1999) Response to novelty after i.c.v. injection of melanin-concentrating hormone (MCH) in rats. Physiol Behav 67:813–817

Monzon ME, de Souza MM, Izquierdo LA, Izquierdo I, Barros DM, de Barioglio SR (1999) Melanin-concentrating hormone (MCH) modifies memory retention in rats. Peptides 20:1517–1519

Nagasaki H, Chung S, Dooley CT, Wang Z, Li C, Saito Y, Clark SD, Houghten RA, Civelli O (2009) The pharmacological properties of a novel MCH1 receptor antagonist isolated from combinatorial libraries. Eur J Pharmacol 602:194–202

Nemeroff CB (1996) The corticotropin-releasing factor (CRF) hypothesis of depression: new findings and new directions. Mol Psychiatry 1:336–342

Okuyama S, Chaki S, Kawashima N, Suzuki Y, Ogawa S, Nakazato A, Kumagai T, Okubo T, Tomisawa K (1999) Receptor binding, behavioral, and electrophysiological profiles of nonpeptide corticotropin-releasing factor subtype 1 receptor antagonists CRA1000 and CRA1001. J Pharmacol Exp Ther 289:926–935

Parks GS, Okumura SM, Gohil K, Civelli O (2010) Mice lacking Melanin Concentrating Hormone 1 receptor are resistant to seizures. Neurosci Lett 484:104–107

Pissios P, Frank L, Kennedy AR, Porter DR, Marino FE, Liu FF, Pothos EN, Maratos-Flier E (2008) Dysregulation of the mesolimbic dopamine system and reward in MCH−/− mice. Biol Psychiatry 64:184–191

Rasmusson AM, Vythilingam M, Morgan CA 3rd (2003) The neuroendocrinology of posttraumatic stress disorder: new directions. CNS Spectr 8:651–656, 665–667

Rocca P, Fonzo V, Scotta M, Zanalda E, Ravizza L (1997) Paroxetine efficacy in the treatment of generalized anxiety disorder. Acta Psychiatr Scand 95:444–450

Roche KW, O'Brien RJ, Mammen AL, Bernhardt J, Huganir RL (1996) Characterization of multiple phosphorylation sites on the AMPA receptor GluR1 subunit. Neuron 16:1179–1188

Rorick-Kehn LM, Witkin JM, Statnick MA, Eberle EL, McKinzie JH, Kahl SD, Forster BM, Wong CJ, Li X, Crile RS, Shaw DB, Sahr AE, Adams BL, Quimby SJ, Diaz N, Jimenez A, Pedregal C, Mitch CH, Knopp KL, Anderson WH, Cramer JW, McKinzie DL (2014) LY2456302 is a novel, potent, orally-bioavailable small molecule kappa-selective antagonist with activity in animal models predictive of efficacy in mood and addictive disorders. Neuropharmacology 77:131–144

Roy M, David NK, Danao JV, Baribault H, Tian H, Giorgetti M (2006) Genetic inactivation of melanin-concentrating hormone receptor subtype 1 (MCHR1) in mice exerts anxiolytic-like behavioral effects. Neuropsychopharmacology 31:112–120

Roy M, David N, Cueva M, Giorgetti M (2007) A study of the involvement of melanin-concentrating hormone receptor 1 (MCHR1) in murine models of depression. Biol Psychiatry 61:174–180

Sailer AW, Sano H, Zeng Z, McDonald TP, Pan J, Pong SS, Feighner SD, Tan CP, Fukami T, Iwaasa H, Hreniuk DL, Morin NR, Sadowski SJ, Ito M, Ito M, Bansal A, Ky B, Figueroa DJ, Jiang Q, Austin CP, MacNeil DJ, Ishihara A, Ihara M, Kanatani A, Van der Ploeg LH, Howard AD, Liu Q (2001) Identification and characterization of a second melanin-concentrating hormone receptor, MCH-2R. Proc Natl Acad Sci USA 98:7564–7569

Saito Y, Nothacker HP, Wang Z, Lin SH, Leslie F, Civelli O (1999) Molecular characterization of the melanin-concentrating-hormone receptor. Nature 400:265–269

Saito Y, Cheng M, Leslie FM, Civelli O (2001) Expression of the melanin-concentrating hormone (MCH) receptor mRNA in the rat brain. J Comp Neurol 435:26–40

Santarelli L, Saxe M, Gross C, Surget A, Battaglia F, Dulawa S, Weisstaub N, Lee J, Duman R, Arancio O, Belzung C, Hen R (2003) Requirement of hippocampal neurogenesis for the behavioral effects of antidepressants. Science 301:805–809

Sears RM, Liu RJ, Narayanan NS, Sharf R, Yeckel MF, Laubach M, Aghajanian GK, DiLeone RJ (2010) Regulation of nucleus accumbens activity by the hypothalamic neuropeptide melanin-concentrating hormone. J Neurosci 30:8263–8273

Shimazaki T, Iijima M, Chaki S (2006a) The pituitary mediates the anxiolytic-like effects of the vasopressin V1B receptor antagonist, SSR149415, in a social interaction test in rats. Eur J Pharmacol 543:63–67

Shimazaki T, Yoshimizu T, Chaki S (2006b) Melanin-concentrating hormone MCH1 receptor antagonists: a potential new approach to the treatment of depression and anxiety disorders. CNS Drugs 20:801–811

Shimomura Y, Mori M, Sugo T, Ishibashi Y, Abe M, Kurokawa T, Onda H, Nishimura O, Sumino Y, Fujino M (1999) Isolation and identification of melanin-concentrating hormone as the endogenous ligand of the SLC-1 receptor. Biochem Biophys Res Commun 261:622–626

Shirayama Y, Chaki S (2006) Neurochemistry of the nucleus accumbens and its relevance to depression and antidepressant action in rodents. Curr Neuropharmacol 4:277–291

Shirayama Y, Ishida H, Iwata M, Hazama GI, Kawahara R, Duman RS (2004) Stress increases dynorphin immunoreactivity in limbic brain regions and dynorphin antagonism produces antidepressant-like effects. J Neurochem 90:1258–1268

Smith DG, Tzavara ET, Shaw J, Luecke S, Wade M, Davis R, Salhoff C, Nomikos GG, Gehlert DR (2005) Mesolimbic dopamine super-sensitivity in melanin-concentrating hormone-1 receptor-deficient mice. J Neurosci 25:914–922

Smith DG, Davis RJ, Rorick-Kehn L, Morin M, Witkin JM, McKinzie DL, Nomikos GG, Gehlert DR (2006) Melanin-concentrating hormone-1 receptor modulates neuroendocrine, behavioral, and corticolimbic neurochemical stress responses in mice. Neuropsychopharmacology 31:1135–1145

Smith DG, Hegde LG, Wolinsky TD, Miller S, Papp M, Ping X, Edwards T, Gerald CP, Craig DA (2009) The effects of stressful stimuli and hypothalamic-pituitary-adrenal axis activation are reversed by the melanin-concentrating hormone 1 receptor antagonist SNAP 94847 in rodents. Behav Brain Res 197:284–291

Spooren WP, Schoeffter P, Gasparini F, Kuhn R, Gentsch C (2002) Pharmacological and endocrinological characterisation of stress-induced hyperthermia in singly housed mice using classical and candidate anxiolytics (LY314582, MPEP and NKP608). Eur J Pharmacol 435:161–170

Takekawa S, Asami A, Ishihara Y, Terauchi J, Kato K, Shimomura Y, Mori M, Murakoshi H, Kato K, Suzuki N, Nishimura O, Fujino M (2002) T-226296: a novel, orally active and selective melanin-concentrating hormone receptor antagonist. Eur J Pharmacol 438:129–135

Tan CP, Sano H, Iwaasa H, Pan J, Sailer AW, Hreniuk DL, Feighner SD, Palyha OC, Pong SS, Figueroa DJ, Austin CP, Jiang MM, Yu H, Ito J, Ito M, Ito M, Guan XM, MacNeil DJ, Kanatani A, Van der Ploeg LH, Howard AD (2002) Melanin-concentrating hormone receptor subtypes 1 and 2: species-specific gene expression. Genomics 79:785–792

Tichomirowa MA, Keck ME, Schneider HJ, Paez-Pereda M, Renner U, Holsboer F, Stalla GK (2005) Endocrine disturbances in depression. J Endocrinol Investig 28:89–99

Tye KM, Mirzabekov JJ, Warden MR, Ferenczi EA, Tsai HC, Finkelstein J, Kim SY, Adhikari A, Thompson KR, Andalman AS, Gunaydin LA, Witten IB, Deisseroth K (2013) Dopamine neurons modulate neural encoding and expression of depression-related behaviour. Nature 493:537–541

Urbanavicius J, Lagos P, Torterolo P, Scorza C (2014) Prodepressive effect induced by microinjections of MCH into the dorsal raphe: time course, dose dependence, effects on anxiety-related behaviors, and reversion by nortriptyline. Behav Pharmacol 25:316–324

Urbanavicius J, Lagos P, Torterolo P, Abin-Carriquiry JA, Scorza C (2016) Melanin-concentrating hormone projections to the dorsal raphe nucleus: an immunofluorescence and in vivo microdialysis study. J Chem Neuroanat 72:16–24

Vale W, Spiess J, Rivier C, Rivier J (1981) Characterization of a 41-residue ovine hypothalamic peptide that stimulates secretion of corticotropin and beta-endorphin. Science 213:1394–1397

Varas MM, Pérez MF, Ramírez OA, de Barioglio SR (2003) Increased susceptibility to LTP generation and changes in NMDA-NR1 and -NR2B subunits mRNA expression in rat hippocampus after MCH administration. Peptides 24:1403–1411

Vasudevan A, Souers AJ, Freeman JC, Verzal MK, Gao J, Mulhern MM, Wodka D, Lynch JK, Engstrom KM, Wagaw SH, Brodjian S, Dayton B, Falls DH, Bush E, Brune M, Shapiro RD, Marsh KC, Hernandez LE, Collins CA, Kym PR (2005) Aminopiperidine indazoles as orally efficacious melanin concentrating hormone receptor-1 antagonists. Bioorg Med Chem Lett 15:5293–5297

Vaughan JM, Fischer WH, Hoeger C, Rivier J, Vale W (1989) Characterization of melanin-concentrating hormone from rat hypothalamus. Endocrinology 125:1660–1665

The manufacturer's authorised representative in the EU is Springer
Nature Customer Service Centre GmbH, Europaplatz 3, 69115 Heidelberg,
Germany. If you have any concerns regarding our products, please
contact ProductSafety@springernature.com

Printed and bound by CPI Group (UK) Ltd, Croydon, CR0 4YY
05/10/2026
02228139-0003